ENVIRONMENTAL CHEMISTRY

Nigel J Bunce
University of Guelph

Wuerz Publishing Ltd
Winnipeg, Canada

Wuerz Publishing Ltd

Environmental Chemistry
Nigel J. Bunce
ISBN 0-920063-46-2 paperback
ISBN 0-920063-47-0 hardcover

Printed in Canada

Front Cover

13th Century stone statuary from Salisbury Cathedral in England, showing the corrosion of the stonework by industrial pollution. See Chapter 6.

Contents

Preface

This text in environmental chemistry has been written for senior undergraduates who have completed at least a full year of general chemistry, and preferably have some knowledge of organic chemistry. It presupposes a knowledge of kinetics, elementary thermodynamics, and equilibria. Environmental chemistry is a broad, interdisciplinary subject, which overlaps with both industrial chemistry and with toxicology when issues of pollution are under discussion. However, the subject is more than simply pollution chemistry. Not only is the chemistry of the unpolluted environment interesting in its own right, but knowledge of the unpolluted environment is essential for understanding the polluted environment. This book was conceived for a one-semester course, and so not all topics of environmental concern could be included—of course the choice of material was inevitably influenced strongly by personal interest and preference.

Since the first appearance of this book in 1990, I have received numerous comments from readers, and have included many of these suggestions in this revision. Recent advances in publishing methods have allowed Steve Wuerz and I to attempt something which, to our knowledge, has not been done before. Instead of writing a new edition of the book after perhaps three or four years, we are printing only enough books for a single year's sales and updating the book annually. This seemed appropriate in such a fast-moving field as environmental chemistry, where new information becomes available almost weekly. The 1991 printing of "Environmental Chemistry" incorporates some new material, as well as minor rewrites based on responses from users, but it is not a new edition. It will not be necessary for all students in a class to buy the 1991 printing. A mixture of second-hand 1990 texts and new 1991 books ought to be fully compatible; the layout is unchanged and all the problems are the same.

A few words about the problems are in order. These are numerical illustrations of concepts presented in the text: in order to give them context, they are cited at the appropriate place within the chapter. Therefore they do not appear in increasing order of difficulty. Complete worked solutions to all problems are available in the Answer Book, available separately.

As they were originally conceived, many of the problems were supposed to be mini-projects, perhaps to be worked by groups of students. Some are conveniently solved with the aid of spreadsheet software. Not all the problems will be relevant to every reader: eg., problems involving kinetic-molecular theory of gases will be inappropriate for those who have not previously studied this topic. A number of previous readers have commented that there is a need for additional, less challenging problems. I intend to provide these in the 1992 printing.

I hope that you, the reader, will derive a measure of the enjoyment from reading these pages that I have had in writing them. Please continue to send suggestions for improving the book.

University of Guelph, August 1991

Smoking Hills
Whitehorse
Swan Hills
Red Deer
Regina
Los Angeles
San Diego

Goose Bay
Dryden
Sudbury
Montréal
Thetford Mines
St Basile le Grand
Guelph
Schenectady
Binghamton
Grand Rapids
New York
Washington
Nitro
Locations cited in text

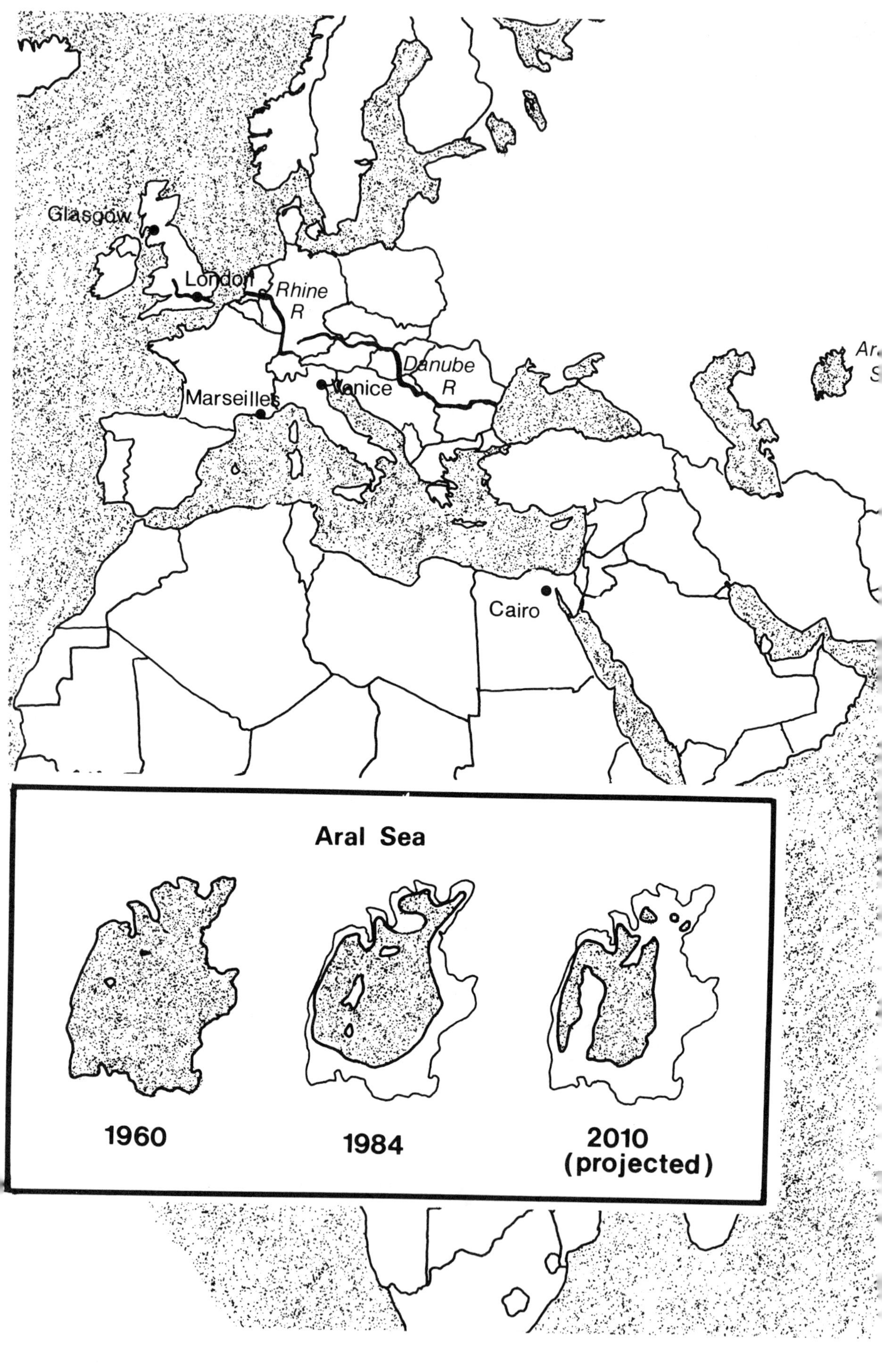
Glasgow
London
Rhine R
Danube R
Marseilles
Venice
Cairo
Aral Sea
1960
1984
2010 (projected)

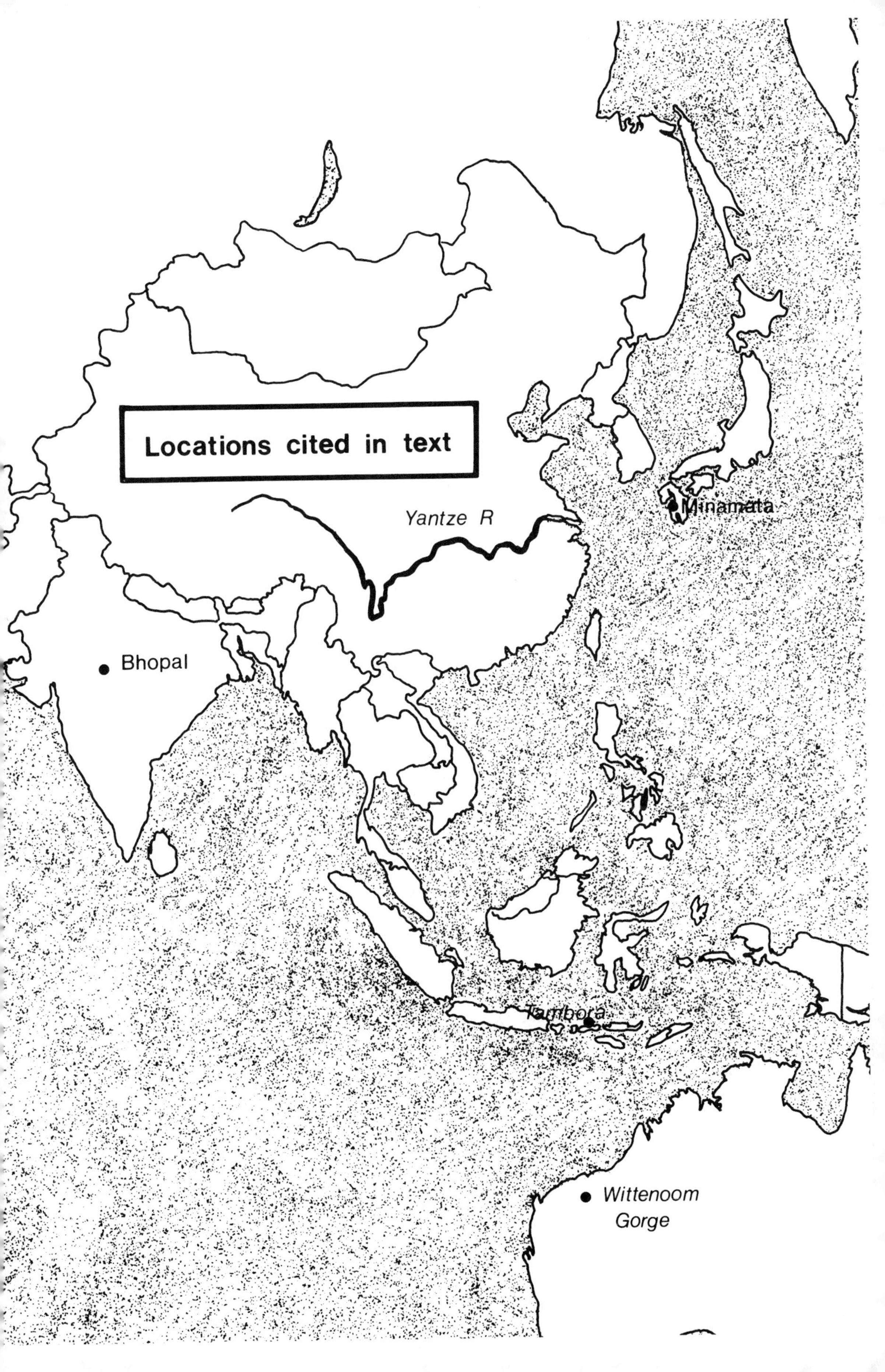

Locations cited in text
Yantze R
Minamata
Bhopal
Tambora
Wittenoom
Gorge

Photo Acknowledgements

Front cover:
By permission of the Dean and Chapter of Salisbury Cathedral.
Photograph by J Proctor, Salisbury Cathedral.
Photo research by Sabine Oppenländer Associates, London, UK.

Page 162:
Courtesy of David Ballard, Park Architect, Gettysburg National Military Park, Gettysburg Pennsylvania (USA).
Photograph by Walt Lane.

Page 169:
Courtesy of Dr Tom Hutchinson, University of Toronto, Toronto, Ontario (Canada).
Photograph by Dr Magda Havas, Trent University, Peterborough, Ontario (Canada).

Page 172:
Courtesy of Dr Ken Mills, Freshwater Institute, Winnipeg, Manitoba (Canada).

Page 175:
Courtesy of Mr Cory McPhee, Public Relations Officer, Inco Ltd, Copper Cliff, Ontario (Canada).
Photograph by Rene T Dionne Ltd.

Page 221:
Courtesy of Dr Ken Mills, Freshwater Institute, Winnipeg, Manitoba (Canada).

Page 244:
Courtesy of Mr Frank Webber, *Medicine Hat (Alberta) News*, Medicine Hat, Alberta (Canada).
Photograph by Paul VanPeenen.

Back Cover:
Courtesy of Ms Kathleen Manscill, Librarian, Great Smoky Mountains National Park, Gatlinburg, Tennessee (USA).

Chapter 1

The Atmosphere

Introduction

In this chapter is presented an overview of the atmosphere: its composition, and the residence times of some of its constituents; the evolution of the present atmosphere and a comparison with the atmospheres around other planets; the temperature profile of the atmosphere and the question of the "greenhouse effect."

1.1 Composition of the atmosphere

Unpolluted dry air at sea level (total pressure = 1 atm) has the composition shown below.

Major constituents (atm)			
N_2	0.781	O_2	0.209
Ar	0.0093	CO_2	0.00034

Minor constituents (ppm)[1]			
Ne	18	He	5.2
CH_4	1.5	N_2O	0.3
H_2	0.5	CO	$\approx$0.1

Even smaller amounts of NH_3, SO_2, Kr, Xe, O_3 and other gases are present. Water content is variable, which is why it has not been included in the table, and ranges from close to zero to about 0.4% (up to 0.004 atm). As a rough guide, think of the atmosphere as being ⅘ nitrogen and ⅕ oxygen.

[1]Note that for gaseous mixtures, 1 ppm (also called 1ppmv, or part per million by volume) corresponds to $10^{-6} \times p(\text{total})$. Therefore at sea level where $p(\text{total}) = 1$ atm, 1 ppm $\equiv 10^{-6}$ atm. However, if $p(\text{total}) = 0.01$ ppm (a value found at about 30 km altitude), 1 ppm $= 10^{-6} \times 0.01$ atm, or 10^{-8} atm. Parts per billion (ppb or ppbv) are defined similarly such that at sea level 1 ppb corresponds to $10^{-9} \times p(\text{total})$.

Be careful to distinguish ppm and ppb for gaseous mixtures (which are parts per.... by **volume**) from the analogous quantities in solution, which are parts per.... by **mass**, see Chapter 5, Natural Waters. They are not the same!

The Earth's gravity causes most of the total mass of the atmosphere ($\approx 5 \times 10^{15}$ tonnes, or $\approx 5 \times 10^{18}$ kg) to be concentrated close to the surface; 99% of the total mass lies below 30 km. There is no outer limit to the atmosphere; the concentration of particles simply decreases almost linearly with altitude (Problem 1) until by 150 km altitude the total pressure is about 10^{-9} atm (which would be considered ultra high vacuum in an Earth-bound laboratory).

Many different regions of the atmosphere may be defined, and each has its own unique characterisics in terms of chemistry and physics. In this book we shall be principally concerned with the **troposphere** (from the surface to about 15 km) and the **stratosphere** (about 15 to 50 km).

Looking ahead briefly to some of the issues in atmospheric chemistry which are of environmental concern today, the table below indicates the phenomena which will be studied in this book, and the region of the atmosphere in which each is most important[2].

Phenomenon	Region of atmosphere	Chapter
Greenhouse effect	troposphere	1
Ozone depletion	stratosphere	2
Photochemical smog	troposphere	3
Smoke and particulates	troposphere	3
Acid rain	troposphere	6

Our knowledge of atmospheric science, especially atmospheric chemistry, has increased tremendously since about 1970, mainly because of research into the different kinds of air pollution just listed. Space probes have permitted glimpses of the atmospheres around some of the other planets and have added new insights into the behaviour of our own planet's atmosphere. One particular view that has changed over the past generation is the importance of biological processes in explaining the chemistry of the Earth's atmosphere. Today the key difference between the atmospheric chemistry of the Earth and that of other planets is recognized to be the major role played on Earth by the biological processes through which the major components of the atmosphere are cycled.

1.2 Residence times, sources, and sinks

Residence times are defined as follows:

$$\text{Residence time} = \frac{\text{amount of substance in the ``reservoir''}}{\text{rate of inflow to, or outflow from, reservoir}}$$

In the present case the reservoir is the atmosphere. Estimates of residence times depend on being able to determine the amount of the substance in

[2]For a brief overview of the interaction between these phenomena, accessible to the non-specialist, see: T.E. Graedel and P.J. Crutzen, "The changing atmosphere," *Sci Am*, **1989**, 261, 58–68.

the atmosphere and the rate of inflow or outflow (Problems 2 and 3)[3]. It will be appreciated that these quantities are not known precisely, and so the values quoted will vary somewhat between different authors. Environmental scientists refer to a "Source" as the origin of a particular substance in a reservoir, and a "Sink" as its destination. A source or sink may be another chemical species within the same reservoir, or it may involve transport from or to a different reservoir.

Residence times are very important in determining whether a substance is widely distributed in the environment. This is true both for naturally occurring substances and for pollutants. For any substance, whether naturally occurring or a pollutant, a long residence time correlates with being well-mixed, i.e., widely distributed, in the environment (and vice versa). Example: chlorofluorocarbons have long residence times and become uniformly mixed in the atmosphere. They represent a global pollution problem. By contrast, acidic gases survive in the atmosphere only for a few days; acid rain is a much more local phenomenon.

1.2.1 Oxygen

The total amount of O_2 in the contemporary atmosphere has been quoted as 3.8×10^{19} mol[4] or 1.2×10^{18} kg[5]. The principal reaction by which oxygen enters the atmosphere is photosynthesis (quoted as 5.0×10^{15} mol yr^{-1}[4] or 4.0×10^{14} kg yr^{-1}[5]). Photosynthesis is almost exactly balanced by processes which consume oxygen: respiration and decay, and the combustion of fossil fuels. Weathering of rocks and ancient sediments also consumes oxygen: weathering includes reactions such as oxidation of low oxidation state metals (e.g., Fe^{2+} to Fe^{3+}). In earlier epochs, weathering was probably a major sink for atmospheric oxygen, but today it consumes less than 0.1% as much oxygen as respiration, decay, and combustion.

The two sets of figures for the amount of oxygen in the atmosphere and its rate of entry to the atmosphere give the residence time of O_2 as 7600 yr and 3300 yr respectively. Both estimates indicate that oxygen has a long residence time in the atmosphere, as a result of which two inferences may be drawn.

- O_2 is well mixed in the atmosphere, i.e., its partial pressure does not vary from place to place. An equivalent statement is that the resi-

[3]The residence time is the same as the reciprocal of the sum of all the first order, or pseudo-first order rate constants for loss from the reservoir. This is reasonable dimensionally in that first order rate constants have the units: time^{-1}. The residence time (also called the lifetime) differs from the half-life in that the concentration falls to $1/e$ (about 37%) of the initial concentration after one lifetime, as opposed to 50% after one half-life.

[4]R.W. Raiswell, P. Brimblecombe, D.L. Dent, and P.S. Liss, *Environmental Chemistry,* Edward Arnold (Publishers), London, England, 1980.

[5]R.P. Wayne, *Chemistry of Atmospheres,* Oxford University Press, Oxford, England, 1985, 23.

dence time is much greater than the time needed for the atmosphere to mix.

- Whatever other problems may be caused by the combustion of fossil fuels, depletion of oxygen is not likely to be one of them, because there is so much oxygen in the atmosphere.

1.2.2 Water

Water is considered next because of the contrast with oxygen. At any time about 7×10^{14} mol of $H_2O(g)$ are present[4], a tiny fraction of the 9.5×10^{19} mol present on the surface as $H_2O(l)$. Evaporation from the oceans (2.2×10^{16} mol yr^{-1}) and from lakes and rivers (3.5×10^{15} mol yr^{-1}) is balanced by precipitation over the land (5.5×10^{15} mol yr^{-1}) and the oceans (1.9×10^{16} mol yr^{-1}), leading to an average residence time of 3×10^{-2} yr (10 days) for water in the atmosphere. Since complete horizontal and vertical mixing of the troposphere requires several years, water is very unevenly distributed in the atmosphere, consistent with the variation in the weather from place to place, and also from time to time. In other words, the atmosphere is poorly mixed as far as water is concerned, and there is great local variability.

The amount of water in the atmosphere is not only variable, but depends on the temperature, since the vapour pressure of water increases strongly with temperature. Some representative values are given below.

t, °C	$p(H_2O)$, atm	t, °C	$p(H_2O)$, atm	t, °C	$p(H_2O)$, atm
−20	0.00102	0	0.00603	20	0.02307
−15	0.00163	5	0.00861	25	0.03126
−10	0.00257	10	0.01188	30	0.04187
−5	0.00396	15	0.01683	35	0.05418

These values are saturated, or **equilibrium** vapour pressures. They represent the maximum attainable values of $p(H_2O)$ at the stated temperature. Under most conditions, the atmosphere is sub-saturated with respect to water vapour. The **relative humidity** is the prevailing $p(H_2O)$ as a percentage of the equilibrium value. Note that a certain value of the relative humidity does not represent a fixed water vapour content; $p(H_2O)$ will vary according to the temperature. Thus a relative humidity of, say, 80% corresponds to $p(H_2O) = 0.0048$ atm at 0°C, but to 0.025 atm at 25°C. Remember also that water vapour can co-exist with ice, so that $p(H_2O)$ is non-zero even below 0°C; this is why snow can evaporate when the temperature is below freezing (Problem 4).

1.2.3 Nitrogen

We are tempted to think of the 0.781 atm of nitrogen in the atmosphere almost as an inert "filler" because in our experience elemental nitrogen is rather unreactive. The biogeochemistry of nitrogen is extremely complex, but leads us to the conclusion that the atmospheric content of nitrogen, like that of oxygen, is regulated principally by biological processes.

The atmosphere contains some 3.9×10^{18} kg of elemental nitrogen[2]. The major natural sinks are biological nitrogen fixation (2×10^{11} kg yr^{-1}) and the production of NO in thunderstorms and through combustion (Equation 1), leading ultimately to deposition of HNO_3 in rainwater (7×10^{10} kg of N per year).

$$N_2(g) + O_2(g) \xrightarrow{\text{high temp.}} 2NO(g) \tag{1}$$

To this must be added perhaps 5×10^{10} kg fixed industrially by the Haber process, Equation 2.

$$N_2(g) + 3H_2(g) \xrightarrow{\text{catalyst, 450°C}} 2NH_3(g) \tag{2}$$

Terrestrial and aquatic nitrogen in the form of NO_3^- or NH_4^+ is cycled through the biosphere to make proteins and nucleic acids. The processes of decay return the nitrogen to the atmosphere as N_2 and as N_2O by the action of denitrifying bacteria[6]. Almost all the nitrogen fixed by the Haber process is used as fertilizer, so that increased fertilizer use also increases the rate of return of nitrogen to the atmosphere through biological denitrification[7]. Indeed, as much as half of all nitrogenous fertilizer applied to crops is denitrified even before the crop takes it up. This effect is reflected in the gradual increase in the atmospheric levels of N_2O, which have risen from 0.29 to 0.31 ppm over the past two decades. Nitrous oxide, N_2O, is rather unreactive and has a residence time of ≈ 20 yr (Problem 5).

Because the rates of transfer of nitrogen in and out of the atmosphere are small in comparison with the size of the reservoir, the calculated residence time of $N_2(g)$ is large ($\approx 10^7$ years).

1.2.4 Carbon dioxide

The amount of carbon dioxide in the atmosphere is 1.4×10^{16} mol. In counterpoint to oxygen, a major source of atmospheric CO_2 is respiration, combustion, and decay, and an important sink is photosynthesis (about 1.5×10^{15} mol yr^{-1} each). Since CO_2 is somewhat soluble in water, exchange with the oceans must also be taken into account, some 7×10^{15} mol yr^{-1} being taken up and 6×10^{15} mol yr^{-1} being released by different regions of the oceans[4]. The result is an atmospheric lifetime of about 2 yr, which

---

[6] W.J. Payne, *Denitrification,* Wiley-Interscience, New York, 1981, Chapters 2, 7, and 8.

[7] Denitrification is usually used to refer to the microbial reduction of NO_3^- to N_2O and N_2. Nitrification is the biological oxidation of NH_4^+ to NO_3^-. Several of the steps in nitrification are the reverse of steps in denitrification, and it has recently been realized that some N_2O and N_2 can be formed during what is usually called nitrification. In the present section, I am using the term denitrification to refer generally to the formation of N_2O and N_2, whether the original nitrogen source was nitrate ion or ammonia.

makes the atmosphere moderately well mixed with respect to carbon dioxide. However, a more recent analysis indicates that terrestrial sinks may in fact be stronger than ocean uptake. Despite extensive research, the authors note that "The global C cycle is not well understood"[8].

An interesting effect is seen in analytical data obtained over many years at the Mauna Loa Observatory in Hawaii, a location far from anthropogenic sources of CO_2 (Figure 1.1). A pronounced one-year cycle in CO_2 concentration is seen, with the peak about April and the trough around October each year. These data show that the atmospheric concentration of CO_2 is not perfectly homogeneous.

Since Hawaii is in the Northern Hemisphere, photosynthetic activity is highest in the period May to October; during the summer CO_2 is removed from the atmosphere a little faster than it is added, while the reverse situation pertains in the winter months. Consistent with this explanation, monitoring stations in the sourthern hemisphere show maximum CO_2 concentrations in October, and minima in April.

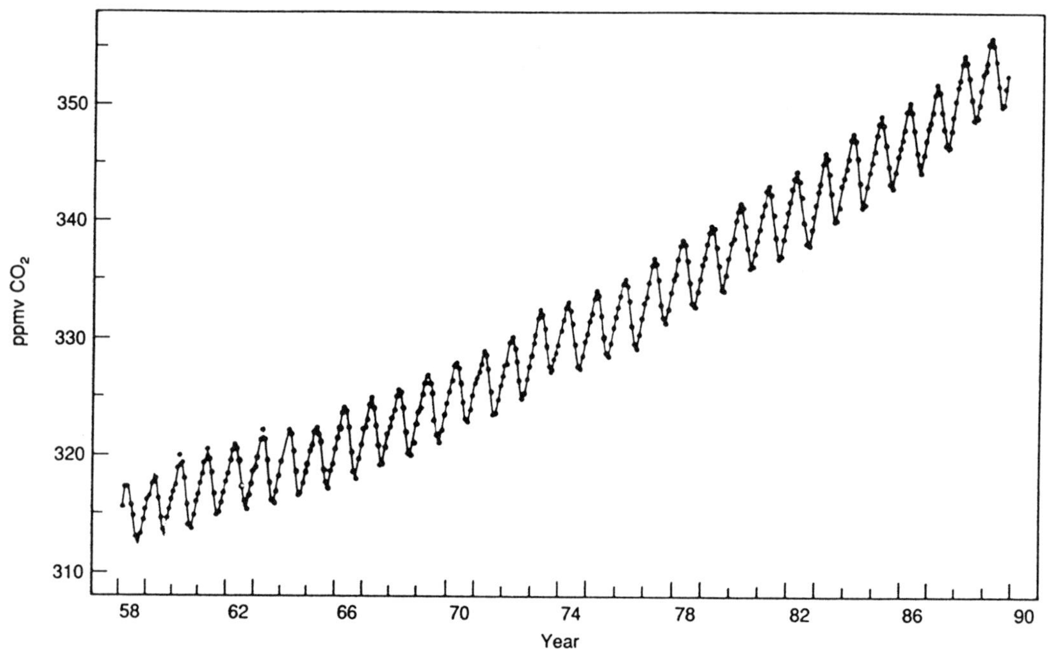

Figure 1.1: Observations of $p(CO_2)$ at the Mauna Loa Observatory for the period 1958–1990.

A second trend that is evident from inspection of Figure 1.1 is the gradual increase in the partial pressure of CO_2 over the years, as $p(CO_2)$ was ca. 315 ppm in 1958, and had reached 350 ppm in 1988. Predictions are for a doubling of $p(CO_2)$ sometime during the latter part of the Twenty-first Century. A rising trend in concentration is also seen for other "greenhouse

[8]P.P. Tans, I.Y. Fung, and T. Takahashi, "Observational constraints on the global atmospheric CO_2 budget," *Science,* **1990,** 247, 1431–1438.

gases" such as CH_4 and N_2O. This will be discussed further in the context of the greenhouse effect, Section 1.5.

1.2.5 Hydrogen[9]

Hydrogen is only a very minor component of the atmosphere (0.5 ppm). A reservoir of 180,000 tonnes of H_2 in the atmosphere, and sinks of about 90,000 tonnes per year afford a lifetime of 2 years. Hydrogen is like CO_2 in that it is moderately well mixed in the atmosphere; its concentration shows an annual cycle with peaks in April and troughs in October. Furthermore, its concentration is rising about 0.6% per year. Unlike CO_2 however the cycle for hydrogen in the northern hemisphere is in phase with that in the southern hemisphere. A likely reason is that the major sink for H_2 is uptake by soil, and the land area of the northern hemisphere is much larger than that of the southern; consequently, the dominant sink is the northern hemisphere. Khalil and Rasmussen point out that a rising trend in $p(H_2)$ will likely lead to an increase in stratospheric $p(H_2O)$, and add to the sink strength for stratospheric ozone.

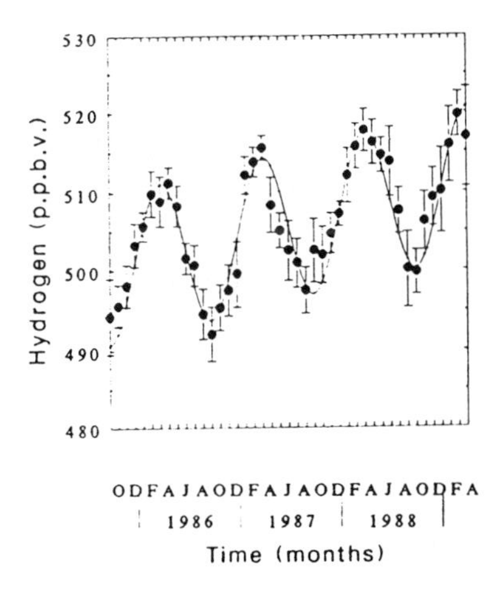

Figure 1.2: Global average concentration of H_2. Unlike CO_2, Figure 1.1, the northern and southern hemispheres are in phase with each other. Reproduced from Reference 9.

[9] M.A.K. Khalil and R.A. Rasmussen, "Global increase of atmospheric molecular hydrogen," *Nature*, **1990**, 347, 743-745

1.3 Evolution of the primitive atmosphere

The discussion above makes clear that the present composition of the atmosphere is regulated by biological processes rather than by inorganic chemistry alone. What must the early atmosphere have been like, and how did the present atmosphere develop?

Outgassing as the early Earth was formed and then cooled probably led to an atmosphere rich in N_2, H_2O, and CO_2, with lesser amounts of NH_3 and CH_4. Such an atmosphere is a reducing environment, in contrast with the present highly oxidizing atmosphere.

The Sun, a star with a surface temperature of 6000 K, emits radiation right across the ultraviolet, visible, and infrared regions of the spectrum, with maximum emission around 475 nm (Problem 6). We do not experience the full spectrum at the Earth's surface because radiation of wavelength shorter than about 300 nm is filtered out (absorbed) by the atmosphere. Also, much of the longer wavelength (infrared) radiation is absorbed by "greenhouse" gases such as H_2O and CO_2: see Section 1.5. As will be discussed in detail in Chapter 3, the active absorbing agents in the ultraviolet region are O_2 and O_3. These were not present in the primitive atmosphere; a much more energetic solar spectrum would have reached the surface, making the surface inimicable to the development of life.

At short wavelengths, the following gas phase reactions[10] would have been possible in the primitive atmosphere.

$$\text{(3)} \qquad H_2O \xrightarrow{h\nu,\ \lambda < 240\,\text{nm}} H + OH$$

$$\text{(4)} \qquad CO_2 \xrightarrow{h\nu,\ \lambda < 240\,\text{nm}} CO + O$$

Reactions 5–8 return the hydrogen and oxygen atoms formed in Equation 3 and 4 back to water; thus the overall scheme of Equation 3–8 converts solar energy into heat, thereby raising the temperature of the atmosphere.

$$\text{(5)} \qquad 2OH \longrightarrow H_2O + O$$

$$\text{(6)} \qquad 2O \xrightarrow{M} O_2$$

$$\text{(7)} \qquad 2H \xrightarrow{M} H_2$$

$$\text{(8)} \qquad 2H_2 + O_2 \longrightarrow 2H_2O$$

In Equations 6 and 7, "M" is a "third body"—any gas phase atom or molecule, which carries away excess energy from the collision of the re-

[10] The following conventions for equations will be used throughout this book. Where, as in the present chapter, the chemistry refers exclusively to the gas phase, the symbol (g) after each substance is omitted, and is to be inferred. Atomic and free radical species, such as H and OH, are given without the "dot" for the free radical (H• or OH•). They are written without charges, and are therefore not to be confused with their charged counterparts such as H^+ and OH^-, which are more familiar in aqueous chemistry. This distinction is extremely important when consulting tables of thermodynamic properties for working the problems at the end of each chapter.

actants, thereby stabilizing the product of the reaction. In numerical work $p(\text{M}) = p(\text{total})$, since M can be any atmospheric constituent.

A factor which has not yet been considered is the possibility of escape of particles into space. The "escape velocity" for leaving the Earth's atmosphere is about 11.2 km s^{-1}. Although the escape velocity is independent of the mass of the particle, the kinetic energy needed ($\frac{1}{2}mv^2$) depends directly upon mass. Only the lightest particles have more than an infinitesimal probability of escaping the Earth's gravitational attraction (Problem 7). For example at 600 K, 1 hydrogen atom in 10^6, but only one oxygen atom in 10^{84}, have sufficient energy to escape[11].

Equations 3–8 represent H_2O in balance chemically with H, H_2, O, and O_2. However, if even a few hydrogen atoms were to escape from the atmosphere, there would be an imbalance between hydrogen and oxygen, and oxygen would begin to build up in the atmosphere. It is calculated that even today nearly 3×10^8 hydrogen atoms per second escape the Earth's gravitational pull for every square centimeter of the Earth's surface (Problem 8).

An excess of oxygen over hydrogen would initially have led to oxidation of reduced metal species in exposed rocks or in aqueous solution (e.g., Fe^{2+} to Fe^{3+}). In the prebiological era, $p(O_2)$ was probably 10^{-10} to 10^{-14} atm. The oxygen content of the atmosphere is thought to have begun its increase with the evolution of photosynthetic bacteria. These microorganisms would likely have had to live under at least 10 m of water (the water would act as a filter to absorb radiation $\lambda \approx 280$ nm which would damage their DNA). As $p(O_2)$ increased, some ozone production would occur (Equation 9) and, with it, some screening of the surface from radiation of wavelength 250–320 nm.

$$\text{(9)} \qquad O + O_2 \xrightarrow{M} O_3$$

As $p(O_3)$ rose to $0.1\times$ its present level, life would be able to migrate first to shallow water and then to dry land.

1.4 Temperature profile of the atmosphere

The pressure of the atmosphere falls steadily with increasing altitude. By contrast, its temperature displays a very complicated profile, Figure 1.2. The temperature of the atmosphere is governed by the following factors: absorption of energy received from the Sun; heat production in the interior of the Earth; and loss of energy by emission from the Earth acting as a blackbody radiator. Whereas the peak intensity of incoming solar energy occurs in the visible, the peak emission by the much cooler Earth occurs in the infrared. The atmosphere plays a vital role in warming the surface of the Earth; in the absence of the atmosphere, that temperature is calculated to be ≈ 245 K, rather than the average of ≈ 288 K actually experienced

[11]Reference 5, p. 58.

today. Since 245 K is well below the freezing point of water, life as we know it could not have developed without the insulating effect of the atmosphere.

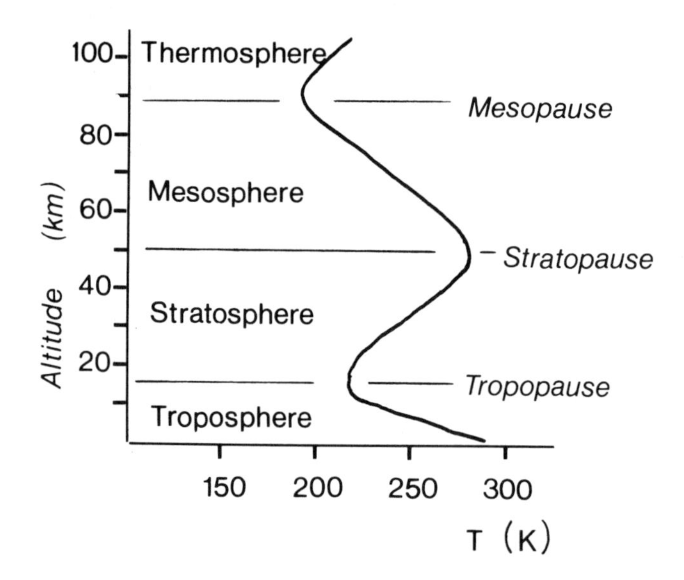

Figure 1.3: Temperature profile of the atmosphere as a function of altitude. Redrawn from Reference 5, p. 56

1.4.1 Temperature regulation in the thermosphere

In the thermosphere (altitude > ca. 90 km) the atmosphere is very thin with $p(\text{total}) < 10^{-8}$ atm. The chemical species present are subjected to the full solar spectrum. Highly energetic photons of wavelengths well below 200 nm (energy > 600 kJ mol^{-1}) are absorbed by both molecules and atoms. These photons have energies sufficient to cleave molecules into atoms, and even to ionize both atoms and molecules (Problem 9).

Typical reactions and minimum energies[12]:

$$N_2 \longrightarrow 2N \qquad \Delta H^\circ = 946 \text{ kJ mol}^{-1}$$
$$N \longrightarrow N^+ + e^- \qquad \Delta H^\circ = 1400 \text{ kJ mol}^{-1}$$
$$O_2 \longrightarrow O_2^+ + e^- \qquad \Delta H^\circ = 1160 \text{ kJ mol}^{-1}$$

The reverse of these processes liberates the equivalent amount of energy, mostly in the form of kinetic energy of the particles. Thus the thermosphere is warmed by the conversion of very short wavelength solar radiation into heat. The thermosphere is the only region of the atmosphere in which the

[12]These energies are approximated as standard enthalpy changes; photon energies are considered as supplying internal energy ΔE, and from the first law of thermodynamics $\Delta H^\circ = \Delta E + \Delta n\,RT$. $\Delta n\,RT$ is almost always negligible compared with ΔH°.

predominant chemical species are atoms and ions rather than molecules; for example, at 80 km altitude 80% of all the oxygen is present as atoms (Problem 10).

At the extremely low pressures of the thermosphere, the mean free paths of the molecules and ions are long, centimeters to kilometers depending on altitude. As a result, the particles are not in true thermal equilibrium, and the concept of temperature has meaning only in the context of the kinetic energies of the particles. The few particles present are highly energetic, but if you placed a thermometer at 150 km altitude, it would register a very low temperature. There are relatively few particles at this altitude, and so the temperature registered by the thermometer at the steady state would be determined by the balance between the energy gained from the relatively infrequent collisions with gas molecules and the radiation of heat by the thermometer.

As this highly energetic radiation penetrates the atmosphere, it encounters an increasing density of particles, and more and more of it gets absorbed. Fewer photons are therefore converted into heat in the lower parts of the thermosphere, and the temperature drops until the mesopause is reached.

1.4.2 Temperature regulation in the stratosphere

In the stratosphere, almost all the chemical species are molecules, unlike the thermosphere, in which atoms and ions predominate. Chemical processes in the stratosphere involve the conversion of solar radiation in the range 200–300 nm into heat. The essential reactions are:

$$O_2 \xrightarrow{\lambda < 240\,\text{nm}} 2O \tag{10}$$

$$O + O_2 \xrightarrow{M} O_3 \tag{11}$$

$$O_3 \xrightarrow{\lambda < 320\,\text{nm}} O_2 + O \tag{12}$$

As altitude decreases from the mesopause, the temperature rises to a maximum near 0°C at the stratopause, and then falls again to a minimum of about −60°C at the tropopause. This is explicable as follows. Near the mesopause, the concentration of molecules is very low, $p(\text{total}) \approx 10^{-5}$ atm. As the concentration of molecules rises at lower altitude, absorption becomes more efficient and so the temperature rises. Maximal conversion of 200–300 nm radiation to heat occurs at the stratopause. Below this level, the rate of conversion of photon energy to heat falls again because most of the photons have been absorbed at higher altitudes, and so fewer are available for absorption (even though there are more molecules available to absorb them).

Figure 1.4 summarizes the last two sections, showing how far into the atmosphere the various wavelengths of incoming solar radiation are able to penetrate.

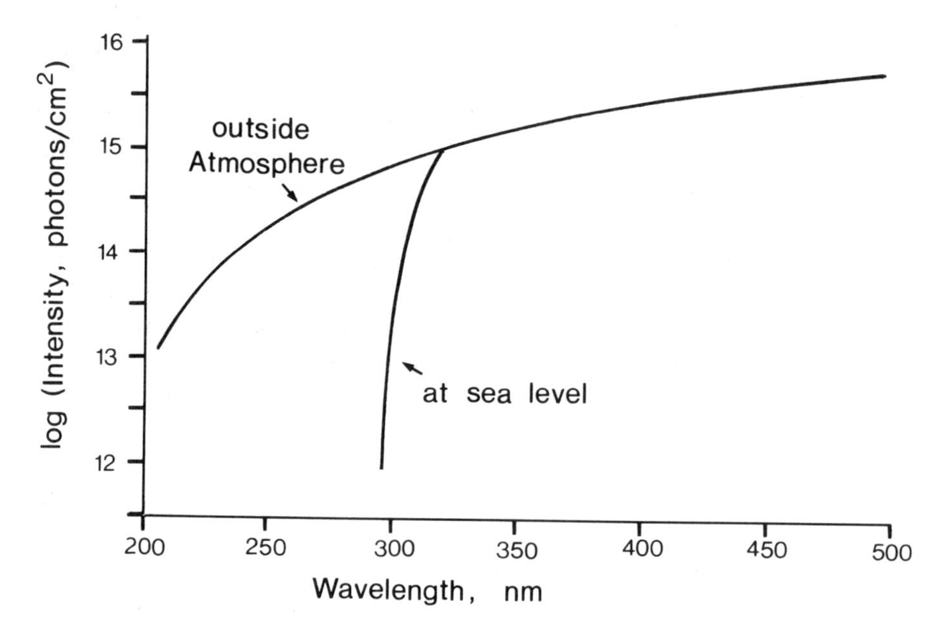

Figure 1.4: Penetration of solar radiation into the Earth's atmosphere.

1.4.3 Temperature regulation in the troposphere

The important process to consider is absorption by the troposphere of infrared radiation emitted by the Earth. Entrapment of infrared radiation is most effective near the surface where the intensity of the radiation is highest. As altitude increases, the temperature in the troposphere decreases. This is because some of the infrared radiation has already been absorbed at a lower altitude, and also because the pressure is decreasing (fewer molecules available for absorption). This trend continues until the tropopause, beyond which altitude the conversion of solar radiation to heat via the O_2/O_3 system outweighs the cooling trend just described.

Not all gases are equally effective at absorbing infrared radiation. The physical process accompanying infrared absorption is the promotion of a molecule from a lower to a higher vibrational state. The requirement for infrared absorption is that the lower and upper vibrational states of the molecule must differ in dipole moment. This condition cannot be fulfilled for atoms such as argon, nor for homonuclear diatomic molecules such as O_2 and N_2, which have zero dipole moment in all their vibrational states. Consequently the major tropospheric constituents, N_2, O_2, and Ar, are ineffective in absorbing infrared radiation, and all the absorption is done by minor atmospheric constituents, of which H_2O and CO_2 are the most important.

1.5 Greenhouse effect and climate change[13]

As just described, the troposphere is warmed by the absorption of infrared radiation emitted from the surface of the Earth, and which would otherwise be lost into space. By analogy, the glass of a greenhouse limits the dissipation of the warmth from inside the greenhouse. (The analogy is really not all that close: in a real greenhouse the major heat-conserving effect is that the glass prevents convectional mixing of the air in the greenhouse with that outside.)

An important question is exactly what is meant by the term "greenhouse effect." As just described, it refers to the process by which minor atmospheric gases such as CO_2 and H_2O trap infrared radiation, as a result of which the Earth's surface temperature is on average +15°C rather than −30°C. Obviously this is not the issue which is of current concern in the news media! In media terms, the greenhouse effect refers to a predicted **increase** in the rate of energy trapping in the troposphere, with a corresponding increase in the temperature of the atmosphere. There are several aspects to the issue, as discussed, in Sections 1.5.1 through 1.5.3.

1.5.1 Increased levels of greenhouse gases

Greenhouse gases (sometimes called radiatively active gases) are those which can absorb infrared radiation. As was shown in Section 1.4.3, atmospheric nitrogen, oxygen, and argon do not absorb infrared radiation.

Carbon dioxide

Carbon dioxide has so far attracted the most attention. The partial pressure of this atmospheric constituent over 30 years (Figure 1.1) shows an increasing trend which shows no sign of levelling off. Currently, $p(CO_2)$ is approximately 350 ppm. The sources and sinks of atmospheric CO_2 described in Section 2.4 are not exactly in balance, with the rate of increase in the 1980's a little less than 0.5% annually[14].

Why carbon dioxide levels should be increasing is a matter of some controversy. A rising trend in $p(CO_2)$ implies that the sources and sinks of this gas are out of balance. Emissions of CO_2 have been increasing due to increased use of fossil fuels (hydrocarbons). What is not known is whether loss of CO_2 from the atmosphere has been decreasing. Recall from Section 1.2.4 that the major sinks for CO_2 are photosynthesis and terrestrial/ocean uptake. Deforestation, especially of the tropical rainforests, simultaneously

[13] For an introduction see K.B. Belton, "Global climate change," *American Chemical Society*, Washington, D.C., **1990,** 12 pp.

[14] H.G. Hengeveld, *Understanding CO_2 and Climate*, Canadian Climate Centre, Atmospheric Environment Service, 1987.

releases CO_2 into the atmosphere through biomass burning[15], and reduces photosynthetic activity.

If fossil fuel consumption increases as little as 1% annually (it was over 5% from 1945–1975), $p(CO_2)$ will still double, to > 600 ppm, within a century[16]. Coal burning for electrical power generation represents a major use of fossil fuels. In this context, Winschel[17] has pointed out that a switch from low rank (i.e., low carbon content) coals to high grade bituminous coals for power generation would significantly reduce CO_2 emissions. This is because high grade coals produce more heat per mole of carbon burned.

Regarding the uptake of tropospheric CO_2 uptake by the oceans, a complex series of equilibria relates $CO_2(g)$ with $CaCO_3(s)$.

$$CO_2(g) \rightleftharpoons H_2CO_3(aq) \overset{-H^+}{\rightleftharpoons} HCO_3^-(aq) \overset{-H^+}{\rightleftharpoons} CO_3^{2-}(aq) \overset{Ca^{2+}}{\rightleftharpoons} CaCO_3(s)$$

The amount of $CO_2(aq)$ in the oceans is sixty times that of $CO_2(g)$ in the atmosphere, suggesting that the oceans can "soak up" most of the additional CO_2 injected into the atmosphere. However, uptake of CO_2 into the surface waters of the oceans is relatively slow (half-life 1.3 years). In addition, the surface waters of the ocean (zero to $\approx$ 100 m depth) mix with the deep waters even more slowly, with a half-life of about 35 years[18]. Thus in the medium term, the surface waters have the capacity to remove only a fraction of any increase in the load of gaseous CO_2 (Problem 11).

Ultimately, carbonate rocks constitute an enormous reservoir of CO_2 which is presently locked up. If conditions changed so that some of this CO_2 began to be released there would be a very strong positive feedback: a "runaway" greenhouse effect in which injection of CO_2 into the atmosphere would raise atmospheric and ocean temperatures, and hence causing more CO_2 to be released into the atmosphere from the oceans[19]. Fortunately, this possibility seems remote (see also Chapter 6, Problem 17). As discussed in Section 6, a runaway greenhouse effect does exist on Venus, where the surface temperatures are > 700 K. No liquid water is present, and most of the total CO_2 is in the atmosphere rather than locked away in rocks.

Water

Global warming, a predicted consequence of an "enhanced" greenhouse effect, would increase the average amount of water in the atmosphere through

[15] Recent information suggests that biomass burning may contribute at least one quarter of CO_2 emissions from all combustion sources: *Chem Eng News,* March 26, **1990,** 4–5.

[16] Current per-capita releases of CO_2 due to fossil fuel burning range from $\approx$ 5 tonnes per year (U.S.; East Germany) through $\approx$ ¼ tonne per year (India, Brazil, Mexico) to < 0.1 tonne per year for the least industrialized Third World countries.

[17] R.A. Winschel, "The relationship of carbon dioxide emissions with coal rank and sulfur content," *J. Air Waste Management Assoc.,* **1990,** 40, 861–865

[18] Data from Reference 4, pp. 34–37.

[19] The reasoning is that the solubility of carbon dioxide in water decreases as the temperature rises: see Chapter 5.

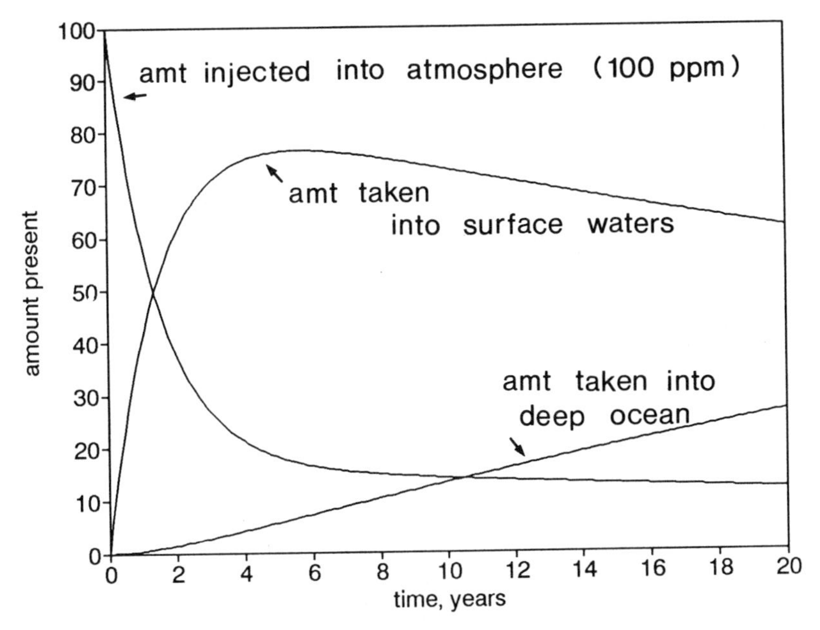

Figure 1.5: Calculated uptake of CO_2 from the atmosphere to the surface and deep oceans. Data on which the diagram is based are found in Problem 11.

evaporation from the oceans; this is because the equilibrium vapour pressure of water rises with temperature. This situation would represent positive feedback, i.e., increased temperature leading to a rise in $p(H_2O)$, causing in turn a further increase in the efficiency of trapping infrared radiation[20].

Trace gases

These include methane, nitrous oxide, ozone, and chlorofluorocarbons. All these gases have atmospheric concentrations which are hundreds or more times less than those of CO_2 and water vapour. This might lead one to suspect that their infrared absorbing potential would be insignificant compared with water and CO_2. Such is not the case, because each greenhouse gas absorbs radiation in its own characteristic region of the infrared; some of these gases absorb radiation in regions of the spectrum in which CO_2 and H_2O are transparent, and from which the radiation would otherwise escape into space. In addition, most of them are more effective infrared absorbers than CO_2 on a molecule-for-molecule basis[21]. Third, their concentrations are increasing faster, on a percentage basis, than that of CO_2.

---

[20] A recent paper by A. Raval and V. Ramanathan, "Observational determination of the greenhouse effect," *Nature,* **1989**, 342, 758–761, suggests that positive feedback due to water vapour is already evident from extraterrestrial satellite data.

[21] D.A. Lashof and D.R. Ahuja, "Relative contributions of greenhouse gas emissions to global warming," *Nature,* **1990**, 344, 529–531

Methane: The concentration of tropospheric methane is currently about 1.7 ppm but is rising at 1–2%[22] annually. Examination of the methane levels in air bubbles trapped in ice cores indicates that the historical level of this gas was ≈ 0.7 ppm until about two centuries ago[23], since when its rate of increase has been accelerating (Figure 1.6). The chief sources of atmospheric methane all involve anaerobic decay. As its common name "marsh gas" implies, wetlands are an important natural source of this gas. Human agricultural activity adds greatly to the natural background. World-wide emissions from cattle almost equal those from wetlands[24], the result of anaerobic fermentation in the animals' rumens, and emissions from rice paddies are two thirds those of wetlands. More intensive agriculture raises the emissions from both these sources.

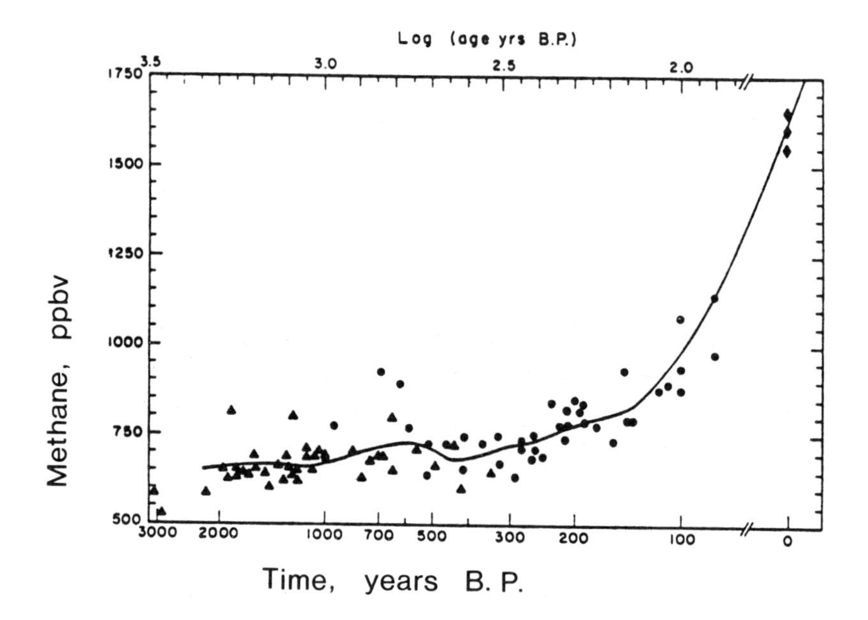

Figure 1.6: Historical trends in the concentration of atmospheric methane. Reproduced from Reference 22. On the axis labels note (abscissa) B.P. = Before Present, and (ordinate) ppbv = parts per billion by volume. Some authors write ppm and ppb; others write ppmv and ppbv. In the gas phase, consider these abbreviations as interchangeable.

[22]M.A.K. Khalil and R.A. Rasmussen, "Atmospheric methane: recent global trends," *Environ Sci Technol,* **1990,** 24, 549-553. Ice core records dating back 160,000 years suggest that $p(CH_4)$ has varied between 350 and 700 ppb over this period: J. Chappellaz *et al, Nature,* **1990,** 345, 127-131.

[23]R.A. Rasmussen and M.A.K. Khalil, "Atmospheric methane in the recent and ancient atmospheres: concentrations, trends, and interhemispheric gradient," *J Geophys Res,* **1984,** 89, 11599–11605.

[24]F. Pearce, "Methane: the hidden greenhouse gas," *New Scientist,* May 6, **1989,** 37–41. See also P.J. Crutzen, "Methane's sources and sinks," *Nature,* **1991,** 350, 380–381.

Rowland *et al*[25] note that the atmospheric lifetime of methane is about 10 years, suggesting a 10% excess of sources over sinks. Therefore, a 10% reduction in methane emissions would bring sources and sinks of this gas back into balance. In this context, a significant source of methane appears to be leakage from natural gas transmission, a source which ought in principle to be controllable.

In addition to these increasing sources of methane, there is a possibility that there is also less efficient removal of this component from the atmosphere. As will be discussed in Chapter 3, the main reaction of CH_4 is hydrogen abstraction by the hydroxyl radical.

$$CH_4 + OH \longrightarrow CH_3 + H_2O \tag{13}$$

There is some evidence that the concentration of OH may be decreasing because of a rise in the concentration of carbon monoxide, which is the chief sink for OH (Problem 12).

$$CO + OH \longrightarrow CO_2 + H \tag{14}$$

Finally, there exists the possibility of a positive feedback cycle for methane. The Arctic tundra regions of Canada and the U.S.S.R. contain vast deposits of methane in the form of a frozen clathrate hydrate having the approximate composition $CH_4.6H_2O$. The methane is physically imprisoned in the framework of the ice structure, and would be released if the temperature rose sufficiently to melt the permafrost (Problem 13).

$$CH_4.6H_2O(s) \longrightarrow CH_4(g) + 6H_2O(l) \tag{15}$$

Nitrous oxide, N_2O: Nitrous oxide, structure N=N=O, has a dipole moment and absorbs infrared radiation effectively. Its present tropospheric concentration of ca. 300 ppb is increasing at the rate of about 0.2% per year. N_2O has no known sinks in the troposphere; it eventually diffuses into the stratosphere where it either decomposes photochemically or reacts with excited state oxygen atoms (Chapter 3). Biological denitrification remains the chief source of atmospheric N_2O, but there is also a contribution from combustion (Chapter 3). The latter probably explains why the concentration is increasing[26].

Chlorofluorocarbons, CFC's: These substances are also of concern in connection with the destruction of stratospheric ozone (Chapter 2). Like N_2O, they have no tropospheric sinks, but are infrared absorbers[27]. Up to 1984, the tropospheric concentrations of three of the major commercial

[25] F.S. Rowland, N.R.P. Harris, and D.R. Blake, "Methane in cities," *Nature,* **1990,** 347.

[26] A recent postulate is that the production of Nylon is a hitherto unrecognized source of nitrous oxide: M.H. Thiemans and W.C. Trogler, *Science,* **1991,** 251, 932–934.

[27] D.A. Fisher *et al,* "Model calculations of the relative effects of CFC's and their replacements on global warming," *Nature,* **1990,** 344, 513–516.

CFC's ($CFCl_3$, CF_2Cl_2, and CHF_2Cl) were each growing at an annual rate of $\approx$ 6%. The 1987 "Montreal Protocol" calls for 50% reductions in the emissions of these compounds before the year 2000, and several countries have now called for a complete ban on those compounds thought to pose the greatest threat to stratospheric ozone. One of the issues in the search for CFC substitutes is that the replacement compounds should not pose a problem as long-lived greenhouse gases. Totally fluorinated compounds have been ruled out on this account, even though they are both non-toxic and pose no threat to stratospheric ozone.

Ozone and carbon monoxide: These substances share the characteristics of very low concentration in the unpolluted troposphere, but much larger concentrations when the atmosphere is polluted (Chapter 3). Ozone is increasingly recognized as one of the more active greenhouse gases.

1.5.2 Climate change

Increased concentrations of greenhouse gases may be anticipated to lead to an increase in the temperature of the troposphere. Data from Antarctica have shown an extremely close correlation between the local temperature and the atmospheric concentration of CO_2 over the past 160,000 years[28]. Present CO_2 levels are now higher than at any time during that period[29].

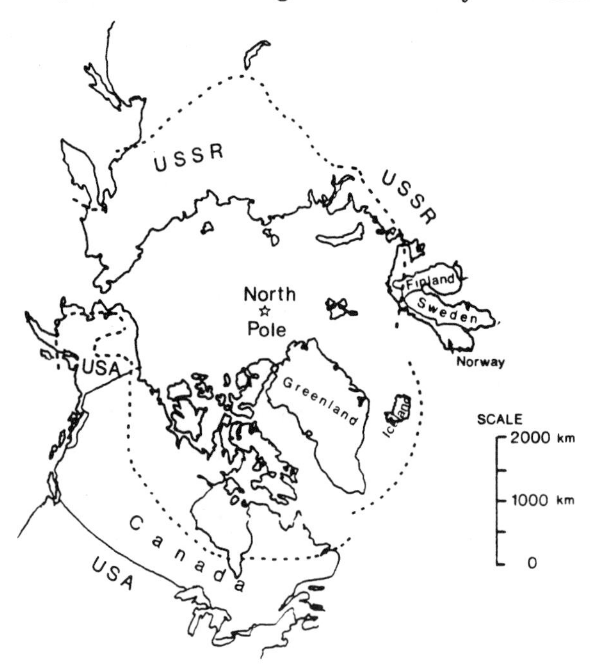

Figure 1.7: Approximate limit of tundra in Canada, Alaska, and USSR.

---

[28] S.H. Schneider, "The changing climate," *Sci Am,* **1989**, 261, 70–79.

[29] B. Hileman, "Global warming," *Chem Eng News,* March 13, **1989**, 25–44.

Until recently, CO_2 has gained almost all the attention in terms of its increased atmospheric concentration and the implications for global warming. However, recent information[30] suggests that the other trace greenhouse gases (methane, ozone, nitrous oxide, CFC's) will likely have a combined effect comparable with that of CO_2: CO_2, 50%; CH_4, 20%; CFC's, 20%; N_2O, 5%. The following factors must all be considered in order to assess the effect of each of these greenhouse gases: absolute levels of emission; efficiency of radiation trapping; and atmospheric lifetime. CFC's contribute almost half as much to greenhouse warming as CO_2, even though they are many orders of magnitude less abundant, because each tonne of CFC emitted is over 6,000 times as effective at trapping infrared radiation as a tonne of CO_2.

News media coverage of the greenhouse effect has, understandably, concentrated on the more sensational of the predictions of a climate change caused by atmospheric warming: melting of the polar ice-caps and consequent inundation of major coastal cities, desertification and massive changes in agriculture in the temperate zones, and wholescale extinctions of species. It must be remembered that such predictions are based on very complex computer models which attempt to extrapolate the behaviour of the climate into the future. Assumptions must be made as to the strengths of all sources and sinks for the greenhouse gases, as to the patterns of air circulation and how they might change with the changing climate, and as to possible changes in the proportions of radiation reflected from or trapped by clouds. Such models must account for mixing of the air masses with latitude, with longitude, and vertically[31].

During the decade of the 1980's, predictions of the magnitude of the global warming to be expected over the next 50–100 years have ranged from 0.5 to 10°C. Most of the more recent estimates have been in the range of 1–3°C. For example, Dr. Michael McElroy of Harvard University has predicted a rise in atmospheric temperatures of about 0.25°C by the year 2010[32]. Over the next century, the atmospheric concentration of CO_2 could rise to between 600 and 1000 ppm, depending on rates of fossil fuel use, but only 30% of the added load of atmospheric CO_2 would be taken up by the oceans.

---

[30]R.A. Houghton, "The global effects of tropical deforestation," *Environ Sci Technol,* **1990,** 24, 414–422. D. Etkin, CO_2 Climate report, Environment Canada report 90–01, p. 5.

[31]For discussion of the relationship between climate modelling and public policy, see P. Rogers, "Climate change and global warming," *Environ Sci Technol,* **1990,** 24, 428–430; S.H. Schneider, "The greenhouse effect: science and policy," *Science,* **1989**, 243, 771–781; S.H. Schnieder, "The global warming debate: science or fallacies?" *Environ Sci Technol,* **1990,** 24, 432–435.

A result of increased infrared trapping in the troposphere would be a lower infrared flux in the stratosphere, and consequent stratospheric cooling: R.J. Cicerone, "Greenhouse cooling up high," *Nature,* **1990,** 344, 104–105.

[32]M. McElroy, "The challenge of global change," *New Scientist,* July 28, 1988, 34–36.

To place these temperature changes in perspective, the difference in global temperatures between today and the last ice age is only some 4–5°C, the same order of magnitude as is under discussion. It is evident that if a drop in temperature of this magnitude can have such dramatic effects on climate, so would a corresponding increase. The major changes seem likely to be these:

- less rainfall in the temperate zones (American Midwest, Canadian Prairies, Russian steppes). These are some of the current "bread-baskets" of the world. Australia is predicted to be relatively unaffected.
- extension of agriculture to higher latitudes.
- increased rainfall in the drier regions of Africa, northern India, and the U.S. Southwest.
- development of a tropical climate in the southern U.S.

Although almost all commentators seem agreed that future tropospheric warming is inevitable, the issue of whether this trend has already begun is still hotly debated. Certainty requires the benefit of hindsight because the weather is naturally so variable. The information upon which these conclusions are based is typified by Figure 1.8[33].

The literature is unhelpful on the issue of whether greenhouse warming is already under way. Three studies have shown that 1990 was an exceptionally warm year, following 1980, 1981, 1983, and 1987 as the four warmest years in North America over the past century[34] However, two retrospective studies of U.S. meteorological records showed no conclusive evidence of global warming[35], and other authors have questioned whether a trend will become evident for several decades[36]. This lack of consensus has made it more difficult to obtain international agreement on limiting emissions of greenhouse gases—especially CO_2—than has been the case with CFC's (Montreal Protocol: Chapter 2).

Finally, a somewhat different perspective on historical climatic data has been provided by Kuo *et al*,[37] who have used statistical methods to show the correlation between $p(CO_2)$ and global temperatures. The two quantities are indeed correlated, but changes in $p(CO_2)$ lag behind the temperature

---

[33] R.A. Kerr, "Global warming continues in 1989," *Science*, **1990**, 247, 521.

[34] R.A. Kerr, "Global temperature hits record again," *Science*, **1991**, 251, 274.

[35] K. Hanson, T.R. Karl, and G.A. Maul, "Are atmospheric greenhouse effects apparent in the climatic record of the contiguous United States (1895–1987)?," *Geophys Res Letters*, **1989**, 16, 49–52. R.W. Spencer and J.R. Christy, "Precise monitoring of global temperature trends from satellites," *Science*, **1990**, 247, 1558–1562.

[36] T.R. Karl, R.R. Heim, and R.G. Quayle, "The greenhouse effect in central North America: if not now, when?," *Science*, **1991**, 251, 1058-1061

[37] C. Kuo, C. Lindberg, and D.J. Thomson, "Coherence established between atmospheric carbon dioxide and global temperature," *Nature*, **1990**, 343, 709–713; for commentary on this paper, see 696–697 in the same issue.

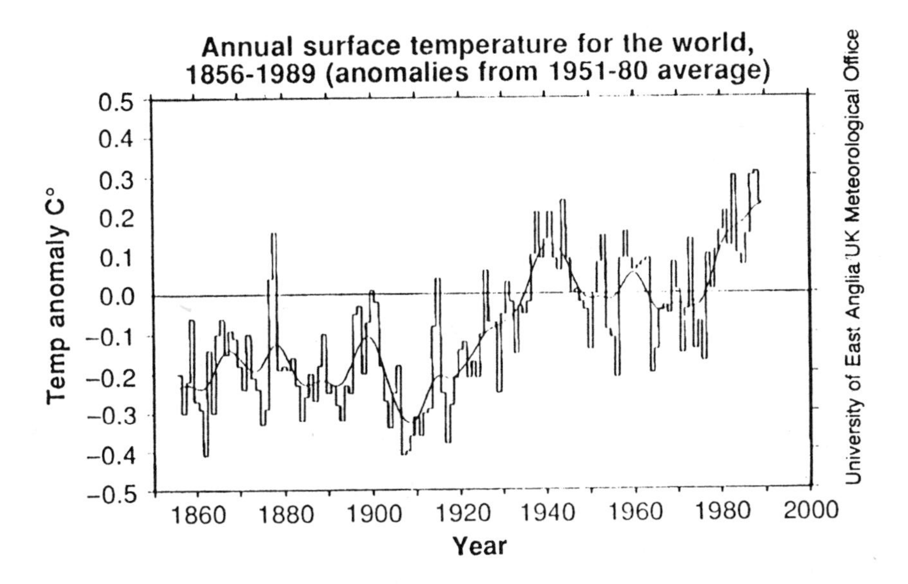

Figure 1.8: Trends in global average temperatures. Redrawn from Reference 28.

change by ca. 5 months. This finding raises the possibility that the temperature change is responsible for changes in $p(CO_2)$, not the other way round. A possible explanation, if this trend is proved correct, would be that natural climatic variability alters the temperature of the oceans, leading to release of CO_2 from the warmer ocean (for explanation, see Chapter 5, Section 1).

1.5.3 Climate change on a geological timescale

Assuming no catastrophe originating on Earth, the ultimate fate of life on this planet is bound up with the natural evolution of the Sun. Over time, the luminosity of the Sun is increasing slowly; eventually the Sun is likely to evolve into a red giant, whose outer fringes will completely engulf the Earth. It seems likely that the near constancy (±10°C) of the Earth's surface temperature has been regulated by the balance between photosynthesis and the greenhouse effect. Over geological time, the increased luminosity of the Sun (warming) has been compensated by increased rates of photosynthesis with the greater solar flux. Since photosynthesis removes CO_2 from the atmosphere, this represents a cooling effect.

Although society's short term concerns involve an increase in $p(CO_2)$, the long term trend in $p(CO_2)$ has been downwards. Eventually, the biosphere will reach a point where $p(CO_2)$ falls so much that the efficiency of photosynthesis is compromised ($\approx$ 150 ppm of CO_2?) and there will be no counterbalance to the Sun's increased luminosity. This point is, however, many million years in the future[38].

1.6 Atmospheres around the other planets

Except for Mercury, all the planets in the solar system have atmospheres, as do some off the more massive satellites such as Saturn's largest moon, Titan[39].

Solar system data.

Body	p(total), atm[a]	Predominant gases (%)
Venus	90	CO_2 (97) N_2 (3)
Earth	1	N_2 (78) O_2 (21) Ar (1)
Mars	0.006	CO_2 (95) N_2 (3) Ar (2)
Jupiter	(Note b)	H_2 (89) He (11)
Saturn	(Note b)	H_2 (94) He (6)
Uranus	5×10^6	Note c
Neptune	8×10^6	Note c
Pluto	10^{-4} (?)	CH_4 and/or N_2[40]
Titan	1.6	N_2(≈80) Ar(≈15) CH_4 (≈5)

[a] i.e., 101.3 kPa

[b] No true surface; the gases probably become increasingly dense, and ultimately liquid or metallic.

[c] Not known; possibly H_2 and He in the same 89:11 ratio found in the Sun and on Jupiter.

Satellite flybys over the past decade have increased our knowledge of these planets to the point that their atmospheric chemistry and geochemistry are becoming understood. The chief difference between Earth's atmosphere and those of Venus, Mars, and Jupiter is the effect of biological processes upon the composition of the atmosphere (compare Section 1.3). Quite probably the primitive Earth had a high proportion of CO_2 in its atmosphere, like Mars and Venus, our nearest planetary neighbours. The chemistry of life has converted most of our carbon into carbonate rocks—the exoskeletons of marine animals—and produced an oxidizing atmosphere through photosynthesis.

Venus' atmosphere is often described as a "runaway greenhouse effect." The surface temperature is over 700 K, explained by closer proximity to the Sun together with highly efficient infrared trapping by the 85 atm of CO_2. The surface of the planet is perpetually covered by clouds of concentrated

[38]Reference 5, Chapter 9.

[39]Data from Reference 5, Chapters 1 and 8.

sulfuric acid, reflection from which explains the brilliance of the night-time Venus. Because the clouds either reflect or absorb all the sunlight incident upon them, photochemically driven processes are only important above the clouds. The chemistry above the clouds is initiated by photodissociation of CO_2.

$$\text{(4)} \qquad CO_2(g) \xrightarrow{h\nu,\ \lambda < 200\,\text{nm}} CO(g) + O(g)$$

Various reaction pathways recycle these products back to CO_2. Below the clouds, the reactions are thermal. Since the surface temperature of Venus is above 700 K, no liquid water is present. In addition, many rocks will be present as oxides rather than as carbonates (Problem 14).

Whereas Venus is characterized by a hot, dense atmosphere, Mars is cold (surface temperature 125 K in winter to 220 K in summer) and its atmosphere is tenuous: $p(\text{total}) = 0.006$ atm. Permanent polar caps of water ice are supplemented in the winter-time by solid CO_2 (Problem 15).

As on Venus, the predominant photochemical process is photocleavage of CO_2. The products are cycled back to CO_2 by reactions very similar to those occurring in the Earth's atmosphere. As on Earth, OH is the active oxidant for carbon monoxide (see Equation 14, above); it is formed by direct photolysis of H_2O (Equation 3), or by the reaction of excited oxygen atoms with H_2O, see Chapter 3.

The chemistry of the Jovian atmosphere is quite different. At very short wavelengths ($\lambda < 100$ nm), hydrogen photodissociates into atoms, and methane is photolyzed below 120 nm. One reaction channel converts CH_4 to free methylene, CH_2, which can insert into the C–H bond of methane, yielding ethane and, ultimately, larger alkanes.

$$\text{(16)} \qquad CH_4 \longrightarrow CH_2 + H_2 \text{ (or 2H)}$$

$$\text{(17)} \qquad CH_2 + CH_4 \longrightarrow CH_3\text{–}CH_3$$

The small amounts of ammonia that are present are susceptible to photolysis, and this reaction leads to the production of a small steady state concentration of hydrazine, N_2H_4.

$$\text{(18)} \qquad NH_3 \xrightarrow{h\nu} NH_2 + H$$

$$\text{(19)} \qquad 2NH_2 \xrightarrow{M} NH_2\text{–}NH_2$$

1.7 Problems

1. (a) Use the diagram below to determine an approximate relationship between the atmospheric pressure and altitude. Comment on why there are deviations from this relationship, especially at altitudes > 100 km.

[40] *New Scientist*, November 19, 1988, 29.

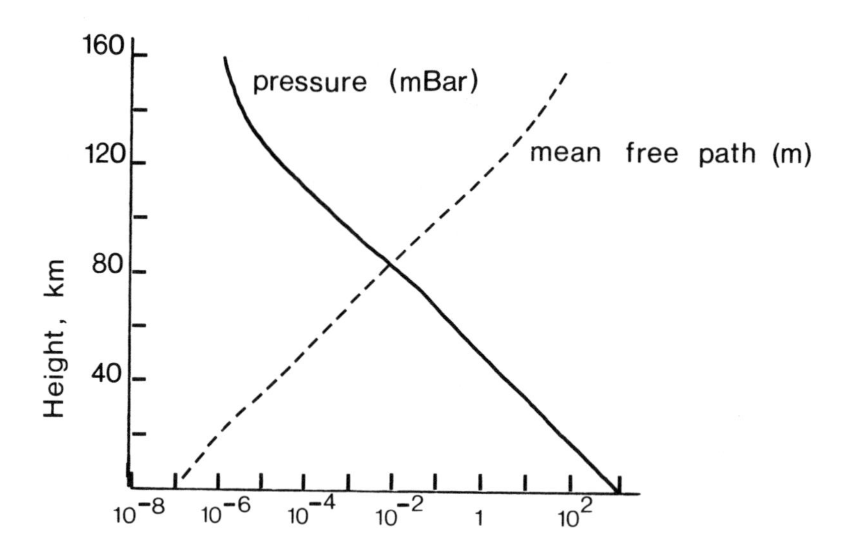

Figure 1.9: Pressure vs. altitude in the Earth's atmosphere. Redrawn from Reference 5

(b) Calculate, using the kinetic-molecular theory of gases, the mean free path of the molecules in the air at 20°C and 1.00 atm, assuming an average molecular diameter of 0.17 nm.

(c) Repeat the calculation of part (b) for the outer atmosphere ($p = 10^{-10}$ atm, T = 1500 K), and assuming that most of the species are atomic, with atomic diameters 75 pm.

2. The mass of the atmosphere is about 5×10^{15} tonnes. COS is present as a trace gas at concentration 0.51 ppb; its major source is from the oceans (6×10^8 kg yr^{-1}).

   (a) Estimate the residence time of COS in the atmosphere

   (b) Calculate the number of molecules of COS in 1 L of air at 1.00 atm pressure and the total mass of COS in the atmosphere.

3. Atmospheric argon consists mainly of the isotope ^{40}Ar, which is formed by radioactive decay of ^{40}K in the Earth's crust. There are no known sinks for argon. Calculate the average rate of emission of argon into the atmosphere over the lifetime of the Earth.

4. (a) On an afternoon when the temperature is 25°C, the relative humidity is 70%. Do you expect that dew will form during the night if the temperature falls to 5°C?

(b) It is −15°C outside and the partial pressure of water vapour is 0.11 mm of mercury. Would you expect frost to be forming, or snow to be evaporating?

(c) Under the conditions of part (a) at 5°C, calculate ΔG for the process

$$H_2O(g) \longrightarrow H_2O(l)$$

5. Use the data in the text to estimate for nitrous oxide, N_2O:

(a) the rate of inflow of N_2O into the atmosphere today, in the units tonnes per year;

(b) the excess of inflow over outflow of N_2O into the atmosphere per year for the past two decades.

6. (a) Wien's Law is used in astronomy to relate the surface temperature of an astronomical body to the wavelength at which radiation intensity is at a maximum:

$$\lambda_{max}(\mathrm{m})\,\mathrm{T(K)} = 2.9 \times 10^{-3}$$

Estimate the wavelength of maximum radiation emission for the Sun (T = 6000 K) and the Earth (T = 288 K).

(b) How much energy must be absorbed by the atmosphere per unit volume to raise the temperature by 1°C? Assume $p = 1.00$ atm and T = 288 K. Express your answer in terms of the number of photons absorbed per liter, both for solar photons and for the Earth's emission, using the average wavelengths calculated in part (a).

7. (a) Verify, using the kinetic-molecular theory of gases, that the escape velocity from the Earth's atmosphere is 11.2 km s^{-1}.

(b) Calculate the average velocity of hydrogen, helium and nitrogen atoms in the outer thermosphere with T = 1500 K.

(c) The Maxwell-Boltzmann speed distribution function allows one to estimate the fraction of molecules ΔN having velocities within the velocity interval Δv:

$$\frac{\Delta \mathrm{N}}{\mathrm{N}} = 4\pi\left[\frac{\mathrm{m}}{2\pi \mathrm{kT}}\right]^{3/2} \bullet e^{-\mathrm{mv}^2/(2\mathrm{kT})} \bullet \mathrm{v}^2\,\Delta\mathrm{v}.$$

Calculate the fraction of hydrogen atoms at 1500 K having velocities in excess of the escape velocity.

8. Estimate the mass of hydrogen lost from the atmosphere each year, at a rate of 3×10^8 atoms per square centimeter per second.

9. (a) Verify the relationship that for photons:

$$\Delta \mathrm{E}\ (\mathrm{kJ\ mol^{-1}}) = 1.2 \times 10^5/\lambda(\mathrm{nm})$$

(b) Calculate the longest wavelength at which each of the following reactions will occur:

$$N_2 \longrightarrow 2N \quad \Delta H^\circ = 946 \text{ kJ mol}^{-1}$$
$$N \longrightarrow N^+ + e^- \quad \Delta H^\circ = 1400 \text{ kJ mol}^{-1}$$
$$O_2 \longrightarrow O_2^+ + e^- \quad \Delta H^\circ = 1160 \text{ kJ mol}^{-1}$$

(c) Calculate the energy associated with the absorption of infrared radiation by greenhouse gases: CO_2 at 2250/cm; H_2O at 1800/cm. What is the physical process which occurs when this energy is absorbed?

10. At an altitude of 170 km, where T = 1100 K, 80% of all oxygen molecules are dissociated into atoms. Assume that p(total) is 3×10^{-10} atm, and that O_2 and N_2 dissociate in equal proportions.

(a) Estimate ΔG for the reaction $2O(g) \longrightarrow O_2(g)$ under these conditions.

(b) Why does the concentration of O(g) stay so high when ΔG is so highly negative?

11. Model, for a period of 20 years, the predicted effect of a sudden, one-time injection of 100 ppm of CO_2 into the atmosphere. Use the following data[4]: CO_2 in atmosphere, 1.4×10^{16} mol; CO_2 in ocean surface waters, 6.1×10^{16} mol; CO_2 in deep ocean waters; 7.5×10^{17} mol; rate constant for transfer of CO_2 from atmosphere to surface waters, 0.54 yr^{-1}; rate constant for transfer of CO_2 from surface waters to deep ocean, 0.02 yr^{-1}; rate constant for transfer of CO_2 from surface waters to atmosphere, 0.10 yr^{-1}.

12. The diurnally and seasonally averaged concentration of OH in the troposphere is 5×10^5 molec cm^{-3}. The two major sinks for OH are these reactions, which consume respectively 70% and 30% of all hydroxyl radicals:

$$OH + CO \longrightarrow CO_2 + H \quad k = 1.5 \times 10^{-13}\ cm^3 molec^{-1} s^{-1}$$
$$OH + CH_4 \longrightarrow CH_3 + H_2O \quad k = 8.0 \times 10^{-15}\ cm^3 molec^{-1} s^{-1}$$

Both rate constants are given at 300 K. The concentration of CH_4 in the atmosphere is currently $\approx$ 1700 ppb.

(a) Estimate the diurnally and seasonally averaged concentration of CO.

(b) Estimate the reduction in the average OH concentration at the new steady state if the concentration of CO were to double.

13. (a) Emissions of methane to the atmosphere are given in the table below, and the current atmospheric concentration is 1.7 ppm. Estimate its residence time.

Sources of atmospheric methane[41]
(millions of tonnes per year)

Wetlands	150
Oceans, lakes etc	35
Cattle	120
Rice paddies	95
Other sources	150

(b) Estimates of the total reserves of methane hydrate in the permafrost and below the ocean floors range up to 10^{14} tonnes. Suppose that 1% of this material were to melt per year, what would be the increase in the amount of methane in the atmosphere per year, assuming no additional sinks for the methane? Give your answer in ppm yr^{-1}.

14. Determine by calculation whether you expect calcium and copper to exist as carbonates or as oxides in the rocks on Venus. Use the following thermodynamic data:

Substance	ΔH_f°, kJ mol^{-1}	S°, J mol^{-1} K^{-1}
$CaCO_3(s)$	-1206.9	92.9
$CaO(s)$	- 635.1	39.7
$CO_2(g)$	- 393.5	213.6
$CuCO_3.Cu(OH)_2(s)$	-1051.1	186.2
$CuO(s)$	- 157.3	42.6
$H_2O(g)$	- 241.8	188.7

The fraction of water in the Venusian atmosphere is 2×10^{-5}.

15. Estimate the temperature on Mars at which you would expect CO_2 to condense to form polar ice caps of solid CO_2. Use the following data which are appropriate to the ordinary melting point of CO_2, $-57°C$:

Fusion: $\Delta H^\circ = 8.37$ kJ mol^{-1} $\Delta S^\circ = 38.64$ J mol^{-1} K^{-1}
Vaporization: $\Delta H^\circ = 16.24$ kJ mol^{-1} $\Delta S^\circ = 75.03$ J mol^{-1} K^{-1}

[41] M.A.K. Khalil and R.A. Rasmussen, "Sources, sinks, and seasonal cycles of atmospheric methane," *J Geophys Res,* **1983,** 88, 5131.

Chapter 2

Stratospheric Ozone

Introduction

As was seen in the last chapter, ozone plays a vitally important role in the atmosphere as the principal absorber of ultraviolet radiation in the range 240–320 nm, thus preventing most of that highly energetic radiation from reaching the Earth's surface. The focus of this chapter is the series of reactions by which ozone is formed and destroyed in the stratosphere. The possible depletion of the "ozone layer" through atmospheric pollution is a matter of great current concern; we shall also examine the chemical reactions involved in ozone depletion and its possible consequences.

2.1 The ozone layer

The term "ozone layer" is misleading, since it implies a distinct region of the atmosphere in which ozone is a major atmospheric constituent. In reality, ozone is found in both the troposphere and the stratosphere; it is never more than a trace constituent, albeit a very important one. The concentration of ozone is at a maximum in the stratosphere, although the actual value depends on both the latitude and the season[1]. Even at this maximum concentration, the absolute amount of ozone is only about 10 ppm. Therefore, even though ozone is so important to the chemistry of the stratosphere, it is a mistake to think that ozone is the major component of this part of the atmosphere. It isn't; the major species in terms of numbers of molecules are still nitrogen (78%), oxygen (21%), and argon (1%).

A rather graphic way of thinking about how little ozone there really is in the "ozone layer" is to imagine compressing all the atmospheric ozone into a single layer at STP. That layer would be just 3 mm thick. The way in which ozone is distributed through the atmosphere from ground level to about 100 km is shown in Figure 2.1, which was obtained using the data of Problem 1.

[1]Problem 13 at the end of this chapter, which is more difficult than the others, involves modelling the distribution of the rate of photolysis of oxygen to give ozone as a function of altitude.

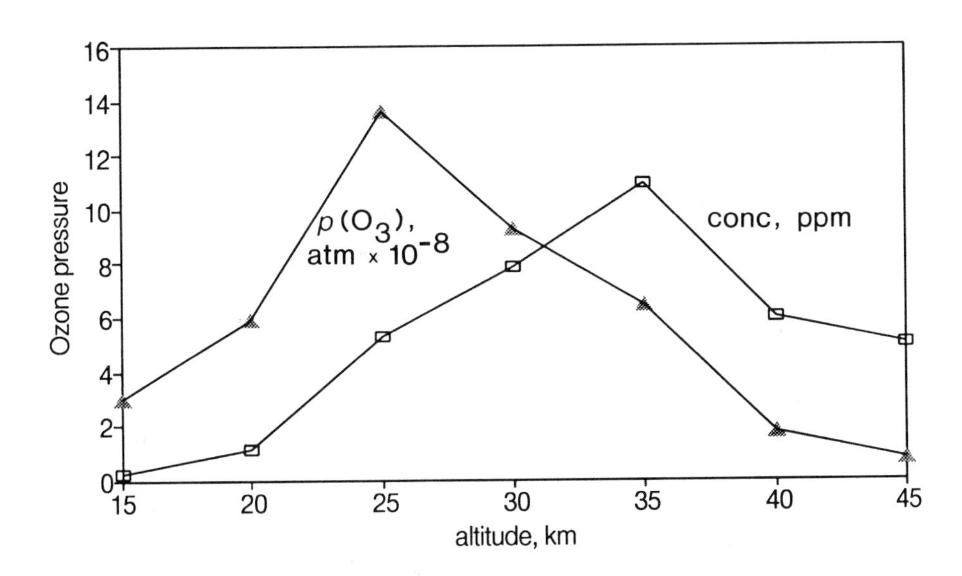

Figure 2.1: Variation of the concentration of ozone with altitude: upper scale $c(O_3)$ in molecules per cm^3; lower scale $c(O_3)$ in ppm.

2.2 Formation and destruction of ozone

Under the influence of sunlight, oxygen (O_2) is continually being changed into ozone (O_3) and ozone is likewise converted back to ordinary oxygen. Each day 350,000 tonnes (3.5×10^8 kg) of ozone are made—and destroyed—in the atmosphere. It would be completely incorrect to think that the natural process is $O_2 \rightarrow O_3$, and that ozone is destroyed because of air pollution. If that were so, all the O_2 in the atmosphere would long since have been changed to O_3! What certain air pollutants can do is to speed up the rate of loss of ozone so that its **steady state concentration** declines. This discussion will be deferred to Section 2.3.

The principal reactions involved in the formation and destruction of ozone are given in Reactions 1–4 below. Taken together, they are known as the Chapman mechanism.

	Reaction	$\Delta H°$, kJ mol^{-1}
(1)	$O_2 \xrightarrow{h\nu,\ \lambda < 240\ \text{nm}} 2O$	495 – E(photon)
(2)	$O + O_2 \xrightarrow{M} O_3$	–105
(3)	$O_3 \xrightarrow{h\nu,\ \lambda < 325\ \text{nm}} O_2 + O$	105 – E′(photon)
(4)	$O + O_3 \longrightarrow 2O_2$	–389

Reactions 1 and 2 represent ozone formation, while 3 and 4 are the reverse process. Notice that oxygen atoms are instrumental in both forming ozone (Equation 2) and destroying it (Equation 4). Oxygen atoms are very reactive, and consequently have a short lifetime in the stratosphere (Problem 2); this means that all four of these reactions come to a halt at sunset, and so the concentration of ozone at night is essentially the same as at the end of the day.

Sunlight drives both ozone formation and removal, but different wavelength ranges are involved. These absorptions and the consequent reactions are responsible for shielding the Earth's surface from two different parts of the spectrum, both of them highly energetic. Ordinary oxygen is the main absorber near 200 nm, while ozone is most important in the range 230–320 nm. The shielding property of ozone is precisely due to its ability to absorb radiation of these wavelengths, which simultaneously converts it back to O_2.

The energetics of Reactions 1–4 have been given as standard enthalpies instead of free energies because a photon is best considered as a source of internal energy (ΔE), rather than free energy. The enthalpy change is closely approximated by ΔE: $\Delta H° = \Delta E + \Delta n\,RT$. When the photon energies are included, Reactions 1 through 4 are all energy releasing. Thus sunlight brings about both ozone formation and destruction, and as a result photon energy is converted to heat (Problem 3). This explains the relatively high temperature in the mid-stratosphere, where ozone formation and destruction are most active.

Because the formation of ozone requires solar energy, the steady state concentration of ozone does not correspond to equilibrium (Problem 4). Recall that the condition for equilibrium is $\Delta G = 0$. The input of solar energy maintains the system $(3O_2)/(2O_3)$ far from equilibrium, "pumping" O_2 uphill energetically until the rate of ozone production is balanced at the steady state by its rate of reversion to O_2.

The cleavage of ozone, Reaction 3, is also driven photochemically. The products of this reaction may be either both formed in their ground states (Equation 3) or both in their excited states (Equation 3a); refer to Problem 5c. Since electron spin must be conserved, it is not possible for one product to be formed in its ground state and the other in its excited state.

$$(3) \qquad O_3 \xrightarrow{h\nu,\, \lambda < 325\text{ nm}} O_2 + O \qquad 105 - E' \text{ (photon)}$$

$$(3a) \qquad O_3 \xrightarrow{h\nu,\, \lambda < 325\text{ nm}} O_2^* + O^* \qquad 383 - E' \text{ (photon)}$$

The difference in energy is accounted for by the excitation energies of O_2 (90 kJ mol^{-1}) and O (188 kJ mol^{-1}). Equation 3a will be encountered again in tropospheric chemistry (Chapter 3), where it is of critical importance in the generation of the hydroxyl radical, OH.

Until the mid-1960's the Chapman mechanism was thought to be a complete description of the chemistry of the O_2/O_3 system in the stratosphere. Research since then has greatly increased our understanding of the chemistry of the stratosphere. With improved estimates of the rates of Equation 1–4, the steady state concentration of ozone was calculated to be two to three times larger than what was observed experimentally. This meant that there must be other natural sinks, which had not then been discovered, by which stratospheric ozone is destroyed. These additional mechanisms were shown to be catalytic processes, each of which is the propagation cycle of a free radical chain reaction. They are shown in generalized format as Reactions A and B.

(A) $X + O_3 \longrightarrow XO + O_2$
(B) $XO + O \longrightarrow X + O_2$

Adding A and B together gives Equation 4:

(4) $O + O_3 \longrightarrow 2O_2$

Therefore the sequence A, B is another way of carrying out Reaction 4, and increases the rate of destruction of ozone.

Four separate catalytic cycles have since been discovered, each with a different catalyst X[2]. X can be a chlorine atom (Cl), or any of NO, OH, or H, but in every case X is an odd electron species. One representative chain sequence is given below.

(5) $Cl + O_3 \longrightarrow ClO + O_2$
(6) $ClO + O \longrightarrow Cl + O_2$

Notice that the chlorine atom consumed in Reaction 5 is regenerated in Reaction 6, allowing the cycle to be repeated over and over. Eventually termination reactions (e.g., Equation 7) remove the "odd electron chlorine" species.

(7) $Cl + HO_2 \longrightarrow HCl + O_2$

These catalytic cycles have now been studied quite thoroughly. For example, in the case of Equation 5 and 6, the concentrations of the reactive intermediates Cl and ClO have been measured experimentally at several different altitudes. In addition, the rate constants for both reactions have been measured in the laboratory at several temperatures. Table 2.1 gives activation energies for the relevant processes.

Under certain conditions[3] these two-step catalytic sequences can be as fast or faster than the uncatalyzed reaction (Problems 6–9). For example, at ≈30 km, the relative rates of the possible reactions for decomposing

[2] R.M. Baum, "Stratospheric science undergoing change," *Chem Eng News*, September 13, **1982**, 21.

[3] Remember that we cannot conclude that the two step cycles will be faster just because of the smaller activation energies: the rate of a chemical reaction depends on the concentrations of the reactants as well as the magnitude of the rate constant. An interesting illustration of this point is that at 30 km altitude, all four catalytic cycles have B as the rate limiting reaction, even though Reaction A has the larger activation energy in each case.

ozone are as follows: NO/NO_2 cycle > uncatalyzed reaction ≈Cl/ClO cycle > OH/HO_2 cycle >> H/OH cycle. The H/OH cycle is not important at 30 km because Reaction 8 is so fast.

$$H + O_2 \xrightarrow{M} HO_2 \tag{8}$$

The H/OH cycle becomes more significant at higher altitude, where the concentrations of O_2 and M are smaller (p(total) is less), and the reaction with O_3 can compete more effectively (Problem 10).

Table 2.1: Activation energies (kJ mol^{-1}) for individual steps of catalytic reactions for ozone destruction[4]. Uncatalyzed reaction, Equation 4: 18.4 kJ mol^{-1}.

	Activation Energies	
X	$X + O_3 \rightarrow XO + O_2$	$XO + O \rightarrow X + O_2$
Cl	2.1	1.1
NO	13.1	≈0
H	3.9	≈0
OH	7.8	≈0

We will digress briefly to discuss how researchers deduce the relative importance of the various sinks for stratospheric ozone. There are three phases to an investigation of this kind.

1. The concentrations of all reactant species must be obtained experimentally as a function of altitude. This is done by sending analytical instruments aloft by means of rockets or balloons, and is very expensive.
2. Rate constants, and their temperature dependence, must be measured in the laboratory for all relevant reactions. Many of these reactions had not been studied before the mid-1970's. The issue of which reactions are relevant is a very complex one, see below.
3. Computer simulations are carried out to try to reproduce the behaviour of the atmosphere. These simulations involve numerical integration of the changes of concentration of all species with time, and make use of the solar photon flux over all wavelengths, the concentrations of all species and the rate constants of all reactions. Very powerful computers are needed for atmospheric modelling because of the complexity of the kinetic models. The simulations can only be as good as the data, and need to be updated as new relevant reactions are discovered, and improved measurements of concentrations and rate constants are made.

[4]From R.P. Wayne, *Chemistry of Atmospheres,* Oxford University Press, Oxford, England, **1985,** 123.

2.2.1 Additional reactions in stratospheric chemistry

The reactions shown in Table 2.1 do not account completely for the chemistry of the stratosphere. Complicating factors include:

1. temporary reservoirs of active species
2. interaction between catalytic cycles
3. null (do nothing) cycles
4. initiation and termination reactions

Temporary reservoirs

Catalytically active species such as NO_X and ClO_X can be converted into substances which reduce their instantaneous concentrations, but from which they can be regenerated. Examples include the following. For NO_2:

$$NO_2 + OH \xrightarrow{M} HNO_3 \tag{9}$$

$HNO_3(g)$ acts as a temporary reservoir from which NO_2 can be regenerated by photolysis. At any moment, as much as half of the "active nitrogen" of the stratosphere may be temporarily inactivated in the form of HNO_3. For Cl:

$$Cl + CH_4 \longrightarrow HCl + CH_3 \tag{10}$$

Here HCl is the temporary reservoir; Cl can be regenerated by the reaction of HCl with OH.

$$HCl + OH \longrightarrow Cl + H_2O \tag{11}$$

Interaction between cycles

The four separate catalytic cycles represented by Reactions (A) and (B), on page 31, all involve a pair of catalysts X and XO. In principle, X from one cycle can react with XO from another cycle. For example, Reaction 12 affects the concentration of reactants in both the NO/NO_2 and the OH/HO_2 cycle.

$$NO + HO_2 \longrightarrow NO_2 + OH \tag{12}$$

Null cycles

Such a cycle results in no net chemical change; generally, sunlight is converted to kinetic energy. An example is:

$$\begin{aligned} NO_2 &\xrightarrow{h\nu,\ \lambda < 400\ \text{nm}} NO + O \\ O + O_2 &\xrightarrow{\quad M \quad} O_3 \\ O_3 + NO &\longrightarrow NO_2 + O_2 \end{aligned}$$

When these equations are added together, all the chemical species drop out on both sides. The previous three reactions may alternatively be considered as the photochemical "pseudo-equilibrium" 13.

$$NO_2 + O_2 \underset{\text{thermal}}{\overset{h\nu}{\rightleftharpoons}} NO + O_3 \qquad (13)$$

Reaction 13 affects the concentrations of reactants in both the NO/NO_2 cycle and the OH/HO_2 cycle.

Initiation and termination reactions

The rates of the chain reactions referred to in Table 2.1 depend on the rates of initiation and termination reactions. These processes determine how many chains are initiated, and how many catalytic cycles are propagated on the average before the cycle is terminated. Initiation reactions generate free radicals from non-radical precursors. They are generally photochemical, hence the rate of initiation depends upon the intensity of sunlight. Examples are given by Equations 14–19.

$$HNO_3 \xrightarrow{h\nu} NO_2 + OH \qquad (14)$$

$$NO_2 \xrightarrow{h\nu} NO + O \qquad (15)$$

$$O_3 \xrightarrow{h\nu} O_2 + O \qquad (16)$$

$$CH_3Cl \xrightarrow{h\nu} CH_3 + Cl \qquad (17)$$

$$O + H_2O \longrightarrow 2OH \qquad (18)$$

$$N_2O + O \longrightarrow 2NO \qquad (19)$$

Reactions 18 and 19 are not photochemical, but the oxygen atom needed as a reactant is formed in Reaction 3, a photochemical process.

Termination reactions remove radicals from the system. Generally, such reactions involve the combination of two radical (i.e., odd electron) species. Some examples are given in Reaction 9 and in 20 to 22.

$$2ClO \longrightarrow ClOOCl \qquad (20)$$

$$2HO_2 \longrightarrow H_2O_2 + O_2 \qquad (21)$$

$$NO_2 + Cl \longrightarrow NO_2Cl \qquad (22)$$

As Equation 17 shows, methyl chloride photolyzes to produce chlorine atoms in the stratosphere. The source of CH_3Cl in the atmosphere is not well understood. Postulated origins for CH_3Cl include volcanoes and biological processes involving Cl^- in the oceans. It appears to be a natural atmospheric constituent with a long residence time rather than a pollutant, as shown by its very uniform distribution in the troposphere. Most of the CH_3Cl is oxidized in the troposphere, but a proportion reaches the stratosphere where it is cleaved by radiation having wavelength < 250 nm. The gas N_2O is also primarily of biological origin; it finds its way to the stratosphere because of a lack of tropospheric sinks.

2.3 Chlorofluorocarbons

Chlorofluorocarbons (CFC's)[5] have been manufactured since the 1930's. CFC-12 (CF_2Cl_2)[6] was introduced originally as the operating fluid in refrigerators, displacing the highly toxic and odorous SO_2 and NH_3 which had hitherto been used for this purpose. The requirements for such a fluid are rigorous; it must be a gas at room temperature, but easily compressible to a liquid. Most successful refrigerants have normal (1 atm) boiling points in a narrow range (CFC-12, −30°C; SO_2, −10°C; NH_3, −33°C). The non-toxic, non-flammable CFC's made the refrigerator safe enough to be operated as a domestic appliance in every home, where previously it had been restricted to industrial use.

Other important uses for CFC's are as blowing agents for expanded foams, as propellants for aerosol sprays, and as cleaning solvents for microelectronic components. CFC-11 ($CFCl_3$) is important in the first two of these applications and CFC-113 ($CF_2ClCFCl_2$) in the third. Again, the lack of toxicity, odour, and flammability, and convenient low boiling points make CFC's ideal for these purposes. In 1988, the Canadian usage of CFC's (20,000 tonnes) was divided among foams (44%), refrigerants (33%), solvents (11%), and aerosols (8%). CFC-11 and CFC-12 are manufactured cheaply from carbon tetrachloride (Equation 23).

$$CCl_4 + 2HF \xrightarrow{\text{catalyst}} CF_2Cl_2 + 2HCl \qquad (23)$$

Their combined worldwide production in 1977 was estimated at 700,000 tonnes. Over 600,000 tonnes were estimated to be released to the atmosphere[7]. Thus almost all the production of CFC's eventually ends up being released to the atmosphere.

CFC's were first discovered in the atmosphere in the early 1970's. Worldwide concentrations of CFC-11 at ground level were ca. 50 ppt (parts per trillion, 10^{12}) in 1971 and had risen to ca. 150 ppt by 1979. In 1974, Molina and Rowland proposed[8] that these seemingly innocuous compounds might

---

[5]CFC's are often known by the trade name Freons. Freon is a DuPont registered trademark. DuPont introduced CFC's to the marketplace in a joint venture with General Motors Corp. and subsequently bought out the GM interest. However, CFC's are now made world-wide by many different companies, as the original patents have long since expired.

[6]The "Rule of 90" enables one to determine the composition of a CFC from the "code number" such as CFC-12. Add 90 to the code number; in the case of CFC-12, this gives 102. The three digits one, zero, and two are the numbers of carbon, hydrogen, and fluorine atoms in the molecule. The rest of the atoms must be chlorine. Another example: CFC-141: $141 + 90 = 231$, so this molecule contains 2 carbons, 3 hydrogens, 1 fluorine, and hence 2 chlorines. You cannot tell from the formula $C_2H_3FCl_2$ which isomer is involved, and letters a, b, etc are used to differentiate between them.

[7]Data from *Atmospheric Chemistry,* B.J. Finlayson-Pitts and J.N. Pitts, Wiley-Interscience, New York, 1986, Chapter 15.

[8]An interesting account of these early proposals is given in "Fluorocarbon File," *New Scientist,* October 2, 1975, 7–17.

cause environmental problems through depletion of stratospheric ozone, thus accounting for the undiscovered sink for ozone which was mentioned in the Section 2.2. CFC-11 and CFC-12 are completely inert in the troposphere, but Molina and Rowland suggested that in the stratosphere they would be susceptible to gradual photolysis, e.g., Equation 24.

$$CF_2Cl_2 \xrightarrow{h\nu,\ \lambda < 250\ \text{nm}} CF_2Cl + Cl \tag{24}$$

The chlorine atom thus released was proposed to initiate a free radical chain reaction, destroying ozone.

$$\begin{aligned} Cl + O_3 &\longrightarrow ClO + O_2 \\ ClO + O &\longrightarrow Cl + O_2 \end{aligned}$$

These reactions are simply Equation 5 and 6 of the previous section.

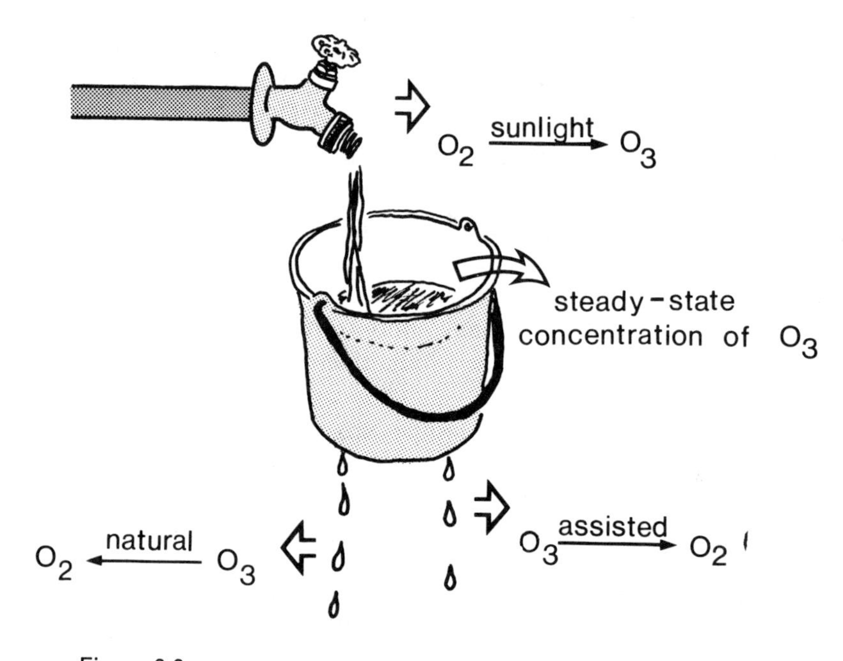

Figure 2.2:

As a point of clarification, it was not known in 1974 that Reactions 5 and 6 occurred naturally in the stratosphere; chlorine atom assisted destruction of ozone was suggested to occur only as a consequence of the release of CFC's to the atmosphere. It is now known that the Cl/ClO chain reaction proceeds even in the unpolluted stratosphere, as also does the NO/NO_2 cycle. The release of CFC's does not introduce a brand-new sink for ozone; rather, it increases the strength of a pre-existing sink. The following analogy might be useful. A bucket is being filled from a faucet at a constant rate; this corresponds to the rate of formation of ozone. The bucket has a hole in it, as a result of which the water drains out; this represents the rates of all natural precesses for ozone removal. The level of water in the bucket at the steady state depends on the rate of inflow and the size of the hole. A new hole is now drilled in the bucket, corresponding to the new sinks for ozone opened up as a result of stratospheric contamination by CFC's; the water escapes more easily, and so the steady state level of water in the bucket is lower than before.

The particular environmental concern about CFC's is that they are predicted to be extremely long-lasting. They are completely unreactive in the troposphere. Diffusion through the troposphere into the stratosphere takes months to years. In the stratosphere, CFC's can break down because they are exposed to radiation $\lambda < 250$ nm. Even here, decomposition is very slow, because O_2 and O_3 absorb this radiation more strongly (Problem 11). The best predictions of the lifetimes of these compounds in the stratosphere range up to and beyond 100 years. Table 2.2 gives some estimated properties of selected commercial CFC's.

Table 2.2: Properties of some chlorinated air pollutants

Compound	(CFC No.)	Lifetime(yrs)	%Increase[a]	Efficiency	$\Delta O_3(\%)$[b]
$CFCl_3$	(11)	70	6	1	2.0
CF_2Cl_2	(12)	110	6	0.86	2.1
CHF_2Cl	(22)	25	> 10	0.05	0.03
$CF_2Cl.CFCl_2$	(113)	90	n.a.[c]	0.8	0.5
$CH_3.CCl_3$	–	< 10	9	0.15	0.5
CCl_4	–	≈ 10	2	1.1	0.6
Other compounds[d]					
N_2O		100			

[a] Percent increase per year in the late 1970's

[b] These values are the calculated reductions in ozone levels at the steady state with release rates at the 1970's levels, calculated component by component. The total percentage loss of ozone would be the sum of all these.

[c] Not available

[d] Greenhouse gases (CO_2 and CH_4) are predicted to cause an increase in the levels of stratospheric ozone. Climatic models predict that warming of the troposphere will be accompanied by cooling of the stratosphere. The ozone depleting reactions in Table 2.1 would therefore be slowed down, especially the direct Reaction 4, which has the largest activation energy.

The ozone depleting potential (ODP) in Table 2.2 is the propensity of the substance to destroy stratospheric ozone, integrated over the life of the compound. Both the total ODP and the time scale of the effect can vary. This is shown in Figure 2.3; the HCFC compounds cited in Figure 2.3 are discussed below in Section 2.6.

The atmospheric concentration of all organochlorine species now stands at ca. 4 ppb, having risen from an estimated < 1 ppb in 1950. The present contributors are[9]:

CH_3Cl (natural)	0.6
$CCl_4 \times 4$	0.6
$CH_3CCl_3 \times 3$	0.5
CFC-11 $\times$ 3	0.8
CFC-12 $\times$ 2	1.0
CFC-113 $\times$ 3	0.2

The key issue for several years was whether the predicted decrease in the levels of stratospheric ozone would be seen in practice. At first, evidence on the point was hard to obtain, because the concentration of ozone varies diurnally and seasonally, and the pattern is not exactly reproducible from year to year. Recent work suggests that some stratospheric ozone loss may now be detectable, the maximum loss at mid-latitudes being seen near 40 km altitude[10].

2.3.1 Polar ozone holes

By the mid-1980's evidence began to accumulate about an "Antarctic ozone hole" which developed in the late winter, with local depletions of stratospheric ozone up to 50%[11]. The Antarctic in winter is a unique location in that the air circulation is entirely circumpolar, with almost no admixture of air from lower latitudes. No ozone is generated during the long dark Antarctic winter while the Sun is below the horizon. In late winter/early spring, when the temperature of the polar stratosphere is at its lowest, crystals of $HNO_3.3H_2O$ (at $T < -70°C$) and of water ice (at $T < -85°C$) are formed. Temporary reservoirs of chlorine such as $ClONO_2$ and HCl are broken down heterogeneously on the surface of these crystals, releasing reactive forms of chlorine[12].

$$HCl + ClONO_2 \longrightarrow Cl_2 + HNO_3 \quad (25)$$

$$H_2O + ClONO_2 \longrightarrow HOCl + HNO_3 \quad (26)$$

Both Cl_2 and HOCl are cleaved by visible or near-UV light after sunrise, releasing chlorine atoms, which vigorously promote the catalytic destruction of ozone by Reactions 5 and 6[13]. Two factors lead to the restoration of

[9]F.S. Rowland, "Stratospheric ozone in the 21st Century," *Environ Sci Technol,* **1991**, 25, 622–628

[10]M.B. McElroy and R.J. Salawitch, "Changing composition of the global stratosphere," *Science,* **1989**, 243, 763–770.

[11]R.S. Stolarski, "The Antarctic ozone hole," *Sci Am,* **1988**, 258, 30.

[12]*Chem Eng News,* January 2, **1989**, 30.

[13]For additional reactions that are important under these conditions, see S.E. Sander, R.R. Friedl, and Y.K. Yung, "Rate of formation of the ClO dimer in the polar stratosphere: implications for ozone loss," *Science,* **1989**, 245, 1095–1098.

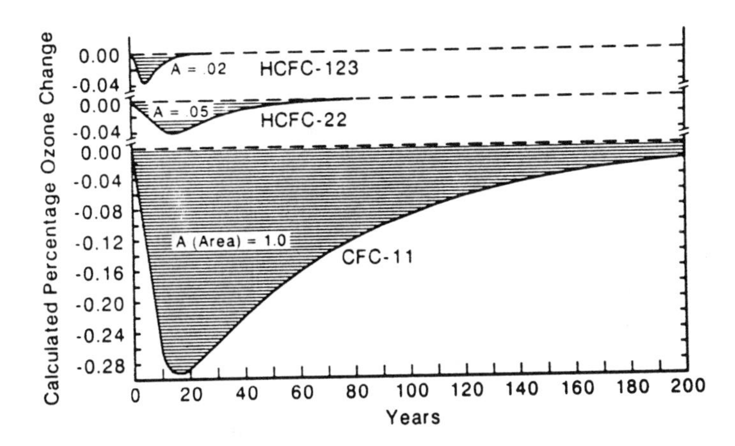

Figure 2.3: The ozone depletion potential (ODP) is a good measure of the relative potential of HCFCs and CFCs to affect stratospheric ozone. Shaded areas for each of the compounds show their cumulative potential to affect ozone over their residence times in the atmosphere. The ratio of the area for one of the compounds to that of CFC-11 is equal to the ODP of that compound. Reproduced from Du Pont Company *Update*, August, 1989, page 4

Antarctic stratospheric ozone as spring returns. First, the intensity of sunlight increases, and more ozone is formed. Second, increasing stratospheric temperatures sublime the polar stratospheric clouds, releasing HNO_3(g), which soon cleaves to NO_2, trapping some of the ClO in a temporary reservoir of chlorine nitrate, $ClONO_2$. Another temporary reservoir of ClO is the dimer ClOOCl, but this is very readily cleaved by sunlight, forming O_2 with the release of two chlorine atoms.

The observation of Antarctic ozone depletion over several successive winters led to efforts in 1989 and 1990 to determine whether similar effects would be seen in the Arctic. Although elevated concentrations of active chlorine species such as ClO were detected and significant amounts of ozone were destroyed, the infusion of air from lower latitudes prevented large scale loss of ozone in the high Arctic[14]. Interestingly, loss of *tropospheric* Artic ozone in early spring has been detected both in Alaska and the Canadian

[14]D.J. Hofmann *et al*, "Stratospheric clouds and ozone depletion in the Arctic during January 1989," *Nature*, **1989,** 340, 117–121. M.H. Proffitt *et al*, "Ozone loss in the Arctic polar vortex inferred from high altitude aircraft measurements," *Nature*, **1990,** 347, 31–36.

Arctic. It appears to involve bromine chemistry analogous to Reactions 5 and 6, though the source of bromine atoms is not yet certain.

2.3.2 Brominated CFC analogs

Brominated analogs of CFC's are in commercial use as fire extinguishers, under the name Halons.[15] These compounds are very valuable in fighting electrical fires, for example at computer installations. Their mode of action, besides smothering the fire with a heavy vapour, involves cleavage by heat of the weak C–Br bond. The bromine atoms thus formed act as terminators for the radical chain reactions which take place in flames. Halons are also sold under the Freon tradename, but are distinguishable by a B (number of bromine atoms) in the code. Examples:

CF_2ClBr	H1211 or F-12B1
CF_3Br	H1301 or F-13B1
$BrCF_2CF_2Br$	H2402 or F-114B2

Concern about the Halons is that they can migrate to the stratosphere like CFC's. Bromo compounds are cleaved more easily by light than the corresponding chlorides, and bromine atoms can initiate radical chains for the decomposition of ozone that are entirely analogous to Equation 5 and 6 written for chlorine atoms. H1211 and H1301 have a greater ozone-depleting potential than CFC-11 and CFC-12.

One cause of controversy at present concerns the testing of fire-fighting systems which "flood" the fire site with Halon. The requirement is for the system to be tested by releasing a full charge of Halon. Consequently, far more Halon is released to the atmosphere in testing fire-prevention systems than is consumed in fighting fires. Firefighters are very unwilling to lose the excellent fire-fighting properties of Halons, for which no substitute has yet been found. A solution would be to use something other than Halon for the system test.

2.4 Consequences of ozone depletion

.

The immediate consequence of a reduction in the total amount of stratospheric ozone would be a greater penetration of short wavelength ultraviolet radiation to the Earth's surface. One consequence of this could be a change in the climate. This would result from changing the circulation of the atmosphere due to lowering the temperature of the stratosphere, and to lowering the altitude of the tropopause. Biological effects could also be anticipated

[15]The Halons are given codes such as H1211. The four digits in order are numbers of carbon, fluorine, chlorine, and bromine in the molecule. Any atoms not accounted for are hydrogen.

if the total "ozone column" above the Earth were reduced. This is because ozone filters out radiation in the range 290–330 nm incompletely, as its absorption cross section is relatively small in this range. A decrease in the amount of ozone would lead to less absorption of the most energetic radiation to reach the surface.

It will be useful in this section to use the terminology of biological scientists to describe different spectral ranges. These are:

400 - 700 nm	visible
320 - 400 nm	UV-A
290 - 320 nm	UV-B
< 290 nm	UV-C

The division between UV-A and UV-B is approximately the cut-off of radiation which respectively cannot, and can, cause sunburn. UV-B also happens to be filtered out by window glass; hence it is virtually impossible to acquire a suntan indoors, even if you sit right by the window (see below). UV-C is not encountered naturally in the troposphere.

We can speculate that many species, animals, plants, and microorganisms, (but perhaps especially the plankton in ocean waters) might be affected by increased doses of high energy radiation. One expected result is an increase in the incidence of skin cancer among humans. Almost all cases of skin cancer can be attributed to over-exposure to sunlight, the exceptions occurring mostly in people suffering from inherited conditions where DNA-repair mechanisms are faulty. Skin cancer is almost exclusively an affliction of light skinned people; dark skin contains a larger quantity of the brown pigment melanin, which is an effective absorber of ultraviolet radiation. Moreover, whites tend to develop skin cancer in areas that are most exposed to the Sun: forehead, nose, around the neck and V of the throat, and on the arms.

There are two main types of skin cancer. Squamous cell carcinoma is common; it is normally localized and spreads slowly. It may usually be removed with complete success, often on a visit to the doctor's office. Malignant melanoma, which is much less common, metastasizes readily and is life threatening. Because of this, better records exist for the incidence of melanoma. Figure 2.4 shows the number of deaths reported in the various states of the U.S.A. in 1960 as a function of the average latitude of the state[16].

The data show a fairly good inverse correlation between the incidence of melanoma and latitude. This is reasonable, since at low latitudes the Sun's rays are more direct. When the Sun is overhead, solar radiation travels through the minimum depth of atmosphere, and filtering by ozone is least effective. When the Sun is low in the sky, radiation reaches the surface

[16] F.S. Rowland, "Chlorofluorocarbons and stratospheric ozone," Chapter 4.6 in *Light, Chemical Change and Life,* Eds. J.D. Coyle, R.R. Hill, and D.R. Roberts, Open University Press, Milton Keynes, England, 1982.

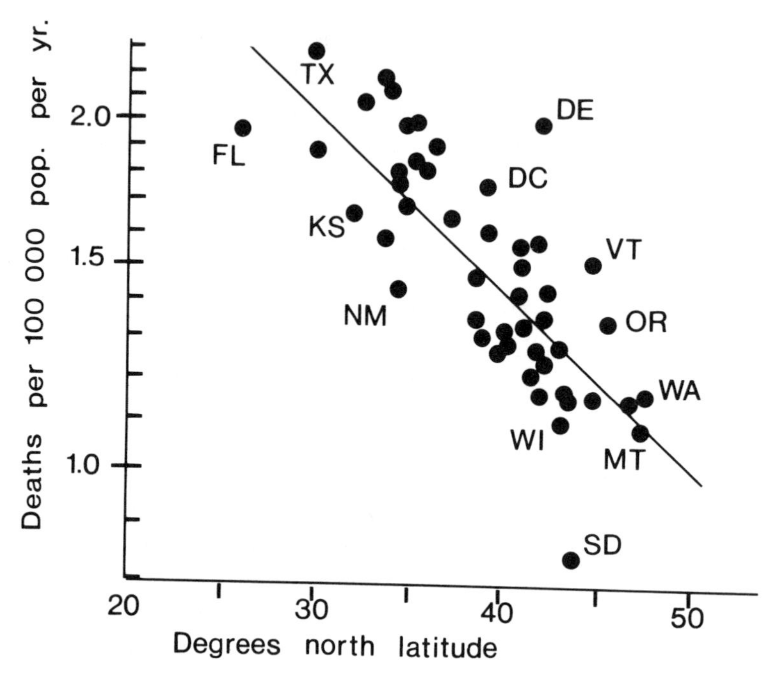

Figure 2.4: Variation of human death with latitude due to skin melanoma among white males in the United States for the year 1960. Reproduced from Reference 15.

obliquely, and travels through a greater depth of atmosphere. This allows more efficient filtering of UV-B radiation by ozone. Consequently, the person who dwells in the tropics (and especially at high altitudes) is exposed to the greatest flux of UV-B, particularly near midday. This is illustrated in Figure 2.5 which shows the variation of the solar zenith angle[17] through the day for several geographical locations and seasons, and the corresponding fluxes of UV-B photons. The same UV-B radiation is responsible for suntanning (and sunburn!), which is also fastest in the tropics and at high altitude. For this reason, fair skinned people are advised to keep out of the Sun in the middle of the day.

A 5% decrease in the amount of stratospheric ozone has been predicted to produce 20% more cases of skin cancer, mostly squamous cell carcinoma, per year in the United States[18]. This projected increase sounds very frightening, but should be kept in perspective: for people living in the middle latitudes, the increased risk of skin cancer due to each 1% decrease in ozone levels is equivalent to that posed by moving 20 km closer to the equator. Probably few of us would turn down a job promotion for fear of skin cancer

[17]The zenith angle is the angle between the direction of the Sun and the vertical direction. When the Sun is directly overhead the zenith angle is zero; at sunrise and sunset, the zenith angle is 90°.

[18]F. Urbach, "Photocarcinogenesis," in *The science of photomedicine*, Eds J.D. Regan and J.A. Parrish, Plenum Press, New York, 1982, Chapter 9.

if it required moving even a few hundred kilometers closer to the equator (e.g., from New York to California, or from England to the Mediterranean!).

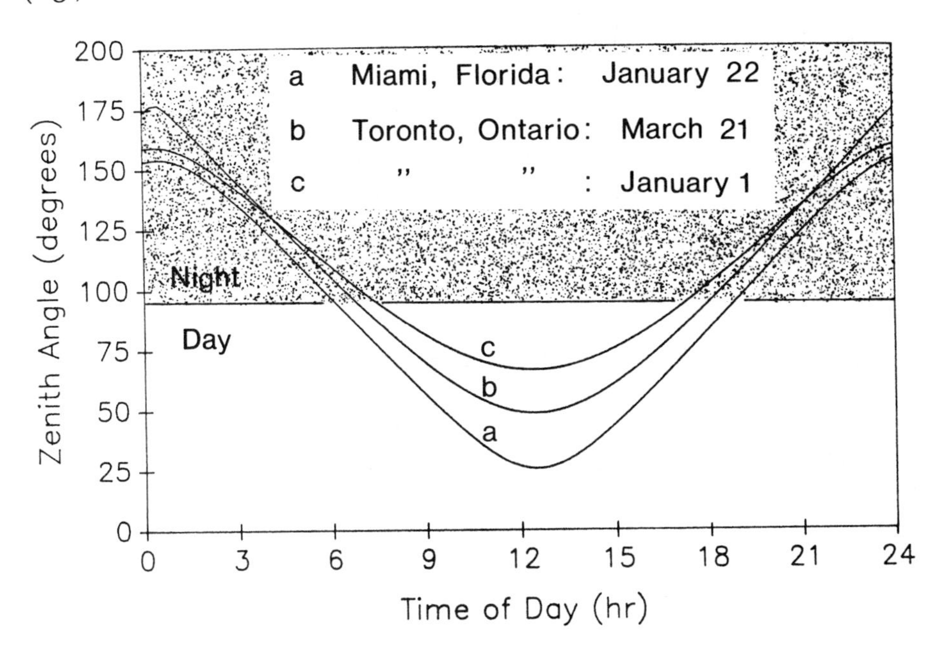

Figure 2.5: Variation of zenith angle at different geographic locations and seasons.

UV-B radiation is harmful because it can be absorbed in the tail of the absorption band of DNA ($\lambda_{max} \approx$ 260–280 nm). The postulate is that photochemical reactions can occur in the DNA, causing the genetic code to be misread at the time of cell division. Such photochemical reactions can be observed experimentally both in aqueous solutions of DNA and in whole cells. While repair mechanisms exist, their efficiency falls with an increase in the number of defects produced per unit time[19]. Patients with the inherited condition **xeroderma pigmentosa** lack these repair mechanisms, and are extremely susceptible to sunburn and to skin cancer.

The different biological activity of radiation of different wavelengths is seen very dramatically in the case of sunburn. The table below shows the energy dose needed to cause perceptible reddening of the skin (erythema) at several wavelengths.

Radiation	Energy Dose J cm^{-2}	Relative Number of photons required
254 nm, UV-C	0.001	1.
300 nm, UV-B	0.02	2.4
310 nm, UV-B	0.2	24.
337 nm, UV-A	15.	2000.

[19] W.L. Carrier, R.D. Snyder, and J.D. Regan, *Ultraviolet-induced damage and its repair in human DNA*, Chapter 4 of Reference 18.

These figures show clearly that it is the energy of the individual photons which matters, not the total energy received. Highly energetic photons are the most damaging biologically.

The view that the UV-B component of sunlight is the main cause of skin cancer in humans has led to the development of commercial tanning booths employing high intensity UV-A radiation. Recent work has shown that these are not without hazard however, since the "action spectra" for pigmentation of human skin and for UV carcinogenesis in mouse skin are almost identical[20].

2.4.1 Sunscreens

Melanin is a natural substance which absorbs radiation in the UV-B range. Melanin is found in specialized cells called melanocytes; its production is stimulated by exposure to UV-B, although it is not produced for some time after the initial exposure. The melanin filters out UV-B radiation before it reaches the living cells underneath. The difference between fair skinned and dark skinned people lies not in the total number of melanocytes which they possess, but rather in the amount of melanin produced by each melanocyte. The pigmentation is found mostly in the surface layers of the skin, which consist entirely of dead, "keratinized" cells.

A sunscreen is a chemical substance which is applied to the skin to carry out the same function as melanin, namely to prevent high energy radiation from reaching the living cells in the lower layers of the skin. Sunscreens have to absorb in the proper region of the spectrum, and to be cosmetically acceptable they should be colourless i.e., their absorption should not extend into the visible. Substances possessing a benzene, or similar aromatic ring as part of their structure are used most commonly, see structures below.

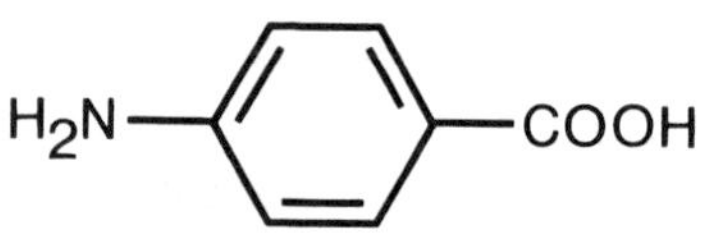

p-Aminobenzoic acid
(PABA)

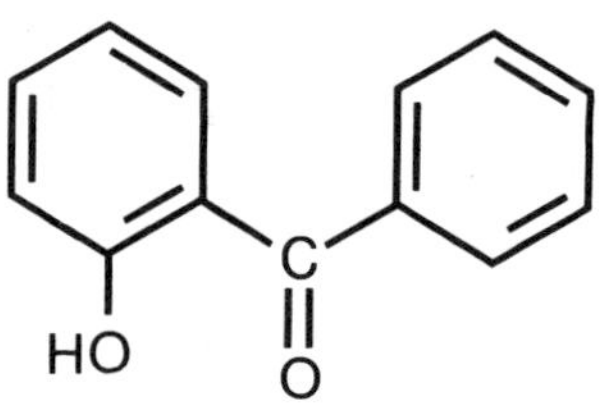

o-Hydroxybenzophenone

Sunscreens must be able to absorb light without undergoing photochemical reactions. There are two reasons for this. First, rapid photodegradation would render the sunscreen ineffective. Second, many substances which are

[20] L. Roza, R.A. Baan, J.C. van der Loon, L. Kligman, and A.R. Young, "UVA hazards in skin associated with the use of tanning equipment," *J Photochem Photobiol(B)*, **1989**, 3, 281–287.

photochemically active while they are in contact with body tissues cause various kinds of "phototoxic" responses: that is, toxic effects which are produced by light exposure. These can include toxicity due to breakdown products of the material, toxicity due to reactive intermediate formed during the photochemical reaction, and photoallergic responses. Cosmetics, including sunscreens, must therefore be carefully tested for phototoxicity problems before they are introduced to the marketplace.

A sunscreen must absorb UV-B radiation (which has an energy of ca. 400 kJ mol^{-1}) in order to be an effective filter, and its molecules are thereby promoted to an excited electronic state. It follows that the successful sunscreen must dispose of its energy in some way other than by chemical reaction. In doing so, the molecule returns to the ground state, and is then available for absorbing another photon. The energy may be dissipated into heat by "internal conversion" of the electronic excitation into vibrational excitation of the molecule (Problem 14). Thermal equilibration of the vibrationally excited molecule with its surroundings then occurs.

Like other aromatic ketones, *o*-hydroxybenzophenone derivatives are reactive photochemically. Most aromatic ketones react photochemically through abstraction by the carbonyl oxygen atom of a hydrogen atom from a suitable hydrogen donor. In the case of *o*-hydroxybenzophenones, the reaction takes place intramolecularly, forming a photoenol.

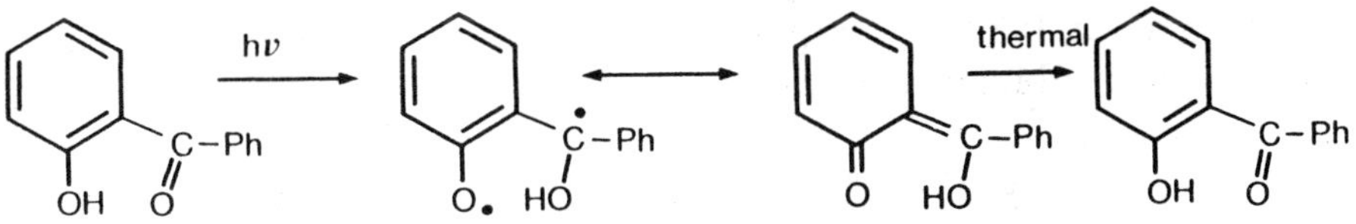

The enol subsequently tautomerises thermally to the more stable ketone starting material; no permanent chemical change occurs, and so the net effect is the degradation of electronic excitation energy to heat.

2.5 The Montreal Protocol

Political acceptance of scientific predictions of stratospheric ozone depletion occurred in North America before any adverse consequences had actually been observed; this stands in contrast to most other environmental problems with which society has had to deal. Thus, the prediction of a threat to stratospheric ozone by CFC's was made by Molina and Rowland in 1974, and as early as 1978 the use of CFC's as aerosol propellants was banned in North America. International consensus on the need to restrict CFC emissions emerged with the discovery that the loss of Antarctic stratospheric ozone (Section 2.3.1) correlated precisely with increases in stratospheric chlorine[21].

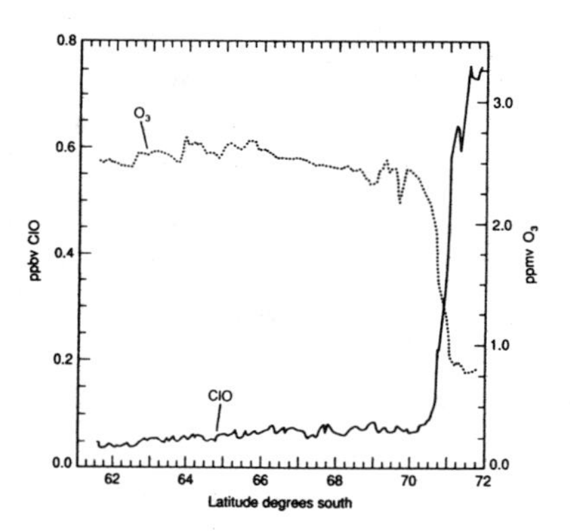

ClO and O_3 concentration over Antarctica at 18 km altitude as a function of latitude on September 21, 1987. Reproduced from Reference 9

A significant event in terms of political awareness of the ozone depletion problem was the signing in 1987 of the "Montreal Protocol on substances that deplete the ozone layer." This is an international treaty which sets targets for CFC production to be cut back to 1986 baseline levels by mid-1989, cut to 80% of baseline by 1993, and to 50% of baseline by 1998. By the end of 1988, the requisite number of signatures by governments adhering to the Montreal Protocol had been obtained, bringing it into force internationally[22].

A NASA study released in 1988 showed experimental evidence of thinning of ozone at locations outside Antarctica[23], particularly in the Arctic, over the period 1969–1986. The NASA data suggested that even the cuts in CFC production proposed in the Montreal Protocol would be insufficient to prevent substantial loss of stratospheric ozone over the next half century (Figure 2.6)[24].

During 1989, governments in both Europe and North America have called for a complete phase-out of all use of CFC's such as CFC-11 and CFC-12, which are the most harmful environmentally. The United States

---

[21] K. Warr, "Ozone: the burden of proof," *New Scientist,* October 27, **1990,** 36–40.

[22] P.S. Zurer, "Producers, users grapple with realities of CFC phaseout," *Chem Eng News,* July 24, **1989,** 7–13; see also *Chem Eng News,* January 2, **1989,** 25.

[23] P. Zurer, "Studies on ozone destruction expand beyond Antarctic," *Chem Eng News,* May 30, **1988,** 16.

[24] M. McFarland, "Chloroflurocarbons and ozone," *Environ Sci Technol,* **1989,** 23, 1203–1207; M.T. Prather and R.T. Watson, "Stratospheric ozone depletion and future levels of atmospheric chlorine and bromine," *Nature,* **1990,** 344, 729–734.

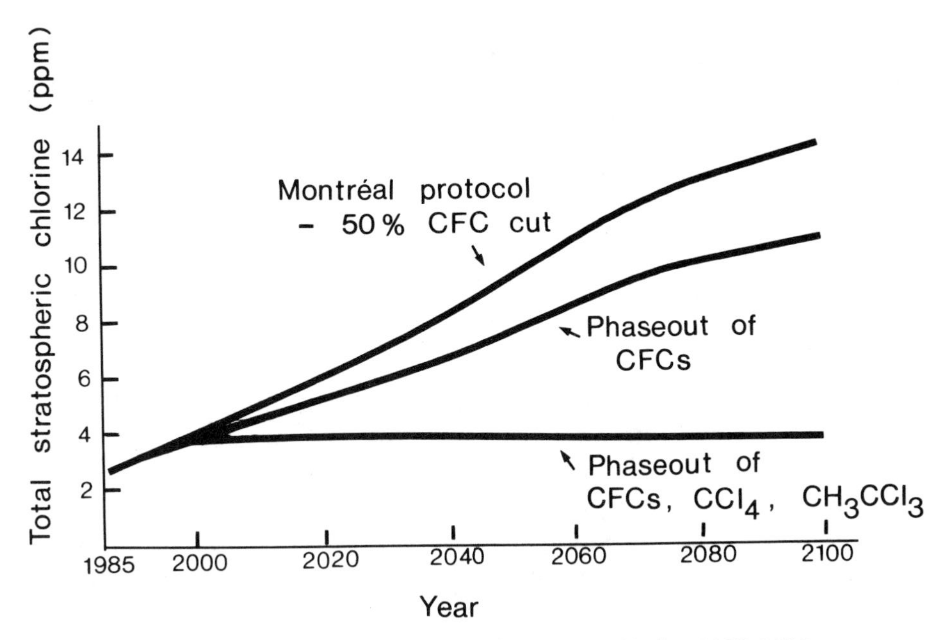

Figure 2.6: Predicted increases of stratospheric chlorine 1985–2100.
Source: Environmental Protection Agency (U.S.A.).

EPA also suggests phasing out CH_3CCl_3, which is used as an industrial degreasing solvent, but which is not covered by the Montreal Protocol. Similar considerations apply to CCl_4. The increasing pace of the movement to phase out CFC's is illustrated by this example from Ontario[25].

1990	remove exemption for CFC propellants in medical aerosols; ban on CFC's in slurries used to prepare aerosols, in insulation foam products (e.g., coffee cups), and in refrigerator and air conditioner home repair kits.
1991	recapture and recycling of CFC's from automobile air conditioners, truck and rail car refrigeration units.
1993	ban on CFC's in rigid foam insulation and flexible foam furniture.

2.6 CFC replacement compounds[26]

It will be clear from Section 2.3 that CFC's have important uses in modern society as the working fluids in refrigerators and air-conditioners, and

[25] "Ontario cutting CFC use in half by 1993," *Environ Sci Eng,* December **1990,** 10.

[26] L.E. Manzer, "The CFC-ozone issue: progress on the development of alternatives to CFCs, *Science,* **1990,** 249, 31–35.

as the blowing agents for plastic foams. In these applications their properties of non-flammability, non-toxicity, and appropriate boiling point are paramount. Other applications do not absolutely require CFC's or comparable substitutes; these include aerosol propellants and degreasing solvents (although even here Japanese companies claim that no substitute yet exists for CFC-113 in cleaning the surfaces of microelectronic parts) and blown foam applications such as fast food containers (the use of CFC-11 in this application has been phased out in North America). The search for CFC replacements has as its goal compounds that will have the desirable properties for the critical applications described, while being more "environmentally friendly."

Recall that the problem with CFC's is that they are *too* stable chemically. Their extreme chemical inertness allows them to survive unchanged in the troposphere for so long that they can eventually migrate to the stratosphere. This leaves two options, short of an outright ban on CFC's (which in view of their usefulness seems impractical). One is to prevent any release to the atmosphere; the second is to substitute a replacement that will be reactive enough in the troposphere that it will not persist long after its release. Realistically, it is impossible to prevent release: blowing agents escape during blown foam manufacture, when the foam "bubbles" are broken in use or when the material breaks down; refrigerators and especially air-conditioners gradually lose their charge of CFC on account of vibrations. Suggestions for a recoverable "deposit" on the CFC in the refrigerator, to be refunded when the unit is taken out of service, have so far not resulted in any action. This leaves a search for a less inert CFC replacement as the only option currently being pursued.

Potential CFC replacement compounds are likely to be small halogenated organic molecules. CFC's owe their extreme chemical inertness to being fully halogenated. At the other end of the structural spectrum, hydrocarbons are rapidly oxidized in the troposphere by reactions involving hydroxyl radicals (see Chapter 3). Unfortunately, hydrocarbons are highly flammable. They are thus totally unsuited to use in refrigerators, and likewise unacceptable in the manufacture of urethane foams for upholstered furniture, since the industry is under pressure to improve the fire resistance of its products. The search therefore narrows to compounds with one or two carbon atoms (to keep the boiling point in the right range) and some appropriate combination of hydrogen, fluorine, and chlorine substituents. Increasing the hydrogen content increases flammability but reduces tropospheric lifetime[27]; increasing the chlorine content of a CFC substance

[27]The rate constant for the reaction of OH with a halogenated alkane is less than that of the unsubstituted compound, even on a per-hydrogen basis. This is because the electronegative halogen substituents decrease the electron density at the adjacent C-H centres, inhibiting attack by the very electrophilic hydroxyl radical. A slower reaction with OH necessarily increases the tropospheric lifetime. Nevertheless hydrogen-containing CFC's have much lower ozone-depleting potential than their fully halogenated counterparts, because of their tropospheric reaction with OH: D.A. Fisher *et al*, "Model

which also contains hydrogen increases its toxicity[28]. Totally fluorinated compounds (i.e., fluorocarbons) are unsuitable; they would not decompose even in the stratosphere, because they do not undergo photochemical breakdown even near 200 nm. They would therefore last almost "forever," and because they absorb infrared radiation (but without chemical breakdown), they would be highly persistent greenhouse gases.

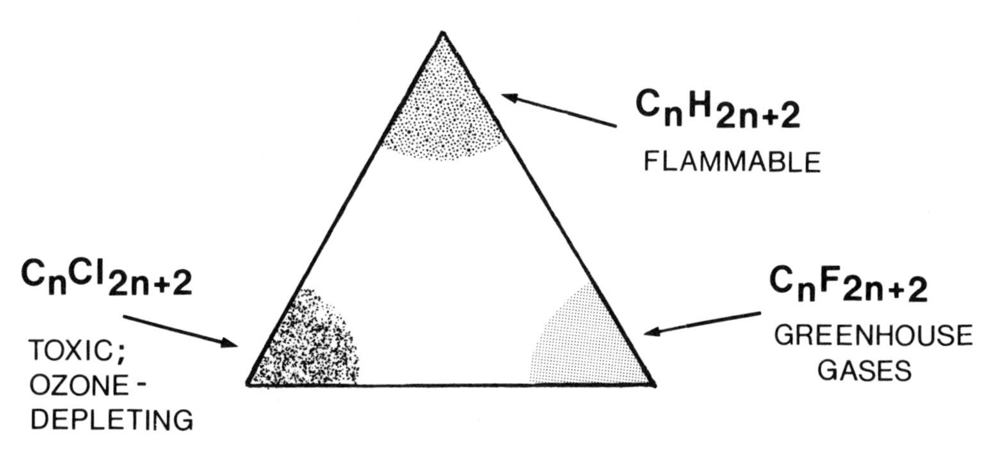

Figure 2.7:

Foams

In this case, the compound to be replaced is CFC-11. CFC-22 is already being used in this application; its use is growing rapidly (> 10% per year). CFC-22 has about 5% of the ozone depleting potential of CFC-12 integrated over its stratospheric lifetime, but presents a substantial medium term threat because its stratospheric lifetime is shorter than that of CFC-12[29]. Either CFC-123 (CF_3CHCl_2) or a blend of CFC-123 and CFC-141b (CH_3CFCl_2) are possible candidates to replace CFC-11 in polyurethane and other rigid foams. CFC-123 is estimated to have a tropospheric lifetime of only 1.5 years, and only 7% of the potential of CFC-11 to deplete stratospheric ozone[30].

calculations of the relative effects of CFCs and their replacements on stratospheric ozone," *Nature,* **1990,** 344, 508–512.

[28]There is not yet agreement on the designation of these partly halogenated alkanes. HCFC (hydrochlorofluorcarbons) is being used in North America; other proposals include HAC (halogenated aliphatic compounds), HFC (hydrofluorocarbons), and HFA (hydrofluoroalkanes). For the moment, I have chosen to stay with CFC, regardless of the presence or absence of H or Cl.

[29]*New Scientist,* September 1, **1988,** 39.

[30]R.G. Prinn and A. Golombek, "Global and atmospheric chemistry of CFC-123," *Nature,* **1990,** 344, 47–49.

Refrigerants

An examination of boiling points suggests the following compounds as possible replacements for CFC-12 as a refrigerant: CFC-134a (CF_3CH_2F, b.p. −26°C), CFC-152a, CH_3CHF_2 (−25), CF_3CF_2Cl (−38), and CFC-22 (CHF_2Cl, −41). Of these, the fully halogenated CF_3CF_2Cl is of comparable ozone depleting ability with CFC-12. Although CH_3CHF_2 is flammable, it can be used as a "drop-in" replacement for CFC-12. CFC-22 is already in service in air-conditioners and as a replacement for CFC-11 in some blown foams. CFC-134a is the top contender to replace CFC-12; it contains no chlorine, and so has no evident ozone depleting potential. CFC-134a contains two hydrogen atoms to facilitate tropospheric attack by hydroxyl radicals.

(27) $$CF_3CH_2F + OH \longrightarrow CF_3CHF + H_2O$$

The CF_3CHF free radical would be susceptible to addition of O_2, and hence further oxidation.

There is an intense race to commercialize the replacements for CFC-11 and CFC-12. At stake is an annual $2 billion CFC market worldwide. Both DuPont (U.S.) and I.C.I. (U.K.) appear to be favouring CFC-134a to replace CFC-12 in refrigerators. DuPont announced the construction of a world-scale manufacturing facility for CFC-134a in 1988, even though final toxicity testing and government approval for use of the compound were not complete[31]. The race for commercialization at present seems to hinge on the cheapest way of making CFC-134a, rather than on which replacement compound to introduce. One unsolved problem with introducing CFC-134a as a refrigerant is that it is incompatible with the lubricants currently used in refrigerators. Alternative lubricants are under development, but are not yet available. Another possible strategy is the used of blends of CFC's as refrigerants. DuPont, for example, has suggested a blend of CFC's 22, 124, and 152a; these types of blend are compatible with present lubricating oils, and so should require minimal modifcation of existing refrigeration systems[32].

Whichever CFC replacements ultimately triumph, they will be more expensive than CFC-12. The latter can be manufactured very cheaply by direct fluorination of CCl_4, whereas the new replacements all require much more complex chemistry. This is shown by one of the proposed routes to CFC-134a below.

(28) $$C_2H_4 \xrightarrow{Cl_2,\,HF,\text{catalyst}} \underbrace{CF_3CFCl_2}_{\text{CFC-114a}}(75\%) + \underbrace{ClCF_2CF_2Cl}_{\text{CFC-114}}(25\%)$$

[31] Short term toxicity testing was completed successfully in 1989 for CFC's 22, 123, 134a, and 141b. Long term inhalation testing will not be completed until at least 1992: *Chem Eng News,* October 4, **1989**.

[32] *Chem in Brit,* April **1990, 314**.

$$(29)\quad CF_3CFCl_2 \xrightarrow{H_2,\ catalyst} \underbrace{CF_3CH_2F}_{CFC\text{-}134a}$$

The byproduct CFC-114 can be converted catalytically to CFC-114a.

Another possible route to CFC-134a starts with tetrachloroethylene.

$$(30)\quad C_2Cl_4 \xrightarrow{HF,\ catalyst} \underbrace{CF_3CHCl_2}_{CFC\text{-}123}(53\%)+\underbrace{CF_3CHFCl}_{CFC\text{-}124}(41\%)+\underbrace{CF_3CHF_2}_{CFC\text{-}125}(2\%)$$

CFC-124 can be catalytically dechlorinated to CFC-134a, while the CFC-123 has been mentioned as a potential replacement for CFC-11 as a blowing agent.

Cleaning solvents for electronic components

The problem of a replacement for CFC-113 as a cleaning agent remains only partly solved. In this application, CFC's are used because of their low surface tension and low viscosity. These properties allow them to penetrate into tiny crevices which would not be wetted by a solvent like water. CFC-132b ($C_2H_2F_2Cl_2$) looked to be a promising substitute, but failed toxicity testing. DuPont is recommending a blend of CFC-22 and methanol for this application, thereby reducing the ozone depleting potential by more than 95% if the same volume of each solvent were used.

Aerosols

In the case of aerosols, North American industry has had a decade to adjust to the ban on CFC's. Hydrocarbons have successfully been substituted for CFC-11 and CFC-12, with methylene chloride added as a flame suppressing agent[33]. CFC-22, CFC-142b (CH_3CF_2Cl) and CFC-152a (CH_3CHF_2) are also in use as less damaging replacements for CFC-11 and CFC-12. Dimethyl ether is also available as a propellant; it can be mixed with up to 50% by weight of water, thus lowering its flammability.

In concluding this section, note that CFC's and their replacements contribute to global warming, a completely different phenomenon from stratospheric ozone depletion. Table 2.3 summarizes this information for some leading CFC compounds. Finally, concerns have been raised that the troposheric chemistry of these CFC replacements has received insufficient study prior to the decision to go ahead with production[34].

---

[33] P.L. Layman, "Aerosols back on road to success," *Chem Eng News,* April 28, **1986,** 29.

[34] *Chem in Brit.* March, **1990,** 217.

Table 2.3: Alternative Fluorocarbon Compounds

Fluorocarbon	Formula	Boiling Point (0°C)	Ozone depletion Potential*	Global Warming Potential*
CFC-22	$CHClF_2$	-40.8	0.05	0.3
CFC-123	CF_3CHCl_2	27.9	0.02	0.02
CFC-124	$CHClFCF_3$	-11.	0.02	0.09
CFC-125	CHF_2CF_3	-48.5	0.	0.5
CFC-134a	CF_3CH_2F	-26.5	0.	0.3
CFC-141b	CH_3CFCl_2	32.0	0.1	0.09
CFC-142b	CH_3CClF_2	-9.8	0.06	0.3
CFC-152a	CH_3CHF_2	-25.0	0.	0.03

* Estimated values relative to CFC-11 as 1.0

Source: Du Pont *Update,* H-07421, August, 1989

2.7 Nitrogen oxides as ozone depleters

N_2O

This substance was encountered in Chapter 1 as a natural trace constituent of the atmosphere. It is produced biologically, as a result of microbial denitrification, and its concentration is now increasing, most likely due to the large increase in the use of nitrogenous fertilizers. Nitrous oxide has no known sinks in the troposphere, and migrates to the stratosphere where it is degraded photochemically (Problem 12).

$$N_2O \xrightarrow{h\nu} N_2 + O^* \qquad (31)$$

While this reaction has little impact on stratospheric ozone depletion, there is also the possibility for conversion of N_2O to nitric oxide by way of the reaction below.

$$O^* + N_2O \longrightarrow 2NO \qquad (32)$$

Another pathway gives $N_2 + O_2$. The excited oxygen atoms that are reactants in Reaction 32 come from photolysis of either O_3 (Reaction 3a) or N_2O.

NO and NO_2

In Section 2.2 it was seen that nitrogen oxides in the stratosphere decompose ozone catalytically:

$$NO + O_3 \longrightarrow NO_2 + O_2 \qquad (33)$$

$$NO_2 + O \longrightarrow NO + O_2 \qquad (34)$$

Nitrogen oxides are naturally present in the stratosphere, but it is reasonable to assume that if their concentrations were to be elevated for any reason the rate of ozone destruction would increase. The argument is identical to that which was used to explain how CFC's lower the levels of stratospheric ozone by increasing the concentration of chlorine atoms.

Several situations have been described in which nitrogen oxides may be found at elevated concentrations in the stratosphere. One of them concerns supersonic transport aircraft (SST's), and similar arguments apply to space rockets, space shuttles, etc. Combustion of any fuel in air produces nitrogen oxides:

(35) $$N_2 + O_2 \longrightarrow 2NO \qquad \Delta H^\circ = +180 \text{ kJ mol}^{-1}$$

Reaction 35 is endothermic, and so its equilibrium constant increases with temperature. This reaction will be considered in more detail in the next chapter; for the moment, note that NO is an inevitable byproduct whenever air is heated. One of the fears expressed when the supersonic "Concorde" aircraft was first put into service was that a large number of SST's flying in the lower part of the stratosphere might produce sufficient NO to deplete stratospheric ozone. (The advantage of flying at maximum possible altitude is to lower air resistance.) In the event, the argument became redundant when the price of oil, and hence aviation fuel, rose substantially in the mid-1970's[35], and development of additional SST aircraft was halted.

The atmospheric nuclear weapons testing of the 1950's and 1960's had a depleting effect on stratospheric ozone. A nuclear explosion would cause hot tropospheric air containing nitrogen oxides to be injected directly into the stratosphere. The levels of stratospheric ozone were found to rise in the early 1970's after atmospheric testing ceased.

Natural events can have the same effect as a nuclear explosion. Major volcanic eruptions can inject hot gases directly into the stratosphere[36], as can large meteorite impacts, the heat being produced by the huge loss of kinetic energy upon impact. One such event has been documented experimentally: the Tunguska meteorite, which landed in a remote region of Siberia in 1908. By chance, the U.S. Geological Survey had recently begun recording atmospheric ozone levels. Reevaluation of these old records nearly seventy years later showed a sharp drop in ozone levels in 1908–9, followed by a gradual recovery to pre-1908 levels over the next 3–4 years[37]. One of the many theories for the Great Extinction (of dinosaurs) is that the Earth suffered a great meteorite impact. Many adverse effects would have resulted, making the planet temporarily less habitable; one of them could have been a catastrophic loss of stratospheric ozone. This could explain the greater loss of land creatures than of ocean species, as apparently occurred.

---

[35] Proposals for an American SST have recently been revived: P.S. Zurer, "NASA paving the way for development of new supersonic plane," *Chem Eng News*, January 1, **1990,** 15–16.

[36] R.B. Symonds, W.I. Rose, and M.H. Reed also report that significant amounts of HCl, a reservoir compound for chlorine atoms, may be injected into the stratosphere in large volcanic eruptions: *Nature,* **1988,** 334, 425.

[37] R.P. Turco, O.B. Toon, C. Park, R.C. Whitten, J.B. Pollack, and P. Noerdlinger, "Tunguska meteorite fall of 1908: effects on stratospheric ozone," *Science,* **1981,** 214, 19.

Further reading

1. R.P. Wayne, *Chemistry of Atmospheres,* Oxford University Press, Oxford, England, 1985, Chapter 4.

2. B.J. Finlayson-Pitts and J.N. Pitts, Jr. *Atmospheric Chemistry,* Wiley-Interscience, New York, **1986,** Chapter 15.

2.8 Problems

1. The data below relate concentrations of all molecules (M) and of ozone (in units molecules per cm^3) with altitude (km) and temperature.

Altitude	Temperature	[M]	[O_3]
15	217	$4.0 \times 10^{+18}$	$1.0 \times 10^{+12}$
20	217	$1.9 \times 10^{+18}$	$2.0 \times 10^{+12}$
25	222	$8.6 \times 10^{+17}$	$4.5 \times 10^{+12}$
30	227	$3.8 \times 10^{+17}$	$3.0 \times 10^{+12}$
35	237	$1.8 \times 10^{+17}$	$2.0 \times 10^{+12}$
40	251	$8.4 \times 10^{+16}$	$5.0 \times 10^{+11}$
45	265	$4.0 \times 10^{+16}$	$2.0 \times 10^{+11}$

Plot the ozone pressure (in ppm and in atm) as a function of altitude to obtain Figure 2.1.

2. The reaction $O(g) + O_2(g) + M \longrightarrow O_3(g) + M$ has $k = 1.1\times\ 10^{-33}$ cm^6 $molecule^{-2}\ s^{-1}$ at 220 K (stratosphere).

   (a) What is the rate of reaction in mol $L^{-1}\ s^{-1}$ if $p(\text{total}) = 0.010$ atm, and $c(O) = 2.1 \times 10^{-4}$ ppm?

   (b) Calculate the pseudo-first order rate constant for this reaction (Hint: which concentrations are assumed to be constant?) Then calculate the half life of O(g) under these conditions. What does your result imply about the concentration of O(g) in the atmosphere once the sun has set?

3. Table 2.4 contains data on flux of photons (per cm^2 of the Earth's surface) for different wavelength ranges. Column A is the flux outside the atmosphere, and Column B is the flux at the Earth's surface when the Sun is directly overhead.

   Calculate the amount of energy absorbed in the atmosphere over the range 200–350 nm. Where in the atmosphere does this absorption occur, and what causes it?

4. (a) Use tabulated data to estimate K_p for the reaction, at −55°C: $\frac{3}{2}O_2(g) \rightleftharpoons O_3(g)$ Include the units of K_p.

Table 2.4: Solar photon intensity vs wavelength (see Problem 3)

Wavelength Range	I_o (Photons cm^{-2}) A	B
202–210	1.2×10^{13}	0
210–220	3.4×10^{13}	0
220–230	5.3×10^{13}	0
230–240	5.6×10^{13}	0
240–250	5.7×10^{13}	0
250–260	8.7×10^{13}	0
260–270	2.7×10^{14}	0
270–280	2.5×10^{14}	0
280–290	4.0×10^{14}	0
290–300	6.9×10^{14}	4.2×10^{11}
300–310	1.0×10^{15}	1.8×10^{14}
310–320	1.1×10^{15}	7.5×10^{14}
320–330	1.4×10^{15}	1.3×10^{15}
330–340	1.7×10^{15}	1.6×10^{15}
340–350	1.7×10^{15}	1.7×10^{15}

Data from *Atmospheric Chemistry*, Finlayson-Pitts and Pitts, Chapter 3.

(b) If $p(O_2) = 2.0 \times 10^{-3}$ atm in the upper atmosphere, what is the equilibrium concentration of O_3?

(c) The actual steady state concentration of O_3 under these conditions is 3.0×10^{15} molecules per litre. Show by calculation whether the system O_2/O_3 is at equilibrium and comment on the significance of your result.

5. This question concerns whether radiation of 300 nm is capable of bringing about the reaction

$$O_3(g) \longrightarrow O_2(g) + O(g)$$

(a) Calculate the energy of 300 nm radiation in kJ mol^{-1}.

(b) Calculate $\Delta H°$ and ΔE for the reaction at −55°C (stratosphere).

(c) Estimate whether 300 nm radiation is capable of bringing about this reaction, and state any assumptions you make.

(d) Estimate the longest wavelength capable of bringing about this reaction.

6. Consider this oversimplified scheme for the destruction of ozone in the atmosphere

(I) $O + O_3 \longrightarrow 2O_2$

$k_I = 1.5 \times 10^{-11} e^{-2218/T}$ cm^3 molecule^{-1} s^{-1}

(IIA) $O_3 + Cl \longrightarrow ClO + O_2$

$k_{IIA} = 8.7 \times 10^{-12}$ cm^3 molecule^{-1} s^{-1} at 220K

(IIB) $O + ClO \longrightarrow Cl + O_2$

$k_{IIB} = 4.3 \times 10^{-11}$ cm^3 molecule^{-1} s^{-1} at 220K

Use the following steady state concentrations, in molecules per cm^3, to answer questions (a)–(e). $O = 5.0 \times 10^7$; $Cl = 1.0 \times 10^5$; $ClO = 6.4 \times 10^7$; $O_3 = 3.2 \times 10^{12}$.

(a) Calculate the activation energy of Reaction I.

(b) Calculate the rates of Reaction I, IIA, and IIB at 220 K, in the units molecules cm^{-3} s^{-1}.

(c) What is the overall rate of the Cycle IIA + IIB? Explain your reasoning.

(d) What fraction of the ozone is destroyed under these conditions by the direct Reaction I, rather than by the Cycle II?

(e) Why is there currently concern that the importance of Cycle II is increasing, and what would be the significance if this concern proved to be well founded?

7. This question follows on from Problem 6. Suppose, which is unrealistic, that Reactions I, IIA, and IIB are the only sinks for O_3, Cl, and ClO, and assume that the concentrations of O_3 and O are maintained at a steady state, which is reasonable. What will happen to the rates of Reactions IIA and IIB, and what will ultimately be the fraction of ozone destroyed by the direct Reaction I?

8. **(a)** Estimate the heat of formation of the ClO(g) radical from bond energy data: $Cl_2 = 243$; $O_2 = 494$; $ClO = 205$ kJ mol^{-1}.

(b) Do you think that this is a very accurate estimate of ΔH_f°? Explain.

(c) What is the importance of the ClO radical in the chemistry of the atmosphere?

(d) Use your estimate of ΔH_f° for ClO to estimate the energetics of the catalytic cycle for the decomposition of ozone, Reactions 5 and 6.

9. The C–Cl bond strength in CFC-12 is 318 kJ mol^{-1}. Estimate the wavelength range over which you would expect this reaction to be possible, and comment on your calculated result.

10. Consider these two reactions involving hydrogen atoms:

(I) $H + O_3 \longrightarrow OH + O_2$ $\quad k_I = 1.4 \times 10^{-10}\ e^{(-470/T)}$

(II) $H + O_2 \xrightarrow{M} HO_2$ $\quad k_{II} = 6.7 \times 10^{-33}\ e^{(+290/T)}$

The units of the rate constants are cm^3 $molec^{-1}$ s^{-1} and cm^6 $molec^{-2}$ s^{-1} respectively.

(a) Calculate the rate of each of these reactions at 15 km and at 50 km, using the data below.

Height	p(total)	$[O_3]$	Temperature
15 km	0.1 atm	1×10^{12} molecules cm^{-3};	215 K
50 km	0.001 atm	8×10^{10} molecules cm^{-3};	280 K

(b) Explain why the catalytic cycle below is more significant at 50 km than at 15 km.

$$H + O_3 \longrightarrow OH + O_2$$
$$OH + O \longrightarrow H + O_2$$

11. Ozone and CFC-11 compete for radiation at wavelengths < 250 nm, but absorption by CFC-11 is **relatively** most favourable near 200 nm. Data on absorption cross sections and solar intensities are given in Table 2.5. The solar intensities strictly refer to those found outside the atmosphere, but should be reasonable in the upper part of the stratosphere. Units of the absorption cross section σ are cm^2 molecule.

Table 2.5:

Wavelength Range	I_o (photons cm^{-2} s^{-1})	σ, O_3	σ, CFC-11
200–210	1.2×10^{13}	2×10^{-19}	4×10^{-19}
210–220	3.4×10^{13}	6×10^{-19}	6×10^{-20}
220–230	5.3×10^{13}	2×10^{-18}	1×10^{-20}
230–240	5.6×10^{13}	6×10^{-18}	1×10^{-21}
240–250	5.7×10^{13}	1×10^{-17}	2×10^{-22}
250–260	8.7×10^{13}	1×10^{-17}	3×10^{-23}
260–270	2.7×10^{14}	9×10^{-18}	1×10^{-23}
270–280	2.5×10^{14}	5×10^{-18}	≈ 0
280–290	4.0×10^{14}	2×10^{-18}	≈ 0
290–300	6.9×10^{14}	8×10^{-19}	≈ 0

Calculate the relative importance of direct photolysis of ozone and the CFC-11 initiated chain reaction under the following assumptions: every photon absorbed causes cleavage of CFC-11 or O_3 as appropriate, and one chlorine atom released by CFC-11 decomposes 2×10^4 molecules of O_3. Assume $p(\text{total}) = 1 \times 10^{-4}$ atm, $p(O_3) = 10$ ppm, and $p(\text{CFC-11}) = 1$ ppb.

12. A natural decomposition route for N_2O is a photochemical reaction

$$N_2O \xrightarrow{h\nu} N_2 + O^*$$

(a) Calculate the maximum wavelength needed to bring about this reaction from the information: ΔH_f° (N_2O, g) = 82 kJ mol^{-1} ΔH_f° (O, g) = 247 kJ mol^{-1}; excitation energy of O^* = 188 kJ mol^{-1}.

(b) Suggest an explanation for the fact that the reaction

$$N_2O \xrightarrow{h\nu} N_2 + O^*$$

does not occur in the troposphere but does occur in the stratosphere.

(c) Calculate ΔG for the thermal reaction

$$N_2(g) + O(g) \longrightarrow N_2O(g)$$

in the stratosphere, assuming $p(O,g) = 2.0 \times 10^{-4}$ ppm, total pressure = 0.010 atm, and $p(N_2O,g) = 1.6 \times 10^{-3}$ ppm. Explain whether the reaction is likely to be an important atmospheric source of N_2O.

(d) The lifetime of N_2O in the atmosphere is reported to be 100 years. What is the apparent first order rate constant for the disappearance of N_2O ?

13. In the stratosphere, the pressure at an altitude of h km can be approximated by $p(\text{total}) = p_o\, e^{(-0.14h)}$, where p_o is the atmospheric pressure at sea level (1 atm). The concentration of O_3 at any altitude is mainly governed by the rate of O_2 photolysis. The absorption of light is governed by the following relationship:

$$I_{\text{transmitted}} = I_{\text{incident}} \bullet \exp(-n(O_2) \bullet \sigma(O_2) \bullet d)$$

$n(O_2)$ is the O_2 concentration (molec cm^{-3}) and σ is in cm^2 molec^{-1}; d is the depth of atmosphere through which the light passes.

(a) Derive an expression to show how the ozone concentration varies with altitude.

(b) Using an average value for $\sigma(O_2)$ of 2×10^{-18} cm^2 molec^{-1} and total incident photon flux of 1.5×10^{14} photons cm^{-2} s^{-1} in the range absorbed by oxygen, plot out the variation of the rate of forming ozone with altitude.

14. One gram of *o*-hydroxybenzophenone is dissolved in 50 cm^3 of ethanol, whose heat capacity is 113 J mol^{-1} K^{-1}, density 0.79 g cm^{-3}. How many photons of wavelength 325 nm must be absorbed to raise the temperature of the solution by 1°C, assuming no heat losses?

Chapter 3

Tropospheric Chemistry

Introduction

The following survey will show that tropospheric chemistry is dominated by the oxidation of trace atmospheric components. As a result of these oxidations, organic compounds such as methane and other hydrocarbons are converted into carbon dioxide and water. Under conditions where the atmosphere contains excessive amounts of oxidizable material, some of the intermediate compounds formed during oxidation can build up to cause pollution problems. One such pollution issue is **photochemical smog,** which will be a major focus of this chapter. We will also consider the effects of particulate matter in the atmosphere, but will defer until Chapter 6 discussion of another major tropospheric pollution concern: acid rain.

3.1 The hydroxyl radical as an oxidant

The hydroxyl radical is central to the chemistry of the troposphere. The purpose of this section is to provide an overview of its reactions and its formation.

The hydroxyl radical OH carries no charge, and is therefore chemically distinct from the hydroxide ion, OH^-. It is extremely reactive, and on this account it has a very short tropospheric half-life (Problem 1). From this we can infer that OH is continually being formed and consumed in the troposphere, just as ozone is continually produced and destroyed in the stratosphere. Parenthetically, the same hydroxyl radical is also very reactive in solution; it is formed by the action of ionizing radiation upon water, and is responsible for the toxicity of ionizing radiation. The two characteristic reactions of the hydroxyl radical are:

1. Abstraction of a hydrogen atom from a suitable substrate. Example:

$$(1) \qquad OH(g) + CH_4(g) \longrightarrow CH_3(g) + H_2O(g)$$

2. Addition to an unsaturated centre. Example:

$$(2) \qquad OH(g) + NO_2(g) \xrightarrow{M} HNO_3(g)$$

Contrast the hydroxide ion, which always acts as a base. Example:

$$OH^-(aq) + HF(aq) \longrightarrow H_2O(l) + F^-(aq) \quad (3)$$

Hydrogen abstraction is the preferred reaction for most substrates which contain hydrogen. This is explained by the very high O–H bond energy in water; the H–OH bond strength is much greater than those of saturated C–H bonds (Problem 2) as in Equation 1. Addition is the normal reaction with unsaturated organic compounds such as benzene derivatives, whose C–H bonds are stronger[1]. Among the few organic substances which are not attacked by hydroxyl radicals are chlorofluorocarbons; this is the reason the latter are so inert in the troposphere (see Chapter 2). Hydrogen-containing CFC derivatives such as CFC-22 and CFC-134a react with OH but only slowly, because of the deactivating influence of the halogen substituents (Table 3.1).

Table 3.1: Rate constants (298 K) for the reaction of the hydroxyl radical with some organic compounds[2].

Substrate	k, cm^3 $molec^{-1}$ s^{-1}	Substrate	k, cm^3 $molec^{-1}$ s^{-1}
CH_4	6.3×10^{-15}	C_2H_6	2.7×10^{-13}
C_6H_6	1.3×10^{-12}	CH_3Cl	4.4×10^{-14}
CFC-22	4.7×10^{-15}	CFC-134a	8.6×10^{-15}
CFC-11, CFC-12[a]	$< 5 \times 10^{-16}$		

[a] Upper limits only

Because hydroxyl is a free radical (i.e., it contains an unpaired electron), its reactions with "even electron" substances such as methane give a free radical as a product. These free radicals, CH_3 in the case of methane, are very reactive towards molecular oxygen, Equation 3.

$$CH_3 + O_2 \xrightarrow{M} CH_3OO \quad (4)$$

The reaction sequence 1, 4 leads to the replacement of one of the C–H bonds in methane by a C–O bond. It is therefore easy to see in principle that successive reactions of this type will eventually cause the complete oxidation of the organic substrate. As an aside, we may note that the mechanism of oxidation of organic compounds in the atmosphere is very similar to the mechanism of combustion in flames.

[1] We must be careful not to over-generalize. Both addition and abstraction occur with such hydrocarbons as 2-butene and toluene, where abstraction takes place at the allylic and benzylic centres respectively.

[2] Data from R. Atkinson, "Kinetics and mechanisms of the gas-phase reactions of the hydroxyl radical with organic compounds under atmospheric conditions," *Chem Rev*, **1986**, 86, 69–201. The value for methane is taken from G.L. Vaghjiani and A.R. Ravishankara, *Nature,* **1991**, 350, 406-409.

The major route for the formation of the hydroxyl radical in the troposphere occurs by a complicated mechanism which is driven by sunlight.

$$NO_2 \xrightarrow{h\nu,\ \lambda < 400\ \mathrm{nm}} NO + O \tag{5}$$

$$O + O_2 \xrightarrow{M} O_3 \tag{6}$$

$$O_3 \xrightarrow{h\nu,\ \lambda < 320\ \mathrm{nm}} O_2^* + O^* \tag{7}$$

$$O^* + H_2O \longrightarrow 2OH \tag{8}$$

Other sources of OH are shown in Equation 9–11, but they contribute little to the overall rate of forming OH under most conditions. An exception is Reaction 10, which can give a "pulse" of OH shortly after sunrise under some conditions of local air pollution.

$$O^* + CH_4 \longrightarrow OH + CH_3OH \tag{9}$$

$$HNO_2 \xrightarrow{h\nu} OH + NO \tag{10}$$

(HNO_3 reacts similarly)

$$H_2O_2 \xrightarrow{h\nu} 2OH \tag{11}$$

The major pathway, Equation 5–8, will now be discussed in detail[3].

Reaction 5: Nitrogen dioxide is a natural trace component of the troposphere. Its immediate precursor is nitric oxide, NO, and because NO and NO_2 are continuously cycled back and forth, it is the sum of their concentrations which is usually reported. This sum is commonly designated NO_X (Problem 3).

Nitric oxide is formed whenever air is heated, through Equilibrium 12.

$$N_2 + O_2 \rightleftharpoons 2NO \qquad \Delta H^\circ = +180\ \mathrm{kJ\ mol^{-1}} \tag{12}$$

Because the reaction is endothermic, the equilibrium concentration of NO increases with temperature (Problem 4). Lightning and combustion both cause NO to be formed, and because the reverse of Reaction 12 is slow, any NO formed at the elevated temperature tends to get "frozen in" when the air cools. That is, suppose NO is produced in an automobile engine: the exhaust gases mix with the cool ambient air, and at equilibrium most of the NO should revert to N_2 and O_2. However, the reaction is slow kinetically, and so a non-equilibrium situation is maintained (Problem 5).

Nitric oxide is oxidized in the troposphere to nitrogen dioxide, so that the "odd nitrogen" species cycle back and forth between NO and NO_2. At the concentrations at which it is found in the troposphere (< 1 ppm) NO is not oxidized by molecular oxygen at an appreciable rate, since the reaction is third order kinetically:

$$\mathrm{rate} = k\,[NO]^2\,[O_2]$$

[3]It will be helpful to memorize Equation 5–8.

This seems peculiar to anyone who has prepared NO in the laboratory, where it reacts "instantaneously" with oxygen (Problem 6). In fact, the slow rate of oxidation in the atmosphere is just a concentration effect, since the rate depends on the second power of the NO concentration.

Since the direct reaction of NO with O_2 is slow in the environment, oxidation takes other routes, involving oxidants such as O_3, HO_2, and RO_2.

$$\text{(13)} \qquad NO + O_3 \longrightarrow NO_2 + O_2$$

$$\text{(14)} \qquad NO + HO_2 \text{ (or } RO_2\text{)} \longrightarrow NO_2 + OH \text{ (or OR)}$$

Nitrogen dioxide is a brown gas, which absorbs in both the visible and the ultraviolet regions of the spectrum. However, visible light is not sufficiently energetic to cause N–O bond cleavage, which takes place only with radiation having wavelength less than about 400 nm (Problem 7). The oxygen atom is formed in its electronic ground state, and is a different species from that produced in Reaction 7.

Reactions 6 and 7: These reactions are familiar from Chapter 2. The formation of ozone, Reaction 6, is faster than in the stratosphere because of the higher concentrations of both O_2 and the third body M (Problem 8). Photochemical cleavage of ozone yields O_2 and O, either both in their ground states or both in their excited states, as was discussed in Chapter 2. Only the shortest wavelengths reaching the troposphere are capable of dissociating ozone into O^* (refer back to Problem 2.5), and the quantum yield for Reaction 7 falls steeply at wavelengths greater than 300 nm, reaching approximately zero by 320 nm[4].

Reaction 8: Although this reaction has a large rate constant at temperatures near 25°C, only a small fraction of all the excited oxygen atoms formed in Reaction 7 actually go on to afford hydroxyl radicals. This is because O^* is rapidly deactivated through collisions with air molecules, the excitation energy of O^* being converted into kinetic energy in the process, Equation 15.

$$\text{(15)} \qquad O^* + M \longrightarrow O + M + \text{kinetic energy}$$

Since M is much more abundant than H_2O, Reaction 15 can predominate over Reaction 8, although the actual proportion of reaction by each route depends on the relative humidity. Ground state oxygen atoms are insufficiently energetic to bring about Reaction 8 (Problem 9).

3.2 Oxidation of carbon monoxide by OH

Carbon monoxide is a natural tropospheric constituent ($\approx$ 0.1 ppm). It is formed biologically, by incomplete combustion (e.g., forest fires and human activities), and as an intermediate in the tropospheric oxidation of

[4]B.J. Finlayson-Pitts and J.N. Pitts, Jr. *Atmospheric Chemistry,* Wiley-Interscience, New York, 1986, Section 3-C-2b.

substances such as methane. Incomplete combustion can lead to local concentrations of CO which are greatly in excess of 0.1 ppm. For example, city air may often contain 2–20 ppm of carbon monoxide, and peak concentrations up to 100 ppm have been recorded in vehicular subways and tunnels. This is on the threshold of representing a health hazard, since carbon monoxide combines irreversibly with the body's hemoglobin, making it unavailable for transporting oxygen. At 100 ppm, up to 15% of a person's hemoglobin would be converted to carboxyhemoglobin at equilibrium. Frequent motor vehicle testing programs and switching to oxygenated fuels have been proposed as ways to reduce urban CO levels, but the value of these strategies has recently been questioned[5].

The tropospheric sinks for carbon monoxide are uptake in soil, followed by microbial oxidation to CO_2, and atmospheric oxidation by OH (Problem 10). The latter reaction (Equation 16) is the only known gas phase tropospheric sink for carbon monoxide. Reaction 16 produces hydrogen atoms, which are the precursors of hydroperoxy radicals HO_2.

$$\text{(16)} \qquad \mathrm{OH + CO \longrightarrow H + CO_2} \qquad k = 2.7 \times 10^{-13} \text{ at } 300\text{ K}$$

$$\text{(17)} \qquad \mathrm{H + O_2 \xrightarrow{M} HO_2}$$

Reaction 16 does not seem to fit into the generalization that OH reacts either by hydrogen abstraction or by addition; however, Reaction 16 has been suggested to be a two-step process rather than an elementary reaction[6].

$$\text{(16a)} \qquad \mathrm{OH + CO \longrightarrow HOCO}$$

$$\text{(16b)} \qquad \mathrm{HOCO \longrightarrow H + CO_2}$$

In the unpolluted troposphere, 70% of all hydroxyl radicals disappear through reaction with CO; most of the remainder react with methane.

3.3 Oxidation of methane

Methane is the simplest hydrocarbon. Even so, its mechanism of oxidation to CO_2 and H_2O is very complex. Much more complicated mechanisms have been written to rationalize the oxidation of larger hydrocarbons, but they are not well established. We have already met the first two steps in this reaction.

$$\text{(1)} \qquad \mathrm{OH + CH_4 \longrightarrow CH_3 + H_2O}$$

$$\text{(4)} \qquad \mathrm{CH_3 + O_2 \xrightarrow{M} CH_3OO}$$

[5] D.H. Stedman, "Automobile carbon monoxide emission," *Environ Sci Technol,* **1989,** 23, 147–149.

[6] R.P. Wayne, *Chemistry of Atmospheres,* Oxford University Press, Oxford, England, **1985,** 186.

In what follows, RH represents any substance (including CH_4) bearing abstractable hydrogen atoms. Once CH_3O_2 has been formed, these reactions occur:

$$(17) \quad CH_3O_2 + HO_2 \longrightarrow CH_3OOH + O_2$$

$$(18) \quad CH_3O_2 + NO \longrightarrow CH_3O + NO_2$$

(HO_2 also oxidizes NO to NO_2)

$$(19) \quad CH_3OOH \xrightarrow{h\nu} CH_3O + OH$$

$$(20) \quad CH_3OOH + OH \longrightarrow CH_3O_2 + HO_2$$

or $H_2O + CH_2O + OH$

$$(21) \quad CH_3O + RH \longrightarrow CH_3OH + R$$

(followed by oxidation of CH_3OH)

$$(22) \quad CH_2O \xrightarrow{h\nu} H + HCO$$

$$(23) \quad CH_2O + OH \longrightarrow H_2O + HCO$$

$$(24) \quad CH_2O \xrightarrow{h\nu} H_2 + CO$$

$$(25) \quad HCO + O_2 \longrightarrow CO + HO_2$$

$$(16) \quad OH + CO \longrightarrow H + CO_2$$

Even this scheme is by no means complete. The overall picture that emerges is one of successive attack on the C–H bonds of methane and their replacement by CO bonds, or the elimination of H_2O. The analogous steps in the oxidation of higher hydrocarbons can be envisioned by replacing CH_3 by R in the reactions shown above. Where alkenes and aromatic compounds are involved as substrates, addition at the unsaturated carbons must be considered in addition to hydrogen abstraction. The reaction schemes are further complicated because the reactive intermediates such as OH, HO_2, CH_3O_2, etc are involved as reactants or products in so many different reactions. The large number of reactions to be considered makes computer modelling tropospheric chemistry extremely difficult. The modeller needs accurate knowledge of all the rate constants and their temperature dependences, and of the ambient concentrations of all substrates and intermediates. A typical recent study in which the tropospheric concentration of OH was modelled employed 39 elementary reactions[7], and even this list was not exhaustive.

3.4 Photochemical smog

3.4.1 Physical description of photochemical smog

At the outset, it must be stressed that the chemistry of photochemical smog is the *same* as that described in the last two sections: the formation

[7]D. Perner, U. Platt, M. Trainer, G. Hubler, J. Drummond, W. Junkermann, J. Rudolph, B. Schubert, A. Volz, and D.H. Ehhalt, *J Atmos Chem,* **1987**, 5, 185.

of OH, and the oxidation of hydrocarbons. Oxidation in the unpolluted and the polluted troposphere differ only in detail, not in principle. As its name implies, photochemical smog is initiated by sunlight, with the photochemical cleavage of NO_2 to NO and O being the most important initiating step.

Physical characteristics of photochemical smog include a yellow-brown haze, which reduces visibility, and the presence of substances which both irritate the respiratory tract and cause eye-watering. The yellowish colour is due to NO_2, while the irritant substances include ozone, aliphatic aldehydes, and organic nitrates. The four conditions necessary before photochemical smog can develop are:

- nitrogen oxides (NO_X),
- hydrocarbons,
- sunlight,
- temperatures above about 18°C.

Nitrogen dioxide is important as the only tropospheric gas with appreciable absorption in the visible region of the spectrum. Recall from Equations 5–8 that light absorption by NO_2 is the first step in the production of ozone and ultimately the hydroxyl radical. Just as in the unpolluted troposphere, the chemical reactions occurring in photochemical smog involve the attack of hydroxyl radicals on organic substrates.

The temperature requirement arises because many of the reactions involved in oxidation have finite activation energies; 18°C is not an absolute cut-off, but gives an idea of the temperature needed for these atmospheric processes to proceed fast enough for the obnoxious byproducts to build up to the levels associated with air pollution.

Photochemical smog was first recognized as a problem in Los Angeles, California in the 1940's, but has since been documented in many other locations in the United States and elsewhere. As long ago as the 1950's the automobile was identified as the leading contributor to photochemical smog. Los Angeles was the first major United States city to build an extensive freeway system and to rely principally on private automobiles rather than public facilities for transportation.

The evidence against the automobile is illustrated by Figure 3.1. Figure 3.1 is interpreted as follows: Early in the morning pollution levels are low. NO and unburned hydrocarbon concentrations rise as people drive to work. As the Sun rises higher in the sky, NO is converted to NO_2 (Problem 11), and subsequently, levels of ozone and aldehydes increase (see formaldehyde CH_2O in the reaction scheme in Section 3.3). The latter maximize towards midday, when the solar intensity is highest. Notice that the concentration of NO_X falls after about 10 a.m. and does not rise again during the evening rush hour—for explanation, see Section 3.4.2. There is no second peak at the evening rush hour, because by then, the free radical chain reactions are already fully under way.

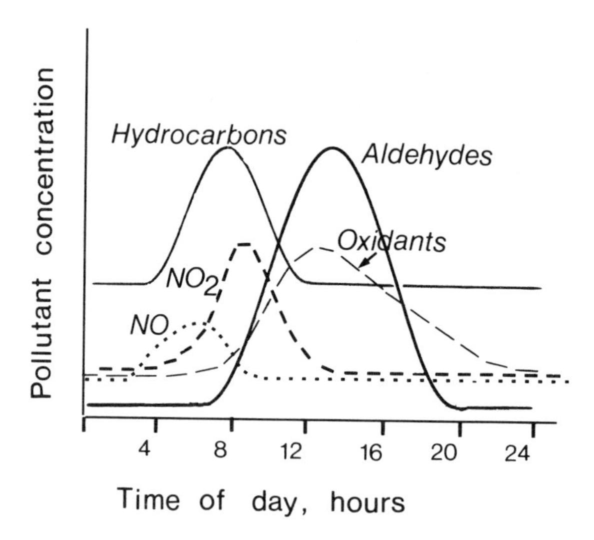

Figure 3.1: Sketch of the diurnal variation in the concentrations of nitrogen oxides, hydrocarbons, ozone and aldehydes under conditions of photochemical smog.

Let us restate the previous paragraph in the context of Equation 5–8. Automobile emissions cause elevated concentrations of NO, which is oxidized to NO_2. Nitrogen dioxide is photolyzed in sunlight, and this reaction proceeds faster the higher the photon intensity. "Ground level" (as opposed to stratospheric) ozone is formed, and its photolysis leads to the formation of OH. Automobile emissions also provide the organic substrates for reaction with OH; intermediates and byproducts—such as aldehydes and organic nitrates—of the oxidation of these substrates to CO_2 and H_2O are the irritating components of the smogs. Many of these reactions are temperature dependent, and so photochemical smog becomes increasingly noticeable the hotter the weather.

All four conditions for photochemical smog must be met simultaneously; consequently, the locations and seasons where this phenomenon is likely to be observed may be predicted. Since automobiles provide the NO_x and hydrocarbons, photochemical smog is a big city phenomenon; however drift of the urban plume can affect neighbouring rural areas. Sunlight and high temperatures are needed; problems may be anticipated near large cities at low latitudes for much of the year, but only in summer at higher latitudes. Readers should consider whether and/or when photochemical smog might be observed in their regions.

Data similar to Figure 3.1 have since been found in many other cities. One observation of significant difference has been made in Toronto, Ontario, where monitoring stations have been set up both at ground level and at 444 m elevation. During an episode of air pollution in July 1988, a pattern similar to Figure 3.1 was seen at the street level sites, but at the elevated site the concentrations of ozone and NO_x showed completely different diurnal behaviour. Ozone showed a baseline level of 0.1 ppm, peaking at 0.15–0.2 ppm in mid-afternoon; NO_X showed baseline levels of 20 ppb, with a peak 0.1–0.2 ppm near noon and a later, smaller peak in the evening. These data may suggest the presence of strong sinks at ground level for these reactive gases. In addition, they suggest that tropospheric ozone may be transported much greater distances than has previously been recognised.

3.4.2 Ground level ozone

Ozone is the chief contributor to "oxidants" in Figure 3.1. It can be a point of confusion that ozone is an undesirable pollutant at ground level, even though it is such a beneficial substance in the stratosphere.

Many countries, and also the World Health Organization, have set standards for the acceptable levels of certain air pollutants. In the United States, seven "criteria pollutants" have been designed (CO, SO_2, O_3, NO_2, non-methane hydrocarbons or "NMHC," suspended particulates, and lead). This section is concerned with CO, O_3, NO_2, and NMHC (Table 3).

Table 3.2: United States and WHO standards for selected air pollutants, in ppm

Pollutant	United States	WHO	Other
CO	35 (1 hour avg)	100 (1 hour avg)	W. Germany, 26 (0.5 hour avg)
	9 (8 hour avg)	30 (8 hour avg)	Canada, 13 (8 hour avg)
O_3	0.12 (1 hour avg)	0.10 (1 hour avg)	Japan, 0.06 Canada, 0.08 (1 hour avg)
NO_2	0.05 (annual avg)		USSR, 0.05 (24 hour avg)
NMHC	0.24 (3 hour avg)†		Canada, 0.24

† Averaged over the period 6 a.m. to 9 a.m.

The concentration of ozone is relatively easy to measure, and correlates well with the severity of an episode of photochemical smog. Consequently, the concentration of ozone is monitored as an indicator of air pollution. In the United States the term "ozone exceedance" is coming into use to describe episodes during which the air quality standard of 0.12 ppm is violated. In many jurisdictions, regulations exist for cutting back or even closing down certain industries when air pollution levels are excessive.

In clean air, ozone concentrations are ca. 20–50 ppb, with both diurnal and seasonal variation. The only tropospheric sources of ozone are from photolysis of natural or anthropogenic NO_2 (Equations 5 and 6) and at some seasons, transport downwards from the stratosphere[8].

Depletion of stratospheric ozone will allow UV radiation to penetrate more deeply into the atmosphere, and will therefore lead to an increase of tropospheric ozone, even outside urban areas. Such a change may already be under way[9] High ozone levels appear to be statistically associated with high summer temperatures, underscoring the importance of temperature as a contributor to photochemical smog, and thus to high ozone levels[10].

Other factors may contribute to photochemical smog. Geographical features may hinder the dispersal of the pollutant plume; this is a factor in the Los Angeles district, where mountains to the east tend to trap the air close to the city. Temperature inversions[11] and lack of wind both serve to localize the pollutant plume and hinder its dispersal.

Ground-level concentrations of ozone appear to be increasing in urban areas[12], and "of the six gaseous 'criteria' pollutants for which air quality standards have been established, ozone remains the most resistant to efforts at abatement."[13]

3.4.3 Chemical aspects of photochemical smog

In chemical terms, the simplest way of considering the interplay between NO_X and ground level ozone is in terms of the reactions below.

$$NO_2 \xrightarrow{h\nu,\ \lambda < 400\,\text{nm}} NO + O \tag{5}$$

$$O + O_2 \xrightarrow{M} O_3 \tag{6}$$

$$NO + O_3 \longrightarrow NO_2 + O_2 \tag{13}$$

The combination of these reaction is Equation 26, a pseudo-equilibrium, which is driven photochemically in one direction and thermally in the other (Problems 11 and 12).

$$NO_2 + O_2 \underset{\text{thermal}}{\overset{h\nu}{\rightleftharpoons}} NO + O_3 \tag{26}$$

[8] J.A. Logan, "Tropospheric ozone: seasonal behavior, trends, and anthropogenic influence," *J Geophys Res,* **1985**, 90, 10643.

[9] A.M. Hough and R.G. Derwent, "Changes in the global concentration of tropospheric ozone due to human activities," *Nature,* **1990,** 344, 645–648.

[10] V. Pagnotti, "Seasonal ozone levels and control by seasonal meteorology," *J. Air Waste Management Assoc.,* **1990,** 40, 206–210.

[11] A temperature inversion occurs when the normal pattern of decreasing temperature with altitude is interrupted by an upper layer of warm air. Since warmer air naturally rises, the warm upper layer acts as a "lid," sealing in the cooler air beneath, and preventing it from dispersing by upward diffusion.

[12] I. Colbeck, "Photochemical ozone pollution in Britain," *Sci Progr,* Oxford, **1988**, 72, 207–226; see also J. Gribbin, "Ozone in smog blocks harmful ultraviolet," *New Scientist,* November 4, 1989, 33.

[13] J.H. Seinfeld, "Urban air pollution: state of the science," *Science,* **1989**, 243, 745–752.

In the absence of other reactions between these gases, the relative concentrations of NO, NO_2 and O_3 should depend upon the solar flux (i.e., upon the season, geographical location, and time of day). Thus at 40°N, the concentration of O_3 is approximately given as $[O_3] = 20 \times [NO_2]/[NO]$ in the absence of urban air pollution.

The peak levels of ground level ozone under conditions of photochemical smog cannot be predicted just from Equilibrium 26. The concentrations of NMHC as well as NO_X must be considered, with maximal $[O_3]$ being produced when the ratio $[NMHC]/[NO_X]$ is near 5. Since the ratio $[NMHC]/[NO_X]$ under conditions of photochemical smog is commonly near 10, it follows that reduction in [NMHC] alone cannot significantly improve urban air quality. The reason is that hydrocarbons are, indirectly, sources of ground level ozone. Hydrocarbons (or, strictly, the RO_2 radicals formed during their oxidation) produce ground level ozone through Reaction 14, since the NO_2 thus formed is the precursor of ozone by Reactions 5 and 6.

(14) $$RO_2 + NO \longrightarrow RO + NO_2$$

Equilibrium 26 is also disturbed by the removal of NO_2, which reacts with OH

(27) $$NO_2 + OH \xrightarrow{M} HNO_3$$

Reaction 27 is important under conditions of urban pollution when both $[NO_2]$ and [OH] tend to be high. An interesting consequence of Reaction 27 is the observation that ground level ozone is frequently present at higher concentration down-wind of an urban area than in the city itself. The clue is that the NO_X concentration follows the opposite trend. We recall that OH is formed from O_3 (Reactions 7 and 8); therefore the production of HNO_3 in Reaction 28 removes material from both the left (NO_2) and right (O_3) hand sides of pseudo-equilibrium 26. However, since one O_3 ultimately yields **two** OH radicals (Equation 7), NO_X is depleted more rapidly than O_3.

The influence of Reaction 27 is also seen in the variation of the concentrations of NO_X and O_3 with time of day under smoggy conditions. In Figure 1, the NO_X concentration peaks about 8 a.m., and the oxidant concentration rises as NO_X falls. This suggests a sink for NO_X, as opposed to the cycling back and forth between NO and NO_X implied by Equilibrium 26. The OH radical, generated from O_3, is the substance which removes NO_2 through Reaction 27. This is evident from the recent work of Dorn[14] *et al* who found that at concentrations of $NO_2 >$ ca. 2 ppb, the concentration of OH remains low, even under strong photolysis (see Section 3.5).

We will now try to identify the differences between normal oxidation processes in the troposphere and photochemical smog formation. As already stated, the basic chemistry is identical. The main difference is that

---

[14] H.P. Dorn, J. Callies, U. Platt, and D.H. Ehhalt, "Measurement of tropospheric OH concentrations by laser long-path absorption spectroscopy," *Tellus,* **1988,** 40B, 437–445

photochemical smog occurs when the concentrations of NO_X and hydrocarbons are higher than normal. This means that photoinitiation through dissociation of NO_2 proceeds faster and that the concentrations of the reactive intermediates are higher than normal. Some of the latter substances are responsible for the adverse effects. Ozone is toxic at low concentrations (< 1 ppm); although stratospheric ozone makes the Earth more hospitable to living organisms, ground level ozone is a toxic pollutant when it occurs at elevated levels[15]. Likewise, the concentrations of aldehydes are higher under conditions of photochemical smog. Table 3.2 gives the concentrations of some of the important pollutants under conditions of photochemical smog.

Table 3.3: Typical pollutant concentrations (ppb) in unpolluted air and under conditions of photochemical smog (from Reference 5, 369)

Pollutant	Unpolluted remote area	Heavily polluted area
CO	< 200	10,000–50,000
NO_X	< 1	1000–3000
O_3	< 50	100–500
Hydrocarbons†	< 65	> 1500
PAN††	< 0.05	20–70
$CH_2{=}O$	< 2	20–75

† Excluding methane (ca. 1500 ppb), most of which is of natural origin
†† Peroxyacetyl nitrate

Aldehydes may be formed by several routes: as intermediates in the oxidation of alkanes and alkenes, initiated by hydroxyl radicals, and by direct attack of ozone upon alkenes. Attack by OH is the dominant sink: e.g., Equation 28.

$$(28) \qquad CH_3CH{=}O + OH \longrightarrow CH_3C{=}O + H_2O$$

One substance which is of particular importance in the present context is peroxyacetyl nitrate (PAN), which is formed from acetaldehyde. Higher homologues such as peroxypropionyl nitrate are also present. Like ozone, peroxyacyl nitrates are toxic and irritant at very low concentrations (< 0.1 ppm). PAN had not been detected in the atmosphere before it was

[15] An interesting historical note is that ozone was formerly believed to be beneficial to health. Ocean-side vacations were perceived to be healthful because ocean air was thought to contain elevated concentrations of ozone. At one time entrepreneurs marketed ozonizers so that people could experience the benefits of a sea side holiday while staying at home. In the 1960's ozonizers enjoyed another brief period of popularity as air fresheners, oxidizing away unwanted smells. In both cases, potentially harmful concentrations of the gas would have been generated. Ozone is formed when an electrical discharge passes through oxygen, and may be smelled around electrical transformers, e.g., those used with model railways.

recognized in photochemical smog, although it presumably occurs in trace amounts.

Figure 3.2 is an outline of the production of PAN from ethane. It follows closely the scheme for oxidation of methane (Section 3.3). It is not intended to imply that PAN is the major product; other reactions lead eventually to CO_2 as the chief product.

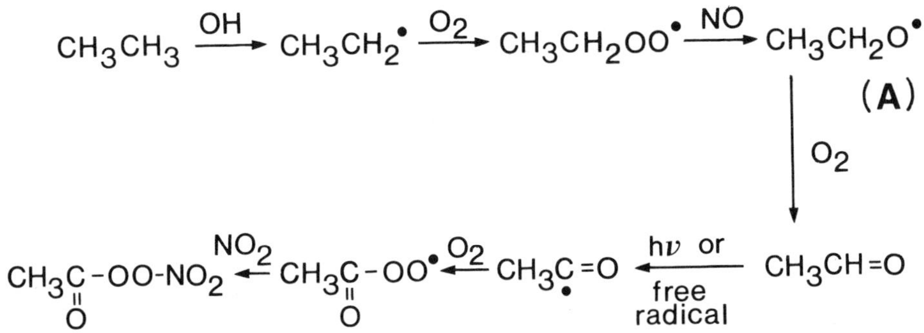

$$CH_3CH_3 \xrightarrow{OH} CH_3CH_2^{\bullet} \xrightarrow{O_2} CH_3CH_2OO^{\bullet} \xrightarrow{NO} CH_3CH_2O^{\bullet}\ (\mathbf{A})$$

$$CH_3CH_2O^{\bullet} \xrightarrow{O_2} CH_3CH{=}O$$

$$CH_3CH{=}O \xrightarrow[\text{free radical}]{h\nu \text{ or}} CH_3\dot{C}{=}O \xrightarrow{O_2} CH_3C({=}O){-}OO^{\bullet} \xrightarrow{NO_2} CH_3C({=}O){-}OO{-}NO_2$$

Figure 3.2: Production of PAN from ethane.

As an aside, note that the formation of CO_2 from larger alkanes requires fragmentation steps. An example of such a reaction is shown in Equation 29, where intermediate (B) is the analog of the ethoxy radical (A) in Figure 3.2. PAN can act as a temporary sink for NO_X. It is thermally labile, and reverts to NO_2 and the acetylperoxy radical with an activation energy of 112 kJ mol^{-1}, thereby re-initiating free radical oxidation of hydrocarbons. If air containing PAN cools quickly, the PAN may be transported to remote locations (Problem 13). This is believed to be the manner in which PAN has come to be found in the Arctic and the mid-Atlantic, far from any industrial source.

$$CH_3CH_2CH(CH_3)CH_3 \xrightarrow[\text{steps}]{\text{several}} \underset{(\mathbf{B})}{CH_3CH_2C(CH_3)(O^{\bullet})CH_3} \longrightarrow CH_3CH_2^{\bullet} + CH_3C({=}O)CH_3 \qquad (29)$$

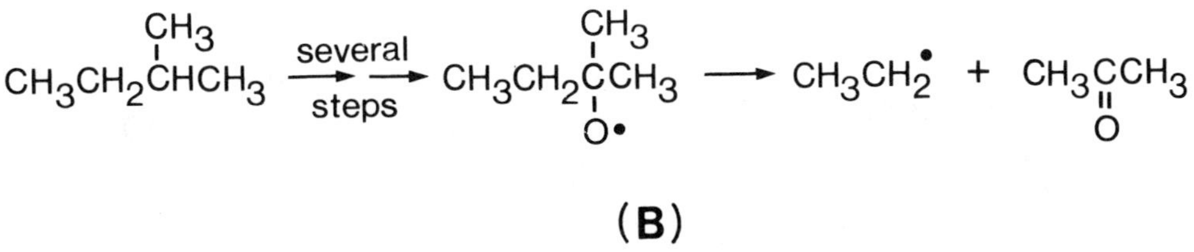

A more important temporary sink for NO_X is HNO_3, which is formed by the reaction of OH with NO_2 (Reaction 27). The HNO_3 either photolyses, releasing NO_2 and OH again, or is removed permanently from the atmosphere by deposition in rain or adsorbed on particles. (This gives a link between photochemical smog and acid rain. As discussed in Chapter 6, acid rain on the west coast of the United States is principally nitric acid, while the eastern part of the continent experiences deposition of sulfuric and nitric acids, typically in a 3:1 ratio.)

The scheme below summarizes the tropospheric chemistry of NO and NO_2.

$$N_2 + O_2 \xrightarrow{\text{heat}} 2NO \underset{h\nu}{\overset{O_3,\ HO_2,\ \text{other oxidants}}{\rightleftharpoons}} NO_2 \underset{h\nu}{\overset{OH}{\rightleftharpoons}} HNO_3 \longrightarrow \text{deposition}$$

While this scheme includes the most important processes, two other nitrogen oxides, N_2O and NO_3, are now known to be active in the troposphere.

N_2O: This substance was discussed in Chapters 1 and 2. It is formed by biological denitrification, and is inert in the troposphere, except as a greenhouse gas. In the stratosphere it undergoes photochemical cleavage, and is potentially a source of ozone depletion. Recently, N_2O has also been shown to be produced anthropogenically as a combustion byproduct, along with NO_X. Typical NO_X:N_2O ratios in combustion products are about 5:1[16].

NO_3: The nitrate free radical is uncharged, and is chemically distinct from the nitrate anion. It is formed in the atmosphere by oxidation of NO_2.

$$NO_2 + O_3 \longrightarrow NO_3 + O_2 \tag{30}$$

Of the various tropospheric intermediates we have discussed, NO_3 is unique in that its reactions are more important at night than during daylight. NO_3 can be formed at night—as well as during the day—through Reaction 30, especially in polluted urban atmospheres[17] Due to rapid photolysis, its daytime concentration is very small.

$$NO_3 \xrightarrow{h\nu,\ \lambda < 670\,\text{nm}} NO_2 + O \tag{31}$$

However, not all the NO_3 is destroyed, because it can also be converted to N_2O_5, which acts as a temporary reservoir.

$$NO_3 + NO_2 \rightleftharpoons N_2O_5 \tag{32}$$

N_2O_5 dissociates thermally (i.e., during both day and night), and releases NO_3 again. In the absence of photodecomposition, major sinks for NO_3 are

[16] W.S. Lanier and S.B. Robinson, "EPA workshop on N_2O emission from combustion," EPA/600/S8-86/035 (1987). For a criticism of this conclusion, see R.K. Lyon, J.C. Kramlich, and J.A. Cole, "Nitrous oxide: sources, sampling, and science policy," *Environ Sci Technol,* **1989**, 23, 392–393.

[17] U. Platt *et al,* "Peroxy radicals from night-time reactions of NO_3 with organic compounds," *Nature,* **1990**, 348, 147–149.

hydrogen abstraction and addition to unsaturated centres. Thus NO_3 plays a similar role in hydrocarbon oxidation at night as the hydroxyl radical does during the day, although NO_3 is intrinsically a less reactive radical [18] than OH.

$$RH + NO_3 \longrightarrow HNO_3 + R \;\; (\xrightarrow{O_2} ROO, \text{ etc.}) \tag{33}$$

Hydrogen abstraction by NO_3 is a significant source of tropospheric HNO_3.

The concentration of NO_3 is reduced at high relative humidity. The lower concentration is probably the result of destruction of the N_2O_5 precursor by water.

$$N_2O_5 + H_2O \longrightarrow 2HNO_3 \tag{34}$$

The reader will appreciate by now that photochemical smog is simply an exaggerated manifestation of the natural oxidation chemistry which occurs in the troposphere. It may reasonably be regarded as the mechanism by which the troposphere cleans itself.

We have stressed the similarity between photochemical smog and normal oxidation in the troposphere. The following example makes the point rather graphically. The Great Smoky Mountains, at the southern end of the Appalachian chain, are well known, and rightly so, for the beautiful bluish haze which hangs over them. In fact, this haze is a kind of natural air pollution. The region is heavily forested with pine trees, which give off the characteristic odour of the pine forest into the atmosphere. This odour is due to hydrocarbons called terpenes, which are secreted by the trees. The haze is an **aerosol**[19] of fine particles and droplets of partly oxidized hydrocarbons, initiated by reactions with OH. The only difference between this and photochemical smog is that the concentration of the other necessary ingredient, NO_X, is normal. Consequently the obnoxious by-products, ozone and PAN, are not formed at abnormal concentrations.

We have not yet discussed the origin of the haze which accompanies photochemical smog, and which is also present in the Great Smoky Mountains. Oxidation of hydrocarbons produces, *en route* to CO_2 and water, oxygenated intermediates such as alcohols, carbonyl compounds, and carboxylic acids, all of which are substantially less volatile than the hydrocarbons from which they were formed (Problem 14). These substances may therefore condense into sub-micrometer sized droplets of liquid. Because of their small size, these droplets remain suspended in the atmosphere and also scatter visible light (see Section 3.7, below). Light scattering by this aerosol is the cause of the reduced visibility.

---

[18] R. Atkinson, S.M. Aschmann, and J.N. Pitts, jr., "Rate constants for the gas-phase reactions of the NO_3 radical with a series of organic compounds at 296 ± 2 K," *J Phys Chem*, **1988**, 92, 3454–3457.

[19] An aerosol includes both the particles and the air in which the particles are suspended. (The word particle is most often used for solids, and aerosol for liquid suspensions. Nevertheless, it is correct to speak of liquid particles or of aerosols containing solid particles.)

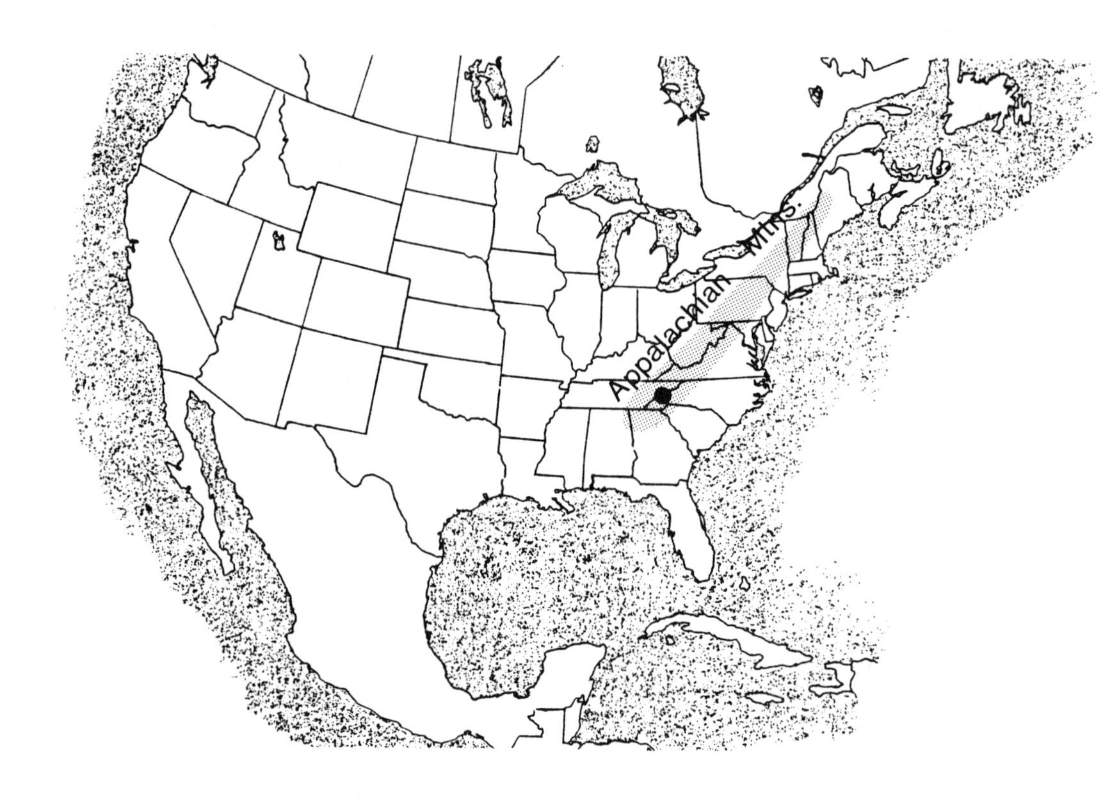

Great Smoky Mountains

To conclude this section it may be helpful to contrast pollution by NO_X and hydrocarbons with that due to CFC's (Chapter 2). Pollution by CFC's is a global problem; CFC's have such long residence times that they become uniformly dispersed around the globe. Their potential problems could be with us for several generations. Photochemical smog is a much more local problem; the atmosphere cleans itself in a few days once the pollution source is removed. Although its effects are both unpleasant and economically damaging locally, photochemical smog is an insignificant global threat to the environment, compared with CFC pollution.

3.4.4 Consequences and remedies of photochemical smog

The consequences of photochemical smog are mainly due to the toxicity of intermediates such as ozone and PAN. Experiments have shown that respiratory impairment results from breathing air containing more than about

0.1 ppm of ozone[20]. Most people have heard of instances in which Los Angeles school-children have been kept in class at recess, to stop them from running around and breathing the polluted air deeply. Fortunately, such episodes are almost a thing of the past in Los Angeles, as pollution control measures have greatly improved air quality in this area. Instead, serious problems now exist in other fast-growing urban regions such as Rio de Janeiro and Mexico City.

Although ozone and PAN are toxic to animals and to people, especially those with respiratory afflictions such as asthma and emphysema, they are much more toxic to plants. At 0.1 ppm of ozone, even sensitive people such as asthmatics, show little overt adverse reaction; in plants however, the rate of photosynthesis is reduced by a factor of more than two. This is a very serious issue in California, where agriculture is such an important industry. The losses in agricultural production resulting from gases such as ozone and PAN in the atmosphere have been estimated in the billions of dollars annually in California alone[21]. Even in Ontario—not a region particularly known for a photochemical smog problem—losses through reduced yields of a single crop, white beans, have been estimated at $20 million (Cdn) annually[22].

The toxicity of ozone can be understood in broad terms because it is such a powerful oxidant. With ΔG_f° for $O_3(g)$ = +163 kJ mol^{-1}, we can appreciate that ozone is a much more powerful oxidant even than elemental oxygen (from which living cells require internal protection mechanisms). Cell membranes are particularly susceptible to attack because they contain unsaturated fatty acid residues; alkenes are well known to suffer oxidation (ozonolysis) by gaseous ozone. In another context, atmospheric ozone substantially accelerates the decomposition of rubber products: e.g., the cracks that are seen in the walls of older automobile tires.

The recognition that the automobile was a major contributor to photochemical smog made possible a partial solution to the problem in the form of emission controls. Not surprisingly, the State of California was the first jurisdiction to institute emission controls on automobiles, beginning in the 1975 model year. Unburned hydrocarbons and nitrogen oxides are the emissions which are regulated. Since that time, emissions have been regulated

---

[20]M. Lippmann, "Health effects of ozone," *J Air Pollut Control Assoc,* **1989**, 39, 672–695; B.E. Tilton, "Health effects of tropospheric ozone," *Environ Sci Techol,* **1989**, 23, 257–263.

[21]The figure $13 billion annually for health expenses, property damage, and agricultural loss has been quoted for southern California: I. Anderson, *New Scientist,* April 1, 1989, 21. Showman (*J Air Waste Management Assoc.,* **1991**, 41, 63–64) quotes $3 billion as the annual loss to U.S. agriculture. Ozone damage is less under conditions of drought, because drought induces stomatal closure, which keeps ozone away from the sensitive interior of the leaf.

[22]These estimates are made by comparing the rates of growth, and of yield, of crop plants grown in environmentally-controlled chambers and exposed to defined concentrations of toxic gases such as ozone. Relatively few crop species have been studied, and extrapolation of the laboratory data to the field is difficult.

by a large number of governments, but California retains the distinction of having the strictest standards (unburned hydrocarbons, 0.25 g km^{-1}; CO, 4.4 g km^{-1}; NO_X, 0.25 g km^{-1})[23]. Unfortunately, advances in technology relating to fuel consumption—CO_2 emissions—and to emissions of CO, NO_X, and NMHC are more than offset by the increasing size of the automobile fleet—and not just in California.

In Section 3.4.2, it was noted that the potential production of ozone and other oxidants depends upon the concentrations of both NMHC and NO_X. Emission control must therefore be directed towards both these pollutants. A problem just recently recognized is that a significant proportion of all NMHC may be of biogenic orgin ($\approx$ 30% in the Los Angeles area, and $\approx$ 50% on the eastern seaboard of the U.S.—compare discussion of the Great Smoky Mountains). Even complete elimination of emissions of vehicular NMHC would therefore not necessarily prevent episodes of photochemical smog.

Emission control devices aim to reduce emissions of hydrocarbons and nitrogen oxides in automobile exhaust to acceptable levels. The functioning of the internal combustion engine allows us to see some of the difficulties which ensue. In the engine, hydrocarbons (gasoline) are burned. There is a stoichiometric relationship between the amount of fuel consumed and the amount of air needed to burn it. For thermodynamic reasons, the operating temperature should be as high as possible in order to maximize the fuel efficiency of the engine. However, the higher the temperature, the greater the potential for forming NO; recall that Reaction 12 is endothermic.

$$(12) \qquad N_2 + O_2 \rightleftharpoons 2NO \qquad \Delta H° = +180 \text{ kJ mol}^{-1}$$

This presents a dilemma. It is difficult to set the fuel:air mixture in an automobile exactly to the stoichiometric ratio, and so most engines are operated with the mixture "lean" (excess air in the mixture) or "rich" (excess fuel). If the mixture is lean, little unburned hydrocarbon should escape in the exhaust gases, but some of the excess oxygen will react with some of the nitrogen in the air used for combustion, forming nitric oxide. As already stated, this is "frozen in" when the hot gases are exhausted to the atmosphere. Alternatively, a "rich" mixture has more than the stoichiometric amount of hydrocarbons, so the excess must inevitably pass out in the exhaust.

[23]United States standards (other than California) show the remarkable improvements which have been made in emission standards over the past two decades: all data in grams per mile. Source: R. Gould, "The exhausting options of modern vehicles," *New Scientist*, May 13, 1989 42–47.

	RH	CO	NO_X
pre 1968	8.8	87	3.6
1975	1.5	15	3.1
1983+	0.41	3.4	1.0

Catalytic converters operate by creating a second zone for combustion outside the engine itself. The canister contains a catalyst which permits the desired chemical reactions to occur at relatively low temperature. The catalysts are finely divided noble metals such as platinum, supported on a matrix of inert oxides. Because the active surfaces of the catalyst are inactivated (poisoned) by lead compounds, leaded gasoline is incompatible with the operation of a catalytic converter, and lead-free gasoline is required[24].

In the early models of catalytic converter, the air:fuel mixture was deliberately made rich, with the aim of limiting the formation of NO. Catalytic oxidation of the excess hydrocarbon was effected in a secondary combustion zone, after admitting additional air. Because the temperature of the secondary combustion was lower, very little NO_X was formed at this stage. The disadvantages of this system were first, that the rich mixture gave poorer fuel economy and tended to foul the spark plugs and valves with soot; second, when the catalytic converter lost its efficiency due to age, even larger amounts of hydrocarbons were emitted than in the absence of any emission control device.

A second generation of catalytic converter, introduced with the 1981 model year, operates in two stages. The mixture entering the engine is approximately in stoichiometric balance between fuel and air. Small amounts of both NO and unburned hydrocarbon therefore leave the primary chamber (the engine). In the first catalytic stage, a ruthenium-based catalyst is used to reduce any NO_X to elemental nitrogen. The reactions involve the production of hydrogen by the reaction of water vapour (a combustion product) with any unburned hydrocarbon, shown here with approximate stoichiometry for methane as an example.

$$2H_2O(g) + 4CH_4 \longrightarrow 2CO + C_2H_2 + 9H_2 \tag{33}$$

The hydrogen thus formed reduces the NO_X over the ruthenium catalyst.

$$H_2 + NO_X \xrightarrow{\text{unbalanced}} N_2 + H_2O \tag{34}$$

The second stage of the converter employs oxidation chemistry; air is admitted, and oxidation proceeds at a sufficiently low temperature that formation of NO_X is minimized. The oxidation step also serves to convert any NO_X which was over-reduced to ammonia (in the first stage) back to molecular nitrogen.

3.5 Tropospheric concentration of OH

In Section 3.3 it was seen that complex kinetic schemes must be used to model tropospheric chemistry. Central to this effort is a good knowledge of

[24]The chemistry of lead compounds in gasoline is covered in Chapter 10. At this point we can note that in North America at least, the introduction of lead-free gasoline was necessitated by legislation requiring the use of catalytic converters, and not by the desire to limit environmental contamination by lead.

the concentration of the hydroxyl radical in the stratosphere, and its diurnal, seasonal, and geographical variation. This is because the hydroxyl radical is the key substance in initiating attack upon the oxidizable materials in both the clean and the polluted troposphere. These reactions are all second order kinetically.

$$\text{rate} = k\,[\text{OH}]\,[\text{oxidizable substrate}] \tag{35}$$

Surprisingly however, the concentration of OH is not well monitored. This is because the OH concentration is difficult to measure. The reasons for this are first that OH has no absorption bands that are well separated from the absorptions of other substances which might interfere, and second, that its concentration is extremely low. In the 1980's, the most often-quoted value for this key concentration has been 5×10^5 molec cm^{-3}, corresponding to 2×10^{-5} ppb, a diurnally and seasonally averaged concentration deduced by Crutzen[25].

The use of an average OH concentration is a reasonable approximation when modelling the chemistry of a tropospheric component having a residence time of several days, but not useful when the substrate remains in the atmosphere only for a few hours, as is true for many of the constituents of photochemical smog. Furthermore, consideration of the mechanism for forming OH in the troposphere (Equation 5–8) makes it clear that the rate of **forming** OH depends on the flux of solar photons (Equation 5 and Equation 7), and on the concentration of NO_2 (Equation 5). The solar photon flux varies with the time of day, season, and latitude, while one of the characteristics of air pollution leading to photochemical smog is an elevated concentration of NO_X. Therefore it is important to know the concentration of OH more accurately. Middle-day estimates of this concentration vary widely: 2.5–25 $\times 10^5$ molec cm^{-3} in rural areas, through values $> 1 \times 10^7$ molec cm^{-3} under highly polluted conditions.

In rural locations, the concentration of OH parallels closely the intensity of sunlight[26] (Problem 15). This is consistent with solar photolysis of ozone being the chief source of OH (Equation 7). Under these conditions, the steady state concentration of OH can be obtained as follows:

$$\begin{aligned} \text{rate of formation} &= \text{rate of disappearance} \\ \text{rate of disappearance} &\doteq \textstyle\sum \text{k[OH] [Substrates]} \\ \text{[OH]} &= \text{rate of photolysis}/\textstyle\sum \text{k[Substrates]} \end{aligned}$$

As noted in Section 3.2, the chief substrates in the unpolluted troposphere are carbon monoxide and methane. However, in polluted urban air, the

[25] P.J. Crutzen, *Atmospheric Chemistry,* Ed. D.G. Goldberg, Ann Arbor Press, Ann Arbor MI, 1982, 313–328; a more recent figure for the globally averaged OH concentration is 7.7×10^5 radicals cm^{-3}: R. Prinn *et al,* "Atmospheric trends in methylchloroform and the global average concentrations for the hydroxyl radical," *Science,* **1987**, 238, 945–950.

[26] U. Platt, M. Rateike, W. Junkermann, J. Rudolph, and D.H. Ehhalt, "New tropospheric OH measurements," *J Geophys Res,* **1988,** 93, 5159–5166.

rapid reaction of OH with NO_2 (Equation 27) is an additional sink for OH. This has two consequences. First, it explains why the concentration of NO_X falls before noon in Figure 3.1. Second, reaction with NO_X keeps the concentration of OH low until most of the NO_X has been removed[14].

3.6 Particles in the atmosphere

The issue of particles in the atmosphere is considered in this chapter, because particles are generally released, at least initially, into the troposphere. There are many different kinds and sources of particles, both natural and anthropogenic, in the atmosphere. Solid particles include smoke, from forest fires, from industrial activity, and from domestic heating; wind-borne materials such as soil particles, pollen, bacterial spores, and sea salt; other industrially derived particles such as fly ash, and those from grinding operations such as cement manufacture; and finely divided rocks from large volcanic explosions. Liquid particles include fog and clouds, and the haze-forming aerosols from photochemical smog. We will defer to Chapter 6 any discussion of the tropospheric chemical reactions which occur in clouds.

The most important characteristic of an atmospheric particle is its size. Stokes' Law gives the rate of sedimentation of a particle in a fluid medium, Equation 36.

$$\text{rate} = \frac{\mathrm{g\,d^2}\,(\Delta\rho)}{18\eta} \tag{36}$$

In this equation, g is the acceleration due to gravity, d is the diameter of the particle, $\Delta\rho$ is the difference in density between the particle and the fluid medium, and η is the viscosity of the medium. Equation 36 indicates that large, dense particles will settle much faster than light, small ones (Problem 16). Clearly, only those particles with long residence times will show more than local effects on the environment. For a density of $\approx$ 2 g cm^{-3}, particles of less than 1 μm diameter will settle so slowly as to remain in the atmosphere for weeks. Strictly speaking, Equation 36 applies only to particles of diameter greater than about a micrometer; the motion of smaller particles begins to resemble the chaotic motion of the air molecules themselves. In practice the chief method of removal of small particles from the atmosphere is usually washout by rain or snow (wet deposition), as opposed to settling (dry deposition). Wet deposition is extremely efficient at scavenging small particles.

Small particles are known to be able to travel long distances in the atmosphere. Particles collected on filters in the Canadian Arctic in late winter, when the flow of air is from the south, are found to contain soot, sulfate, vanadium and lead (from combustion; acid precipitation; coal; and gasoline respectively). Isotope ratio analysis of the lead indicates that the origin of these particles is Europe and the USSR. North American particles are carried out over the Atlantic, where they are washed out of the atmosphere by rain. The interpretation is that in late winter/early spring, there is little

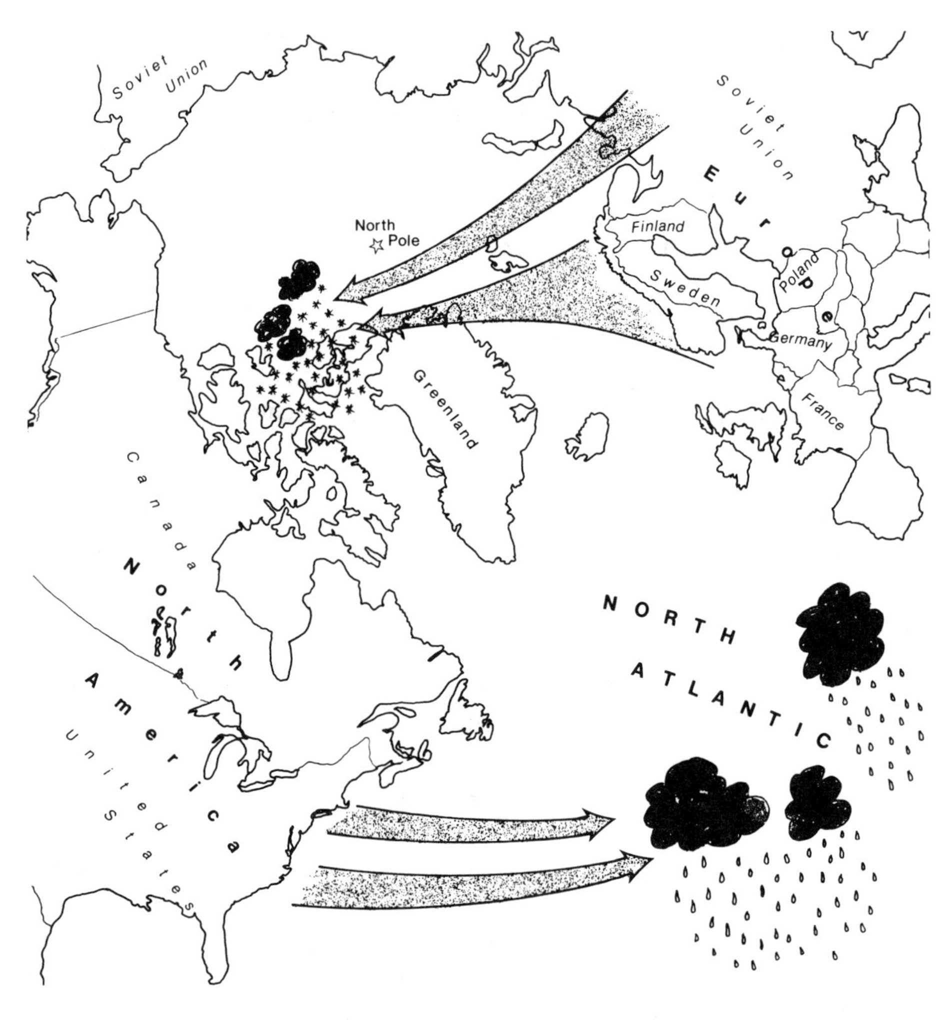

precipitation, and so the particulate matter may be carried for thousands of kilometers. This implies that the particles reside in the atmosphere for at least a week or so (168 h $\times$ 20 km h^{-1}).

Sub-micrometer particles have two other characteristics. First, they scatter light, because their diameter is comparable with the wavelength of visible light (0.4–0.7 μm). The efficiency of scattering depends both on the diameter of the particle and on its refractive index, but the relationship is very complex. Light scattering is responsible for the opacity of clouds. Particles can therefore alter the heat balance of the atmosphere. Second, they are respirable deep into the lungs. Large (diameter > several μm) particles are trapped high in the respiratory tract, where they are transported upwards and into the digestive tract by the action of cilia lining the bronchial passages. The alveoli deep in the lungs have no cilia, and so there is no mechanism to remove particles which lodge there. Silicosis in miners, emphysema in smokers, and lung cancer among asbestos workers are all statistically associated with prolonged breathing of these sub-micrometer particles. Such particles also have a very high surface-to-volume ratio, so that those that are chemically, biologically, or catalytically reactive will exert a near-maximum effect. One example is the catalytic oxidation of SO_2 to SO_3 by particles containing vanadium compounds as part of the chemistry of acid rain (Chapter 6). In another context, it is the small size of the ice particles in the polar Antarctic clouds which makes them so active catalytically in decomposing stratospheric ozone: see Chapter 2.

Some particles of chemical or toxicological interest are examined next.

Sea salt: Particulate matter collected over the open ocean is largely sodium chloride, and is formed when wind entrains surface sea-water. At $> 75\%$ relative humidity, the particles are deliquescent, and so they grow in size. Rainwater and mists near the coast may be appreciably corrosive due to their salt content.

An interesting phenomenon is the partial replacement of chloride in these particles by sulfate or by nitrate. The relevant reactions are:

$$H_2SO_4(g) + NaCl(\text{“s”}) \longrightarrow HCl(g) + NaHSO_4(\text{“s”})$$
$$HNO_3(g) + NaCl(\text{“s”}) \longrightarrow HCl(g) + NaNO_3(\text{“s”})$$

The "s" refers to the particle, although the reaction may be taking place in an adsorbed aqueous phase. These reactions have been postulated to be a significant natural source of atmospheric chlorine, but recent work has shown that more facile reactions occur with NO_2 and N_2O_5, releasing $ClNO_2$ and ClNO respectively. These latter compounds are easily cleaved by visible light, and may act as photoinitiators for photochemical smog in polluted marine areas[27].

Fly ash: This material is a very fine dust which is formed when coal is burned, and in municipal and industrial incinerators. Fly ash particles are small enough to be respirable. The particles are composed of metallic and non-metallic oxides, principally SiO_2, Al_2O_3, Fe_2O_3, and CaO. Their composition varies with the source of the material combusted and the combustion temperature. Small amounts of catalytically active metals such as vanadium, manganese, chromium and cobalt are present, together with more toxic elements such as arsenic, copper, silver, and lead. Part of the oxidation of SO_2 to SO_3 associated with acid rain is believed to occur on the surfaces of such particles, since they have high surface-to-volume ratio. In addition, fly ash may adsorb potentially hazardous materials such as chlorinated dibenzodioxins and dibenzofurans (Chapter 9); much of the controversial emissions of the latter substances from incinerators are adsorbed on the fly ash.

Soot: Incomplete combustion of organic materials produces soot, which is an impure form of elemental carbon (graphite). Soot particles are roughly spherical, whereas pure graphite has a layered structure (Figure 3.3). Soot is believed to form through accretion of graphite-like precursors, but these

[27] L. Ember, "Solid NaCl can affect atmospheric chemistry," *Chem Eng News*, October 24, 1988, 22–23.

precursors contain many structural defects, and incorporate other elements at their peripheries[28]. Because soot is black, it is especially effective at reducing visibility.

Figure 3.3: Structure of graphite

Important environmental sources of soot particles are burning coal, petroleum products, and wood. Coal is discussed below in Section 3.7. In the case of petroleum products, diesel engines are overall much dirtier than gasoline engines (3 g of carbon per kg of diesel fuel vs. 0.1 g per kg of gasoline). The destruction of Kuwaiti oil wells in early 1991 caused a severe regional pollution problem due to particulate and acidic emissions - reduced visibility and oily acidic precipitation[29]. Particulates from wood burning comprise both forest fires and residential wood stoves. The recent resurgence in the use of wood-stoves is the source of considerable winter-time air pollution in, for instance, parts of New England and Whitehorse, Y.T. Hardwood typically produces 0.4 g soot per kg as against 1.3 g from the same amount of softwood.

Soot (graphite) is a highly condensed structure of benzene rings fused together. If soot is extracted with an organic solvent, smaller molecules of similar structure can be isolated. Some of these substances, which are called **polycyclic aromatic hydrocarbons** (PAH's) are shown below.

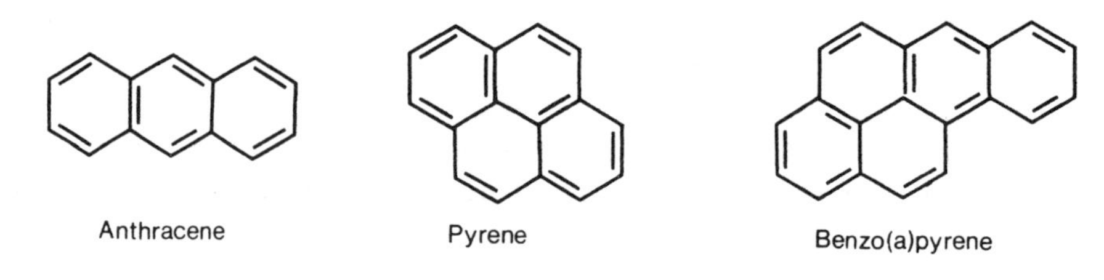

Structures of PAH's.

In polluted air those PAH's with the smallest molar masses (naphthalene, phenanthrene, biphenyl, etc.) will be completely in the gas phase. The

[28] Controversy has recently erupted as to whether soot really has the layered structure of graphite, or whether the soot particles grow as overlapping spirals: for discussion and references, see *Chem Eng News,* February 5, **1990,** 30–32.

[29] R.D. Small, "Environmental impact of fires in Kuwait," *Nature,* **1991,** 350, 11–12

larger molecules (more than 5 fused rings) are almost completely associated as solids with the soot particles. Although diesel vehicles discharge more total particulates, they emit mostly low molecular weight members of the family, whereas the emission from gasoline powered vehicles is mostly high molecular weight PAH's, which tend to be more carcinogenic[30].

PAH's have attracted wide attention because some of them are carcinogenic. They are the compounds formerly responsible for scrotal cancer in young chimney-sweeps in Victorian England, and for causing skin cancer upon prolonged exposure to coal tar. In fact, they were first isolated from coal tar, and their structures confirmed by independent synthesis. They were the first pure chemicals shown to be carcinogenic. Interest in them continues because of their presence in various kinds of smoke, including tobacco smoke and the smoke which rises from the back-yard barbecue[31]. Heating organic compounds to high temperatures without sufficient air for complete oxidation always produces PAH's; formation of these unsaturated compounds is thermodynamically favoured under such conditions.

In recent years, the **Ames assay** has been used to show that the sooty emissions from auto and truck exhausts, from wood-stoves etc. are mutagenic, and by implication carcinogenic. The Ames assay is a much more rapid screening procedure than analyzing for the individual PAH's, which requires elaborate sample clean-up followed by gas chromatography/mass spectrometry. The Ames assay involves the treatment of mutant forms of *Salmonella typhimurium* with the test chemicals. The bacteria are grown in a nutrient-deficient medium which lacks at least one nutrient for the growth of the mutant strain. If the test chemical causes mutation back to the wild type (this is called reversion), a colony of bacteria can establish itself, since the wild type does not require this nutrient. The mutagenic activity of the test chemical or environmental sample is determined by counting the number of revertant colonies.

Gas phase PAH's undergo facile chemical transformation in the atmosphere: for example, the tropospheric lifetime of naphthalene under conditions of photochemical smog is only a few hours. The chief reaction is addition of OH to an aromatic ring; although phenols are among the resulting products, the product mixture is complex and not fully identified. Addition of NO_3 to the aromatic ring occurs at night, but direct photolysis is unimportant, even though PAH's absorb in the ultraviolet. Those PAH which are adsorbed on particles appear to be unreactive, and survive unchanged until they are washed out in rain.

---

[30] B.A. Benner, G.E. Gordon, and S.A. Wise, "Mobile sources of atmospheric polycyclic aromatic hydrocarbons: a roadway tunnel study," *Environ Sci Technol,* **1989**, 23, 1269–1278.

[31] In the case of barbecued meat, the PAH's are formed when the fat melts, drips on to the coals, and pyrolyzes. Meat cooked over a barbecue contains PAH's in easily measured amounts. In contrast, almost no PAH's are detectable in meat cooked under an electric griddle where the heat source is above the meat.

Among the minor products of PAH's are nitro derivatives, which have gained attention because they are mutagenic in the Ames assay. Nitro PAH are mainly formed by atmospheric nitration of PAH as opposed to being directly emitted to the atmosphere, and hence are referred to as secondary pollutants. The mechanism of formation of these compounds in the atmosphere is not well understood, but is most probably *not* electrophilic nitration involving HO_3. Laboratory studies[32] with naphthalene suggest a thermal reaction involving N_2O_5; the reaction with biphenyl is complex, apparently comprising:

1. addition of OH at positions 2 and 4;
2. addition of NO_2 at position 3;
3. elimination of H_2O.

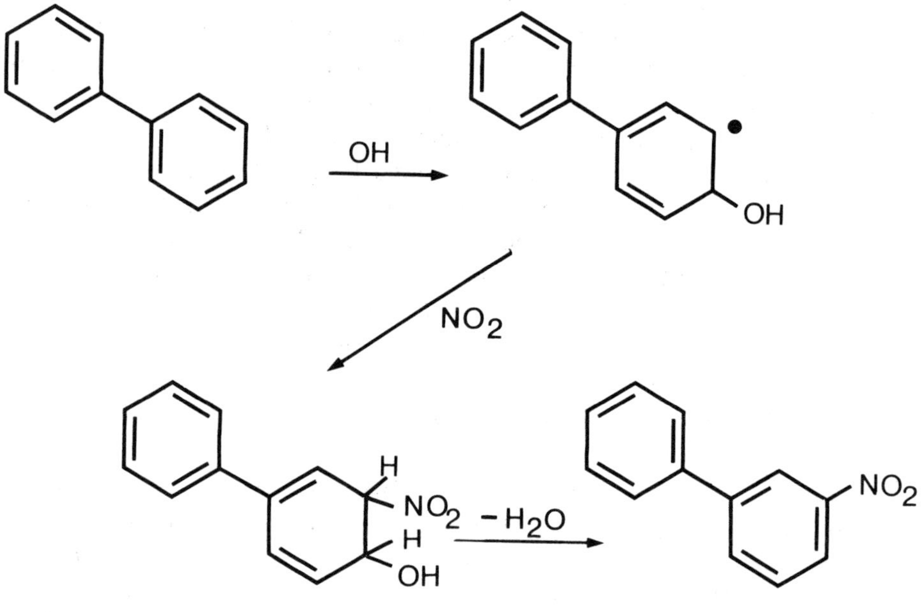

Nitration of biphenyl

3.7 London smog

No Dickensian drama or Sherlock Holmes movie set in Victorian London would have the right atmosphere unless the barely recognizable actors were

[32] R. Atkinson, J. Arey, B. Zielinska, and S.M. Aschmann, "Kinetics and products of the gas-phase reactions of OH radicals and N_2O_5 with naphthalene and biphenyl," *Environ Sci Technol,* **1987**, 21, 1014–1022. For later work on this topic by the same group, see *Environ Sci Technol,* **1989**, 23, 321–327 and *Atmos Environ,* **1989**, 23, 223–229.

viewed through a nearly impenetrable murk. Smogs were common to all the industrialized countries of the 19th and early 20th centuries (the Ruhr valley in Germany, the industrial northeastern United States etc.). The causative agent in this kind of smog is coal, upon which the Industrial Revolution was founded, and indeed still is founded in the still-emerging economies of eastern Europe, the Soviet Union, India, and the Peoples' Republic of China. Coal burning produces large amounts of soot (particles), but also emissions of the acidic gas sulfur dioxide. Only the particles are discussed in this chapter; the acidic aspects of the problem will be deferred to Chapter 6.

London-type smog is quite different from photochemical, Los Angeles-type smog. This is shown in the following comparison.

	London-type smog	Los Angeles-type smog
Pollution source	coal	oil
Cause of reduced visibility	smoke, soot	liquid aerosol
Chemistry of the atmosphere	reducing (SO_2)	oxidizing (O_3, PAN)

Air pollution due to burning coal has been recorded in London since the Middle Ages[33]. The incidence of the nuisance probably peaked in Victorian times, but it was only in the 1950's that concerns about health finally forced remedial action in the form of clean air legislation. On one particularly smoggy day in January 1955, the light intensity fell for a time to 0.1% of what it would have been under sunny conditions, leading to a period of almost total darkness. Under such conditions, solid particulate loadings higher than 4000 μg m^{-3} were recorded. During the smog of December 1952, which lasted for nearly a week, the death rate in London was over 4000 more than normal. These deaths, mainly among the elderly, mostly affected persons with respiratory ailments. Cleaner air was only achieved with a ban on burning coal in domestic fireplaces.

3.8 Particles and climate

Fine particles can remain in the atmosphere for a long period, especially if they have a long way to fall to Earth. Such particles have the effect of scattering solar radiation back into space, and thereby leading to climatic cooling. Volcanic events of especially great magnitude can cause injection of fine particles directly into the stratosphere, from which their return to Earth may take up to a year. Reviews of historical records support a link between major volcanic events and climate changes in the months following. For example, the great Krakatoa eruption of 1893 was followed by a year of especially beautiful sunsets. Particles scatter blue light more than red light, so that at sunset, when the Sun is low in the sky, the blue light is scattered

[33]P. Brimblecombe, *The big smoke: a history of air pollution in London since medieval times,* Methuen, London, 1987.

off into space, leaving a red sunset. Droplets of water in the atmosphere also lead to red sunsets and sunrises.

A volcanic eruption in Iceland about 210 B.C. was followed by a series of particularly cold summers in China[34], resulting in successive crop failures and massive starvation. Probably the largest volcanic explosion in modern times occurred on the island of Tambora, Indonesia on April 5, 1815. So much dust was injected into the atmosphere that complete darkness was experienced for three days up to 500 km away. The next year, 1816, was described as "the year without a summer." Crops failed to ripen, and in New England frost was experienced in every month of the year[35].

A related scenario is the "Nuclear Winter" which, it has been suggested, might follow a nuclear war[36]. Nuclear explosions would inject a great amount of dust and smoke high into the atmosphere. The fires, including forest fires, that these explosions would ignite would add substantially to particulate loads in the atmosphere. According to the nuclear winter scenario, this smoke and dust might substantially obscure the Sun for weeks or months, leading to a worldwide failure of agriculture, and thus to the starvation of the people and animals which survived the original explosions. As evidence in favour of this scenario, Crutzen and Birks cite the severe forest fires in Alberta in 1950, which reduced the intensity of sunlight in Washington D.C. to the extent that the Sun could be observed with the unshielded eye. However, a contrary point may also be noted, namely that black soot particles are strong infrared absorbers, and this might offset the cooling caused by radiation scattering.

Climatic alteration due to particles, including clouds, is in the direction of global cooling. As discussed in Chapter 1, increased concentrations of greenhouse gases are predicted to cause the opposite trend. There is thus a legitimate difference of scientific opinion about the long-term effects of increased industrialization, given that increases in the amounts of greenhouse gases are likely to be accompanied by greater emissions of particles. An imponderable question is the extent to which an unfolding climate change would alter cloud cover, and the scope and direction of any such effect[37].

3.9 Control of particles

Control of particulate emissions is most practical for industrial facilities such as coal-burning power plants, cement manufacturing, metal smelters, and incinerators. As we have seen already, emission control in the domestic

[34] The way in which historical Chinese records were matched up with analysis of Icelandic ice cores is told in *New Scientist,* December 17, 1987, 12.

[35] F.M. Bullard, *Volcanoes of the Earth,* University of Texas Press, Austin, TX, 1984, 511–520.

[36] P.J. Crutzen and J.W. Birks, "The atmosphere after a nuclear war: twilight at noon" *Ambio,* **1982**, 11, 114–125.

[37] V. Ramanathan *et al,* "Cloud radiative forcing and climate: results from the Earth radiation budget experiment," *Science,* **1988**, 243, 57–63.

sphere has been handled by prohibiting the use of certain fuels, rather than by retro-fitting smoke control devices on every chimney.

The two leading technologies for removing particles from effluent gas streams are woven fabric bags and electrostatic precipitators. The former method is similar to the system used to collect dust in a household vacuum cleaner. Finely woven fabric containers are used to trap the particles while allowing the passage of air. The fabric is made to rigid specifications to allow maximum retention of dust while minimizing clogging. In the baghouse of such a facility, the effluent stream can be automatically switched from one set of bags to another to permit removal of the accumulated dust.

Electrostatic precipitators work on the principle of charge neutralization. When particles are formed, and when they rub together, their surfaces acquire an electrostatic charge. Chemically similar particles gain charges of like sign, and hence will not coagulate. In the electrostatic precipitator, the gas stream travels down a tube containing a central rod. The central rod and the outside of the tube are maintained at a potential difference of 50,000 to 100,000 V. Charged particles are attracted to the electrode of unlike relative charge, where they discharge. Once neutral, the particles aggregate spontaneously: their enthalpy of association is more negative than the loss of entropy accompanying this process. The aggregated particles are large enough to settle, and to be collected.

In finishing this section we note one industry in which dust control is of crucial importance. Grain storage elevators and flour mills inevitably contain dust. However, suspensions of these particular dusts, which comprise mainly starch, are highly combustible on account of their very high surface-to-volume ratio. Oxidation by atmospheric oxygen, once initiated, can therefore proceed with extreme rapidity, as numerous devastating explosions in these facilities attest.

Additional Reading

1. B.J. Finlayson-Pitts and J.N. Pitts, Jr., *Atmospheric Chemistry,* Wiley-Interscience, New York, 1986, Chapters 3, 9, 10, 12, and 13.

2. R.P. Wayne, *Chemistry of Atmospheres,* Oxford University Press, Oxford, England, Chapter 5.

3.10 Problems

1. Using the rate constants in the text, calculate the lifetime of the hydroxyl radical in the troposphere under conditions where $p(\mathrm{CO})$ = 4.6 ppm, $p(\mathrm{CH_4})$ = 1700 ppb, and reactions with these two substances are the major sinks for OH.

2. **(a)** Use tabulated bond energy data to estimate the enthalpy of the reaction of OH with methane.

(b) Look up the entropies of the reactants and products of this reaction to calculate $\Delta S°$. Is this reaction driven mostly by enthalpic or entropic considerations?

3. **(a)** Explain why nitrogen oxide concentrations are usually reported as "NO_X," rather than separately as NO, NO_2, etc.

(b) Close to an urban freeway the concentration of NO_X is 60 μg of N per m^3. Express this in atm, ppm, and mol L^{-1}.

(c) Suppose that a city is 25 km across. What is the number of moles of "NO_X" to be found if the average NO_X concentration is 0.04 ppm and this is uniformly mixed to an altitude of 1.0 km?

4. **(a)** Look up the thermodynamic constants of N_2, O_2, and NO. Use them to calculate K for the following reaction at 15 °C and at 800 K.

$$N_2(g) + O_2(g) \rightleftharpoons 2NO(g)$$

(b) Calculate the equilibrium concentration of NO

(i) in air at 15°C;

(ii) in an automobile engine at 800 K with $p(N_2) = 3.0$ atm, $p(O_2) = 0.01$ atm.

(c) The NO concentration over a town is found to be 0.30 ppm, when the temperature is 15°C. Is the system $N_2/O_2/NO$ at equilibrium?

5. The rate constant for the reaction

$$2NO(g) \longrightarrow N_2(g) + O_2(g)$$

is given by the relationship $k = 2.6 \times 10^{12} e^{(-63{,}800/RT)}$ $cm^3\ mol^{-1}\ s^{-1}$, when R is given in cal/mol K. Calculate the half-life of NO with respect to reversion to N_2 and O_2 at 20°C when the concentration of NO is 45 ppb. Use the result to explain what is meant by the NO formed during combustion being "frozen in."

6. Assume that $2NO(g) + O_2(g) \longrightarrow 2NO_2(g)$ is an elementary reaction.

(a) Write the rate law for this reaction.

(b) A sample of air at 290 K is contaminated with 1.0 ppm of NO. Under these conditions can the rate law be simplified? Explain, and if simplification is possible, write the simplified rate law.

(c) Under the conditions described in (b), the half-life of NO has been estimated as 100 h. What would the half-life be if the initial NO concentration were 12 ppm?

(d) Suppose that in the laboratory 0.1 L of pure NO were mixed with 5 L of air, both at 1.00 atm and 290 K, what would be the half-life of NO under these conditions?

7. **(a)** Calculate the maximum wavelength of sunlight capable of dissociating NO_2 into NO and O.

(b) Could photodissociation to NO and excited state O occur with conservation of spin? If so, calculate the maximum wavelength of light needed to bring about this reaction also. Comment on your results.

8. **(a)** What is the lifetime of atomic oxygen in the troposphere if its major sink is the reaction

$$O_2 + O \xrightarrow{M} O_3$$

assuming 15°C, 1.00 atm, given that:
$k = 6.0 \times 10^{-34}\,(T/300)^{-2.3}$ cm^6 molec^{-2} s^{-1}?

(b) Compare your answer with that obtained in Problem 2.2 for a similar calculation under stratospheric conditions.

9. Excited state oxygen atoms undergo deactivation in competition with reaction with water vapour.

$$O^* \xrightarrow{M} O + \text{kinetic energy} \qquad k_1 = 2.9 \times 10^{-11}$$

$$O^* + H_2O \longrightarrow 2OH \qquad k_2 = 2.2 \times 10^{-10}$$

The rate constants are given for 25°C, in the units cm^3 molec^{-1} s^{-1}.

(a) At 25°C, $p(H_2O) = 3.2$ kPa; calculate the fraction of excited oxygen atoms which react with water vapour at 25°C as a function of the relative humidity (0–100%) when $p(\text{total}) = 1.00$ atm. Assume no other sinks for O^*.

(b) Show by calculation whether ground state oxygen atoms are able to convert H_2O to OH radicals.

(c) Calculate the collision rate between excited oxygen atoms and M at 25°C, $[O^*] = 2 \times 10^3$ atom cm^{-3}, and $p(\text{total}) = 1.00$ atm. Compare this rate with the actual rate of deactivation of O^* under these conditions.

10. The reaction

$$OH(g) + CO(g) \longrightarrow CO_2 + H(g)$$

has $k = 2.7 \times 10^{-13}$ cm^3 molecule^{-1} s^{-1} at 300 K.

(a) If the hydroxyl radical concentration is maintained at a steady state of 5.2×10^6 molecules cm^{-3}, calculate the initial rate of the reaction in mol L^{-1} s^{-1} when the CO concentration is 8.5 ppm.

(b) Explain briefly whether or not this reaction could be treated as a pseudo first order process.

(c) What is the residence time of CO under these conditions (same steady state concentration of OH)?

11. The rate constant for the reaction

$$O_3(g) + NO(g) \longrightarrow O_2(g) + NO_2(g)$$

is $2.3 \times 10^{-12} e^{(-1450/T)}$ cm^3 molecule^{-1} s^{-1}.

(a) Calculate the rate constant at 295 K, and the activation energy.

(b) Calculate the initial rate of the reaction if the initial concentrations of O_3 and NO are 1.0 and 5.4 ppm, respectively.

(c) Could this reaction be treated as a pseudo-first order process?

(d) Calculate ΔG°_{295} for this reaction, and use the information to calculate the rate constant at 295 K for the reverse reaction:

$$NO_2 + O_2 \longrightarrow O_3 + NO$$

12. Following Problem 11, consider this oversimplified scheme for the formation and removal of O_3 under conditions of photochemical smog

$$NO_2 \xrightarrow{h\nu} NO + O$$

$$O + O_2 + M \xrightarrow{k_1} O_3 + M$$

$$O_3 + NO \xrightarrow{k_2} O_2 + NO_2$$

where, k_1 is given by $1.1 \times 10^{-34} e^{(+510/T)}$ cm^6 molecule^{-2} s^{-1}; k_2 was calculated in Problem 11. Assume steady state concentrations of O and O_3, and calculate, at 295 K

(a) the steady state concentration of O(g) if $c(O_3) = 0.08$ ppm and $c(NO) = 0.04$ ppm;

(b) the rate of photochemical dissociation of NO_2;

(c) the rate of conversion of solar energy to kinetic energy if the average photon causing the dissociation of NO_2 has a wavelength of 360 nm.

(d) Would you expect photochemical smog to develop (explain why or why not):

(i) in New York at noon in January

(ii) in Delhi, India at noon in July

(iii) in Mexico City at 2 p.m. in March

(iv) in the central Sahara Desert in June?

13. Peroxyacetyl nitrate (PAN) decomposes thermally with a rate constant $1.95 \times 10^{16} e^{(-13540/T)}$ s^{-1}

(a) With what chemical process can you associate the activation energy?

(b) Calculate the half-life of PAN in the atmosphere at 25°C and at −10°C.

(c) Warm air containing 20 ppb of PAN rises and cools to −10°C. What assumptions would you have to make if this air mass was to be the origin of a concentration of 1.5 ppb of PAN measured at a rural location 2000 km away two weeks later?

14. The least volatile oxidation products of hydrocarbons are usually carboxylic acids. Experimentally, the following transformations have been detected:

(i) 1-hexene to pentanoic acid (vapour pressure 0.25 torr at 25°C)
(ii) 1-decene to nonanoic acid (vapour pressure 6×10^{-4} torr, 25°C)
(iii) cyclohexene to adipic acid (vapour pressure 6×10^{-8} torr, 25°C)

Suppose at any moment that 1% of the original hydrocarbon has been converted to the carboxylic acid. What concentration (ppm) of each of the three hydrocarbons (separately) would be needed to cause the formation of haze? Are these concentrations likely to occur in polluted air?

15. From the work of Platt *et al*,[26] it is possible to deduce the relationship:

$$[\mathrm{OH}] = 4.30 \times 10^9 (\mathrm{I_o}\,\sigma \mathrm{O_3})\,\phi(\mathrm{O_3}))\,[\mathrm{O_3}]$$

where I_o is the photon flux absorbed by ozone (units: photons $cm^{-2}\,s^{-1}$), $\sigma(O_3)$ is the absorption cross section for ozone, $\phi(O_3)$ is the quantum yield for photolysis of O_3 to O^*, and the ozone concentration is given in ppb. At midday on March 21 the product $(I_o\,\sigma\,\phi)$, summed over all wavelengths absorbed by ozone, has the value 3.19×10^{-5} in Miami compared with 1.70×10^{-5} in Montreal.

(a) Calculate the steady state concentration of OH in Miami ($[O_3]$ = 45 ppb) and Montreal ($[O_3]$ = 15 ppb) under these conditions.

(b) The hydroxyl radical reacts with toluene with a rate constant 6.2×10^{-12} cm^3 $molec^{-1}$ s^{-1} assumed, for this question, to be independent of temperature. If the concentration of toluene is 1 ng m^{-3}, calculate its percent of reaction per hour in the air over each city under these conditions.

16. **(a)** Soot particles have a density close to 2.2 g cm^{-3}. Use Stokes Law to estimate the rate of settling of particles having diameter

(i) 15 μm
(ii) 0.3 μm

taking the viscosity of air as 182 μp (1 poise (p) = 1 g cm^{-1} s^{-1}).

(b) How long will it take particles of these sizes to settle out of the atmosphere from a height of 5 km assuming that the air is still?

(c) Under highly polluted conditions, concentrations of particulates up to 4000 μg m^{-3} have been recorded. Assuming the density given above, and an average particle diameter of 1 μm, calculate the number of particles per liter. Estimate the number of such particles respired by a person breathing this air for a day.

17. An oil-fired power station consumes 1,000,000 L of oil daily. Assume the oil has an average composition of $C_{15}H_{32}$ and density 0.80 g cm^{-3}. The gas emitted from the stack contains 75 ppm of nitric oxide.

(a) Calculate the mass of NO emitted per day.

(b) Assuming that the stack gases become uniformly mixed to an altitude of 2 km over a city 20 km across, what concentration of NO_X (in ppm) would be added to this air?

18. The rate constant for the reaction of OH with ethane fits the equation:

$$k = 1.37 \times 10^{-17} T^2 e^{(-444/T)} \text{cm}^3 \text{molec}^{-1} \text{s}^{-1}$$

over a wide temperature range. Give the best Arrhenius form of this rate constant,

(i) over the temperature range 200–240 K;

(ii) over the range 300–330 K.

Why are the Arrhenius parameters different?

Chapter 4

Indoor Air Quality

Introduction

The three previous chapters have dealt with various aspects of the outdoor environment. However, indoor air quality is a topic of growing concern. In this chapter we concern ourselves with two different facets of indoor air quality: air quality standards for chemical substances used in work places, and the emerging issue of air quality in offices and homes.

4.1 Air quality in the workplace

Workplace air quality is regulated in order to protect the health of workers. A century ago, little thought was given to any adverse effects that the workplace environment might cause. Better knowledge of the biological effects of workplace chemicals, coupled with pressure from labour unions, have led to the introduction of government legislation to protect the health of the workers, although in many less developed countries the legislation remains inadequate. In parts of Europe and North America, workers have the right to refuse to work in a workplace which they believe to be unsafe, without risk of reprisals from the management. Stronger unions and the desire for a good corporate image mean that almost all large employers in the developed countries provide excellent working environments, and even toxic or obnoxious chemicals such as chlorine, styrene, and hydrogen cyanide are manufactured on a large scale in clean, safe, and almost odour-free conditions.

The control of toxic and other nuisance substances in the workplace can be effected by reducing emissions and/or increasing the rate of ventilation so as to reduce worker exposure. However, increased ventilation may not be possible if it causes emissions of the substance in question outside the plant site at levels above those which are permitted.

Although each jurisdiction sets its own standards, the "Threshold Limit Values" (TLV's) of the American Conference of Governmental Industrial Hygienists are typical both in design and in the choice of the actual permissible concentrations of workplace pollutants. They will be the focus of this section.

The philosophy behind TLV's is seen in these excerpts from the Introduction to the booklet "TLV's: Threshold limit values and biological exposure indices[1]."

> Threshold limit values refer to airborne concentrations of substances and represent conditions under which it is believed that nearly all workers may be repeatedly exposed day after day without adverse effect....
>
> Threshold limit values are based on the best available information from industrial experience, from experimental human and animal studies, and, when possible, from a combination of the three....
>
> These limits are intended for use in the practice of industrial hygiene as guidelines or recommendations in the control of potential health hazards.... (They) are **not** fine lines between safe and dangerous concentration nor are they a relative index of toxicity....

Note in the first sentence quoted that TLV's do not offer a guarantee of freedom from problems for all workers. Some people may become allergic, or sensitized, to very tiny concentrations of specific substances, or they may have preexisting medical conditions which make them more susceptible to the substances in question.

The American Conference of Governmental Industrial Hygienists recommends several categories of concentration limits. The Time Weighted Average TLV (the one most often quoted) is based on a continuous exposure for an 8-hour day, 40–hour working week. The Short Term Exposure Limit TLV is a higher level to which workers may be exposed for no more than four 15–minute periods during the 8–hour workday, provided that the Time Weighted Average is not exceeded over the whole day. When no identifiable Short Term Exposure Limit is given, short term exposures up to three times the Time Weighted Average TLV are permitted but for no more than 30 minutes per day, again provided that the Time Weighted Average is not exceeded over the whole day. Under no circumstances should a level of five times the Time Weighted Average be exceeded. The Ceiling TLV is used for irritant gases, or those that are toxic even upon very short exposure; it should not be exceeded at any time.

TLV data are available for between seven and eight hundred individual substances, and even these do not include all the chemicals and solvents

---

[1] "TLV's: Threshold limit values and biological exposure indices," American Conference of Governmental Industrial Hygienists, Cincinnati, OH, 1987. Further information on specific compounds is available in the "Documentation of the Threshold Limit Values and Biological Exposure Indices," 5th Ed., American Conference of Governmental Industrial Hygienists, Cincinnati, OH, 1986.

that are encountered in the workplace. Several thousand new chemicals are introduced into commerce every year. Complete toxicological information is available for relatively few substances; in the past the impetus for gathering such information has often been the incidence of industrial health problems, though this is changing. What follows is a short list of TLV's for a few common substances (Problems 1 and 2). In order to compare the liquids and solids directly, all values are given in mg m^{-3}, thus explaining the apparently curious choice of values (e.g., 1780 mg m^{-3} for acetone, equivalent to 750 ppm). Among the solids, no distinction has been made in the above table between total dust (particles of all sizes) and respirable dust (particles which are deposited deep in the lungs in the gas exchange zone). Unless noted, the values are Time Weighted Averages.

Generally, the substances with the lowest TLV's are either severely irritating (HCl, NH_3, SO_2) or extremely toxic (Hg, O_3). The organic compounds with the largest TLV's, mostly common solvents such as acetone and ether, act as narcotics in high concentrations. The following comments refer to specific entries in the table.

Benzene is a suspected human carcinogen, and in some jurisdictions much lower limits for benzene are required. Benzene is a component of gasoline; service stations have, up to now, been specifically exempted from legislation controlling the atmospheric concentration of this substance. Note the much higher TLV for toluene: toluene is much less toxic than benzene because the methyl group can be oxidized *in vivo*, giving benzoic acid which can be excreted.

Bis-(chloromethyl) ether $(ClCH_2)_2O$ can be formed by the action of HCl on formaldehyde, and is one reason for concern about indoor emissions of formaldehyde (see later).

Methyl isocyanate $CH_3{-}N{=}C{=}O$ is the substance which gained notoriety in the accident in Bhopal, India in December 1984; some 2000–3000 deaths were attributed to this incident. In addition, thousands of people were left blinded or with respiratory impairment. The accident is thought to have occurred when water entered a holding tank containing the volatile liquid methyl isocyanate, b.p. 39°C. The heat released caused a build-up in pressure in the vessel and ultimately to the venting of the material to the outside. Methyl isocyanate is highly toxic at even a few ppm[2]. Toluene-2,4-diisocyanate is used as the cross-linking agent in polyurethane foams used, for example, as blown foam insulation; its chemical properties are similar to those of methyl isocyanate, but it is much less hazardous to handle since it is a solid at room temperature.

Asbestos[3] is the common name used for a family of naturally occurring silicate minerals which can exist in a fibrous form. In the past, asbestos has been widely used commercially, but these uses have recently been drasti-

---

[2]The Bhopal disaster is the subject of a *Chem Eng News Special Issue*, February 11, **1985.**

[3]P. Holt, "Asbestos dust," *Chem in Britain*, **1988**, 24, 903

Table 4.1: TLV's for a few common substances.

Chemical	TLV mg m^{-3}	Comments
Liquids and Gases		
acetone	1780	
ammonia	18	
benzene	30	suspected human carcinogen
1-butanol	150[a]	absorbed through skin
carbon dioxide	9000	short term limit 54,000
carbon monoxide	55	short term limit 440
chlorine	1.5	
chloroform	50	suspected human carcinogen
bis(chloromethyl) ether	0.005	confirmed human carcinogen
CFC-11	5600[a]	
CFC-12	4950	
diethyl ether	1200	
formaldehyde	1.5	suspected human carcinogen
hydrogen chloride	7[a]	
hydrogen cyanide	10[a]	
hydrogen sulfide	14	
mercury (elemental)	0.05	
methyl isocyanate	0.05	absorbed through skin
nitrobenzene	5	absorbed through skin
ozone	0.2[a]	
sulfur dioxide	5	
toluene	375	
toluene diisocyanate	0.04	
vinyl chloride	10	confirmed human carcinogen
Solids		
asbestos	0.2–2	fibers/cm^3, depending on form; carcinogen
cement	10	
coal dust	2	
grain dust	4	
graphite	2.5	
lead salts	0.15	except lead chromate (0.05)
silica	0.05–0.1	depending on form

[a] ceiling value.

cally curtailed with the finding that asbestos can represent a serious health hazard.

The commonest crystalline modification of asbestos, chrysotile $3MgO\,2SiO_2\,2H_2O$, has long white fibers which can be woven into fabrics. Protective suits for firemen, and even household items such as oven mitts were made from chrysotile until recently. Today the synthetic organic polymer polybenzimidazole is replacing chrysotile in fire protective equipment. Chrysotile asbestos was widely used as a spray-on insulation material in public buildings such as hospitals and schools in the 1950's and 1960's. A matter of current concern is that poorly maintained insulation may flake off the walls or ceilings, releasing the fibers into the air. It is a very difficult problem to decide whether to remove this material, because disturbing it releases far more fibers into the air than leaving it be[4].

Chrysotile is the commonest form of "serpentine" asbestos, whose fibers occur in bundles, and which are relatively easily intercepted in the upper airways when inhaled. One of the largest deposits of chrysotile in North America occurs near Thetford Mines, Quebec. The other class of asbestos comprises the "amphiboles," which are rod-like. Crocidolite (also called blue asbestos) $Na_2O\,Fe_2O_3\,3FeO\,8SiO_2\,H_2O$ is the commonest amphibole. Crocidolite has been used as a filter pad because it is chemically very inert, even to strong acids and bases. At one time asbestos filter pads were used to clarify beer, but this is no longer done. Crocidolite fibers are particularly dangerous to health: they are small, and hence very likely to penetrate deep into the lungs. It has been concluded from a recent study of some 7000 workers at a crocidolite mine in Western Australia between 1943 and 1966, that 2000 either have developed asbestos-related diseases or will contract them by the year 2020. The danger to health from the mine tailings has been rated so high that recreation in the scenic Wittenoom Gorge, the site of the former mine, has been restricted by the state government[5].

Asbestos is now well known as a human carcinogen, from epidemiological studies on asbestos mineworkers. The several different crystalline forms of asbestos all appear to be carcinogenic, although crocidolite is the most dangerous. It is the physical characteristics of the fibers which cause cancer rather than their chemical properties, and so the physical dimensions of the particles determine the degree of hazard. For this reason, the TLV's for the various kinds of asbestos are given in terms of number of fibers per cm^3 rather than in $mg\ m^{-3}$. Mossman *et al*[4] present evidence that much of the

---

[4]B.T. Mossman *et al*, "Asbestos: scientific developments and implications for public policy," *Science,* **1990**, 247, 294–301. Risk assessment suggests that there is a smaller overall risk involved when workers or even schoolchildren are exposed to very low levels of asbestos in well-maintained buildings insulated with spray-on asbestos, compared with the risk associated with removing the material. For subsequent correspondence, see *Science,* **1990,** 248, 795–801. See also M. Reisch, "More workers at risk from asbestos exposure" *Chem Eng News,* July 2, **1990,** 10.

[5]I. Anderson, "Employment records reveal the detail of asbestos danger," *New Scientist,* January 13, **1990,** 29.

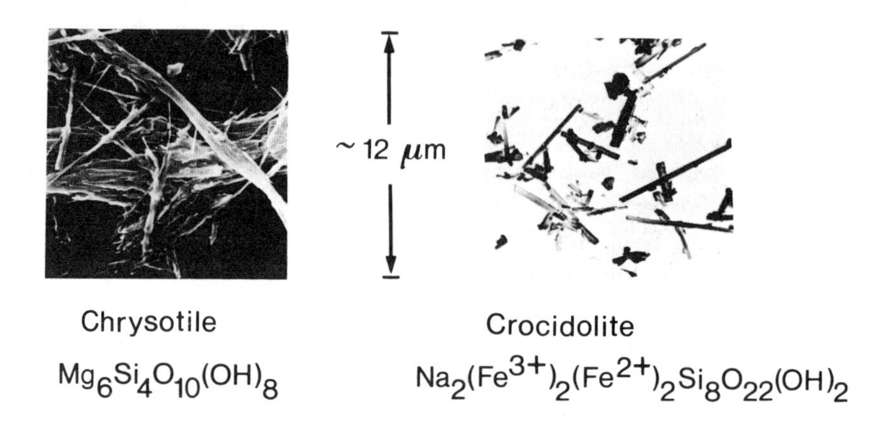

Figure 4.1: Micrographs showing asbestos fibres, see Reference 4.

danger associated with chrysotile asbestos may actually be due to the small amounts of amphiboles, such as crocidolite, that it contains.

Asbestos is associated with a particular form of cancer called mesothelioma, which occurs when an asbestos fiber penetrates the tissue of the gas exchange region of the lung and enters the pleural cavity. Today asbestos is limited by regulation to 0.2–2 fibers cm^{-3}, depending on the form of asbestos; earlier in this century it was not uncommon for industrial workers to be exposed to thousands of particles per cm^3, and it is said that workers in asbestos weaving mills were unable to see 20 feet, so thick was the dust.

The latent period for asbestos related disease is very long, at least 20 years, so that even with recent very stringent limits on the allowable concentration of asbestos fibers in the air, deaths from this cause will continue into the next century. Ironically, mesothelioma is becoming more prevalent since the concentration of asbestos dust has been more carefully controlled in the workplace, and fewer cases of injury to the lung itself occur. Holt[3] quotes a United States prediction of 8000–10,000 deaths per year from this cause, and a comparable figure of 1500–2000 in the United Kingdon In some jurisdictions, asbestos-related diseases are grounds for allowing "Workmens' Compensation" payments.

In the United States, the use of asbestos is to be phased out completely during this decade[6]. Spray-on asbestos insulation was banned in the early

---

[6] *Chem Eng News*, July 10, **1989**, 14.

1970's, asbestos was eliminated from roofing felt in 1990, and will be removed from automotive brake linings and similar components in 1994, and asbestos-cement products such as water pipes and roof shingles will no longer be sold after 1997.

Among other solids in Table 4.1, lead chromate (which is carcinogenic) is familiar as the yellow pigment used in the paint on North American school buses. Silica, which like asbestos exists in several different crystalline forms, is responsible for the disease called silicosis among miners (also called miners' lung). Cement dust is strongly alkaline (see Chapter 8). Grain dust represents a hazard from the pollen (allergen) and spores which it contains, and is a problem to farmers ("farmer's lung") and workers in grain elevators. Other "nuisance dusts" not listed have no specific adverse effects, but their concentrations should not exceed 10 mg m^{-3}.

Mixtures of toxic agents: Many workplaces contain more than one airborne toxic substance. When humans or animals are exposed to mixtures of such agents, their combined outcome may be **additive, synergistic,** or **antagonistic.** As mentioned earlier, detailed toxicological information is available for very few single substances. Even less is known about mixtures. Two different approaches can be taken, depending on the nature of the substances, and their modes of action, if these are known. Of course, the worth of these methods is only as good as the appropriateness of the decision as to whether the modes of action of the substances are similar or not (Problems 3 and 4).

1. Dissimilar compounds: these are substances that act by different mechanisms; usually they will be identified as being chemically different. Examples:

   - CO (ties up hemoglobin) and HCl (irritant acidic gas)
   - diethyl ether (narcotic solvent) and toluenediisocyanate (reacts with OH and NH_2 groups)
   - O_3 (powerful oxidant) and Hg (neurotoxin)

   The method to be taken here is to determine separately whether the TLV is exceeded for each of the components of the mixture. The overall TLV is exceeded if any one of the concentrations is greater than the corresponding TLV.

2. Similar compounds: these act by similar mechanisms. Examples:

   - HCl and HBr (both irritant gases)
   - acetone and ethyl acetate (both narcotic solvents)
   - ammonia and ethylamine (both bases)
   - CO and HCN (both interfere with oxygen uptake by the tissues: CO ties up hemoglobin; HCN inhibits cytochrome oxidase, and hence prevents the reduction of O_2)

For similar compounds the approach is to add the fractions obtained by dividing the actual concentration c by the corresponding TLV. The TLV is taken as exceeded if the sum of these fractions is greater than unity.

$$c_1/\mathrm{TLV}_1 + c_2/\mathrm{TLV}_2 + c_3/\mathrm{TLV}_3 + \cdots > 1$$

By an extension of this approach, we can designate an effective TLV for a defined mixture, such as a mixed solvent with several components. If the fractions f by weight of the components of the mixture are known, the effective TLV of the mixture is calculated as follows (Equation 1).

$$\mathrm{TLV}_{\mathrm{mixture}} = f_1/\mathrm{TLV}_1 + f_2/\mathrm{TLV}_2 + f_3/\mathrm{TLV}_3 + \ldots\}^{-1} \quad (1)$$

Biological exposure indices

For some twenty compounds, exposure limits have been established or proposed based on an analytical measurement on the worker (as opposed to the air he or she breathes). Such measurements are made on air exhaled by the worker, or on a sample of the worker's blood or urine. At first sight, biological testing seems like an ideal way to monitor workplace exposure to chemicals; in fact, there are several possible confounding issues. For example:

- the worker may be exposed away from the work environment (e.g., CO by smoking);
- the general health status of the worker will affect how rapidly the chemical is metabolized;
- the test may not be completely specific for the chemical in question (e.g., one method for assessing exposure to benzene involves analyzing for its metabolite phenol in the urine. Exposure to phenol will invalidate the assay).

In addition, the timing of the test is important. For substances such as lead, which have long residence times in the body, the timing of the test is immaterial. Short-lived chemicals are best assayed either at the end of the shift or prior to beginning the shift.

4.2 Air quality in offices and homes

Many environmental issues have come to public attention in the past couple of decades as a result of improved analytical capability. An excellent example is the detection in the early 1970's of sub-ppb concentrations of CFC's in the troposphere; this could not have been an issue a generation ago because these concentrations could not have been detected. The matter of indoor air quality has surfaced for a different reason, namely the drive to increase energy efficiency in public and private buildings following the steep rise in energy prices in the mid-1970's. Quoting from a Canadian report[7], "it should be quite clear that the major contributing factors to poor indoor air quality are the attempts at energy conservation." A significant finding is that people may be exposed to higher concentrations of potentially toxic air pollutants indoors, in their homes and offices, than outside, even in heavily industrialized cities[8]. New buildings show especially high concentrations of volatile organic compounds. In Scandinavia, 100% outdoor air must be used for ventilation during the first six months of the life of an office building, in order to counteract the release of organic compounds from the new materials used in its construction.

At the heart of this discussion is the topic of the number of air changes in the building per hour (air changes per hour is often abbreviated "ach"). In an energy efficient building, the goal is to minimize the rate of exchange of indoor with outdoor air. Indoor air is maintained at a temperature and humidity consistent with human comfort, while that outside may be too cold, too hot, too dry, or too humid. Every air change requires energy to be supplied by the heating or air-conditioning system. The homeowner seeks to reduce energy loss by sealing drafts and by installing better insulation; in a commercial building, these measures are augmented by partial recirculation of the air in the ventilation system. The result can be that annoying, or even harmful, gases can build up to levels much greater than those in a drafty older building. These gases include carbon dioxide, from respiration and combustion; carbon monoxide, from combustion, notably smoking; formaldehyde and other organics, by emission from a whole range of synthetic materials; and radon, from the ground. Particles may also be present from smoking and from the use of wood burning stoves.

The reason for pollutant build-up can be understood in simple kinetic terms[9]. For any gaseous indoor pollutant, let R be the rate of emission, in some appropriate units such as mg h^{-1}, and let k_1 be the number of air changes per hour. Denoting the indoor concentration of the pollutant as $[c_i]$ in units such as mg m^{-3} and the volume of the building as V m^3, we have, at the steady state:

$$\mathrm{R} = k_1\,[\mathrm{c}_i]\,\mathrm{V} \tag{2}$$

In this equation it is assumed that the outdoor concentration of the pollutant is zero. If, as is more likely, the pollutant is also present outdoors at concentration c_o, the following equation is obtained.

$$\mathrm{R} + k_1\,[\mathrm{c}_o]\,\mathrm{V} = k_1\,[\mathrm{c}_i]\,\mathrm{V} \tag{3}$$

[7] R.J. Milko, "Indoor air quality," Background paper for parliamentarians, Library of Parliament, Ottawa, Canada, August 1985.

[8] L.R. Ember, "Survey finds high indoor levels of volatile organic chemicals," *Chem Eng News,* December 5, **1988,** 23–25.

[9] W.J. Fisk, F.J. Offermann, R.K. Spencer, B. Petersen, D.T. Grimsrud, and R. Sextro, *Indoor air quality control techniques,* Noyes Data Corp., Park Ridge, NJ, 1987, Chapter 2.

Therefore the steady state concentration of the pollutant indoors is given by Equation 4.

$$[c_i] = [c_o] + R/(k_1 V) \tag{4}$$

Figure 4.2 shows some sample curves for the variation of the steady state indoor concentration of pollutant with the air exchange rate (Problem 5). Because of the reciprocal dependence upon k_1, the indoor concentration is predicted to rise steeply as k_1 approaches and then falls below one air change per hour. This is exactly the range of k_1 in a modern energy efficient home or office, values of k_1 between 0.2 and 0.6 being common for such structures[10].

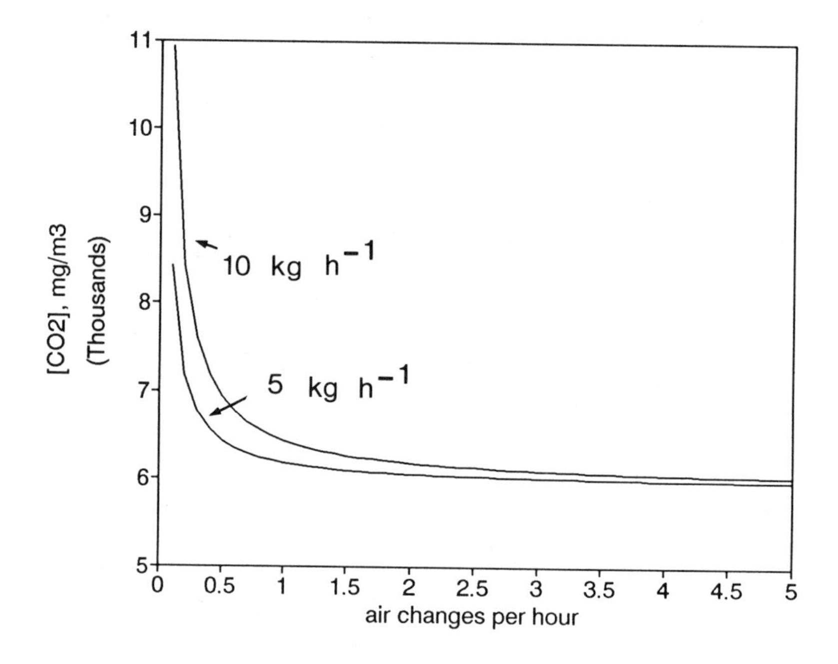

Figure 4.2: Indoor pollutant concentration as a function of air exchange rate. The example is CO_2 under the conditions of Problem 5. The outdoor concentration of CO_2 is 5930 mg m^{-3}.

Before discussing individual substances in the context of indoor air quality, we can add further comments about Equation 4. First, the rate constant k_1 may be a composite of the natural rate of exchange with the air outdoors (this is called infiltration) and any forced ventilation by means of mechanical systems. Part of the natural infiltration is due to the so-called "stack effect." This arises when the outdoor temperature is lower than that indoors. Warm air tends to escape from the top of the building and be replaced

[10] D.T. Grimrud, R.D. Lipschutz, and J.R. Girman, "Indoor air quality in energy efficient residences," Chapter 5 in *Indoor air quality,* Eds. P.J. Walsh, C.S. Dudney, and E.D. Copenhaver, CRC Press, Boca Raton, Fla., 1984.

by cold air at the bottom. In commercial buildings ventilation stacks may be incorporated into the structure to increase natural ventilation through the stack effect. Second, the whole building has been treated as a single compartment; more complex equations can be derived to describe the air quality in different parts of the building (basement, kitchen, living area etc). The rates of emission, ventilation, and exchange from and between compartments may all be different (Problem 6).

Traditionally, we have always thought of home as a very safe environment. In episodes of severe outdoor pollution, such as during episodes of photochemical or "London" type smog (Chapter 3), it is good advice to stay indoors and keep all windows and doors shut. Under these conditions, for pollutants such as SO_2, O_3, PAN etc, there are few or no indoor sources, so $c_o > c_i$. Minimizing the air exchange rate k_1 by keeping doors and windows closed delays the build-up of the pollutant concentration indoors (Problem 7). This effect has recently been shown experimentally for the indoor ozone levels in a series of office-laboratory buildings in New Jersey[11]. Indoor ozone concentrations closely tracked the outdoor levels, while the ratio (indoor level)/(outdoor level) increased in parallel with the ventilation rate (Figure 3). The results showed clearly that ozone in offices originates almost entirely from outside, rather than from electrical equipment such as photocopiers inside the building. A similar phenomenon was seen among southern California museums, where the ingress of outdoor nitric acid aerosol was of concern to the curators. Again, the indoor concentration of the outdoor pollutant was strongly influenced by the design of the ventilation system[12] The reverse effect was seen for the case of chlorinated pesticides in United States homes[13], where indoor use of these substances for pest control was deduced to be their source. For the rest of this chapter we shall be concerned with the latter situation, where the concentration of the contaminant is higher in the indoor air than it is outside[14].

A final comment before covering individual indoor air contaminants: the solutions to these problems are all the same, namely increased ventilation. There are no generally applicable chemical or physical devices to clean up indoor air. Devices of limited applicability include range hood filters in kitchens; these contain activated charcoal (which must be replaced periodically) to absorb odorous compounds from cooking. Extreme cases of indoor pollution include the "sick building syndrome[15]," in which the occupants

---

[11] C.J. Weschler, H.C. Shields, and D.V. Naik, "Indoor ozone exposures," *J Air Pollut Control Assoc,* **1989**, 39, 1562–1568.

[12] L.G. Salmon, W.W. Nazaroff, M.P. Ligocki, M.C. Jones, and G.R. Cass, "Nitric acid concentrations in southern California museums," *Environ Sci Technol,* **1990**, 24, 1004–1013.

[13] D.J. Anderson and R.A. Hites, "Chlorinated pesticides in indoor air," *Environ Sci Technol,* **1988**, 22, 717–720.

[14] For comparison of concentrations of organics between outdoor and indoor air see: J.J. Shah and H.B. Singh, "Distribution of volatile organic chemicals in outdoor and indoor air," *Environ Sci Technol,* **1988**, 22, 1381–1388.

[15] For a brief review, see W. Jones, "Sick building syndrome and its elusive causes," *At*

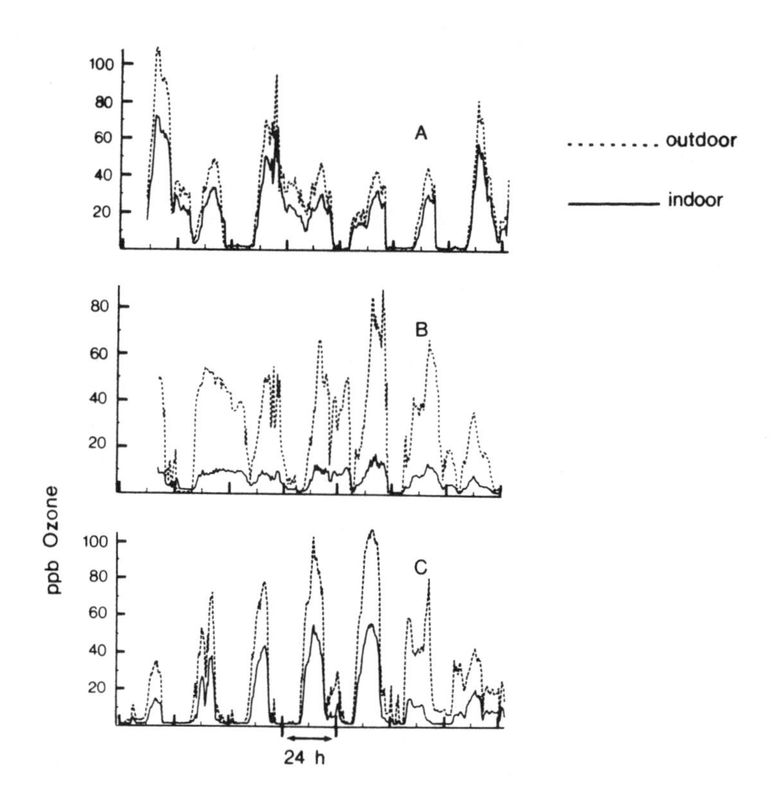

Figure 4.3: Indoor and outdoor ozone concentrations at three different office buildings over several days. Building A, 8.2 ach, Building B, 0.6 ach, Building C, 4.0 ach. Reproduced from Reference 11.

complain of all sorts of unspecific symptoms: headaches, fatigue, inability to concentrate, and sometimes nausea. Most often these symptoms can be alleviated by increasing the ventilation rate or, if the building uses recirculated air, by increasing the proportion of fresh air (see also Section 4.2.3 below). Sometimes, improper setting of air intakes and air exhausts is the problem; poor air quality and excessive odours are often encountered in research laboratories when air exhausted from the fume hoods is drawn into the fresh air intakes.

The Centre (publication of the Canadian Centre for Occupational Health and Safety), June **1989,** 10–11; see also page 20 of the same issue.

4.2.1 Radon[16]

Contamination of indoor air by radon emerged in the late 1980's as a cause for considerable public concern. The United States Environmental Protection Agency (EPA) has estimated that radon may be responsible for between 5000 and 20,000 lung cancer deaths in the United States per year. The British government makes a comparable estimate of 1500 additional lung cancer deaths from this cause while the Swedish government has attributed up to 25% of lung cancer deaths to radon exposure[17].

Considerable misunderstanding and controversy surrounds the radon issue. Over 50 years ago, radon was implicated as a cause of abnormally high rates of lung cancer among miners. By the 1950's it was established that the hazard was not mainly due to radon itself, but rather to the radioactive disintegration products of radon, known collectively as radon daughters. Radon, element 88, is a colourless noble gas with boiling point of $-62°C$ and density 9.73 g L^{-1}. Radon has twenty known isotopes, all of which are radioactive[18]. Of these, the longest lived is ^{222}Rn, with a half-life of 3.8 days. Virtually all the radon isotopes decay by ejection of an α-particle (a bare $_4He$ nucleus), affording isotopes of $_{86}Po$, and ultimately $_{84}Pb$.

The following is just one part of one of the radioactive decay chains. The numbers on the arrows represent the energy involved in the decay, and its half-life.

$$^{226}\text{Ra} \xrightarrow{-\alpha,\ 4.8\,\text{MeV},\ 1600\ \text{yr}} {}^{222}\text{Rn}$$
$$^{222}\text{Rn} \xrightarrow{-\alpha,\ 5.6\,\text{MeV},\ 3.8\ \text{d}} {}^{218}\text{Po}$$
$$^{218}\text{Po} \xrightarrow{-\alpha,\ 6.1\,\text{MeV},\ 3.1\ \text{min}} {}^{214}\text{Pb}$$
$$^{214}\text{Pb} \xrightarrow{-\beta,\ 1.0\,\text{MeV},\ 27\ \text{min}} {}^{214}\text{Bi}$$
$$^{214}\text{Bi} \xrightarrow{-\beta,\ 3.3\,\text{MeV},\ 20\ \text{min}} {}^{214}\text{Po}$$
$$^{214}\text{Po} \xrightarrow{-\alpha,\ 7.8\,\text{MeV},\ 0.0002\ \text{s}} {}^{210}\text{Pb}$$
$$^{214}\text{Bi} \xrightarrow{-\alpha,\ 5.6\,\text{MeV},\ 20\ \text{min}} {}^{210}\text{Tl}$$
$$^{210}\text{Tl} \xrightarrow{-\beta,\ 5.5\,\text{MeV},\ 1.3\ \text{min}} {}^{210}\text{Pb}$$

Lead-210 is quite long-lived, half-life 21 years.

Radon is unreactive and non-polar, and so it is unlikely to lodge in the lungs if inhaled. The main hazard from radon itself is from those few atoms which happen to disintegrate whilst in the lungs. The radon daughters are

---

[16] D.J. Hanson, "Radon tagged as cancer hazard by most studies, researchers," *Chem Eng News*, February 6, **1989**, 7–13. P.J. Walsh and W.M. Lowder, Chapter 8 in *Indoor air quality*, Eds P.J. Walsh, C.S. Dudney, and E.D. Copenhaver, CRC Press, Boca Raton, FL, 1984.

[17] *New Scientist*, September 22, **1988**, 24–25.

[18] From *Handbook of Chemistry and Physics*, CRC Press, 64th Edition.

a different matter. At the moment of their formation, they are produced as gaseous ions, and are both reactive and readily adsorbed onto particles. As metallic elements, they react rapidly to afford oxides, initially in a gaseous form. These oxides are thought to be scavenged with high efficiency, both by bronchial tissue and by airborne solid particles, which may also lodge in the lungs. In either case, this brings the radon daughters into close proximity with the bronchial tissue where they may later disintegrate. For the rest of this section, "radon" will refer to both radon itself and the radon daughters, unless specified otherwise.

As already noted, radon itself is the immediate decay product of radium, an intermediate in the radioactive decay of elements such as uranium and thorium. Exposure to radon was first recognized several decades ago as a hazard among miners, particularly uranium miners, who were found to suffer an abnormally high incidence of lung cancer. In the 1970's, there were reports of high radon levels in homes constructed on land reclaimed from mining tailings. Thus until the late 1970's exposure to radon appeared to be the result of industrial exposure, either in mines or in homes built over mine tailings. Only in the last few years has radon come to be recognized as a problem in homes in which there is no connection with mining activity.

Two factors have led to radon being found in these latter homes. First, people have only recently begun to look for it. In the United States, public awareness of radon was triggered by the case of a worker at a nuclear power plant, who was found to set off the plant's nuclear monitors when he arrived at work. The source of the radioactivity was radon in his home. The second factor is the construction of energy efficient, airtight homes. Any radon released into the basement will tend to remain in the house. Older draftier homes have more air changes per hour. The concentration of radon in the outdoor air is extremely low, so that in the context of Equation 4, c_o can be taken as zero.

In retrospect, it is not surprising that radon should be so pervasive. Radon accounts for over half the natural background radioactivity to which we are all exposed[19]. On a global average, over 80% of all radon emissions come from soil. All soil contains tiny amounts of radium, decay of which releases radon into the atmosphere. Radon enters homes through the basement from soil. However, the extent of contamination varies greatly depending on the composition of the soil. Boyle[20] suggests that soil testing for radon might prevent future home-building in areas of high radon release.

The other main source of radon is ground water, which equilibrates with the underground rock. Radon is mobile because it is a gas; its precursors such as uranium, thorium, and radium are all metals which are immobilized in the rocks in a chemically combined form. There is evidence that in some

[19] Other major sources of ionizing radiation are cosmic rays, natural radionuclides other than radon, and medical exposures. Nuclear fallout, and radiation from nuclear power generating stations are insignificant: R.H. Clarke and T.R.E. Southwood, "Risks from ionizing radiation," *Nature,* **1989**, 338, 197–198.

[20] M. Boyle, "Radon testing of soils," *Environ Sci Technol,* **1988**, 22, 1397–1399.

homes a large fraction of the radon present comes from the water supply, for example from faucets and shower heads.

Two kinds of radon monitors are in use to gauge the concentrations of radon in homes. The short term monitor employs a canister of activated charcoal through which air is drawn for a few days, and the radon daughters (but not radon itself) adsorb to the charcoal. The radioactivity is then counted at a laboratory specializing in this work. Since, as was seen above in the partial radioactive decay chain, the immediate daughters of radon are all very short lived, it is the longer-lived decay products such as ^{210}Pb that are counted. The amount of radon in the home can be computed from the time of exposure of the canister and the efficiency of absorption. A problem with the short term monitor is that radon levels fluctuate substantially, and so a single reading may not give an accurate assessment of the long term average radon concentration. Radon emission from the ground changes with the temperature and also with the humidity. In addition, the amount in the home changes according to the adequacy of ventilation; levels tend to be higher in winter, when windows and doors are kept closed[21].

The longer-term track etch monitor is a passive device which operates by recording the "tracks" which an α particle produces when it ionizes the molecules in a special polymer film. The track begins when the α particle is formed by decay of a precursor. As the α particle moves through the polymer, it ionizes the molecules, leaving a track of damage to the film. The end of the track occurs when the α particle has given up all its energy and can no longer cause ionization. The tracks are later visualized and counted by etching the film with base. Exposures of several months are needed to obtain accurate data because the formation of an α particle right in the vicinity of the film is such a chance event.

The United States EPA has set an "action level" of 4 picoCuries (pCi) per liter for radon in homes. Above this concentration, EPA recommends that the homeowner take steps to reduce the level (see below). In Canada, the action level has been set at 20 pCi L^{-1} (800 Bq m^{-3})[22]. while the British action level has been set at 400 Bq m^{-3}. It is clear that no international agreement exists as to the appropriate action level.

In terms of risk assessment, it was claimed in the British study that a lifetime exposure to 1000 Bq m^{-3} brings a 5% risk of death from lung cancer. The fact that some of the radioactivity comes from radon and some from the daughters presents a problem, because the α particles from radon, polonium and further decay products are not isoenergetic. For this reason, a "working level" has been developed in the United States. This corresponds

[21] See T.J. Bierma, K.G. Croke, and D. Swartzman, "Accuracy and precision of home radon monitoring and the effectiveness of EPA monitoring guidelines," *J Air Pollut Control Assoc,* **1989**, 39, 953–959.

[22] The S.I. unit of radioactivity, the becquerel (Bq) corresponds to 1 disintegration per second. By definition, 1 curie (Ci) = 3.7×10^{10} Bq. Note that the concentration of radon is expressed in terms of its radioactivity. The absolute concentrations are so low that no other method of analysis is feasible (see Problem 8).

to an energy release of 2.08×10^{-5} J per cubic meter of air if all the short-lived daughters were to disintegrate (Problem 9). Health effects of radon exposure are often reported in "working level-months." Since the concept of the working level was originally developed for use in the mining industry, a working level month corresponds to this degree of exposure for 170 h (i.e., a 40 h work week $\times$ 4¼ weeks per month).

The mechanism by which radon daughters are thought to cause cancer is through radioactive disintegration, which releases an α particle (a bare helium nucleus)[23]. Alpha particles have only a very short range, since they interact with matter, removing electrons (hence the term ionizing radiation) and being converted ultimately to a neutral helium atom. In lung tissue, this range (ca. 60 μm) is sufficient to allow damage to the stem cells, alteration of which can lead directly to cancer. Retrospective studies on miners suggest that there is no threshold dose below which there is no risk of cancer. EPA statistics[24] suggest that the risk of lung cancer over 70 years for individuals spending 75% of their time in the home is equivalent to having 200 chest x-rays per year at the 4 pCi L^{-1} action level, and equivalent to smoking 2 packs of cigarettes daily at 20 pCi L^{-1}.

There is controversy in the United States as to the extent of danger which radon presents to the general public[25]. One factor is that the EPA recommends placing the radon monitor in the basement. This will maximize the calculated exposure to radon, because the gas is dense and tends to accumulate in the basement, where most people spend little of their time. Another factor in the public controversy is a recent study in which lung cancer deaths among the general population were related to typical radon concentrations in homes in over 400 United States counties. Fewer, not more, lung cancer deaths were recorded in counties where the radon levels were higher[26].

One possible way of reconciling this study with that involving the miners is that many of the miners who died of lung cancer were smokers. There appears to be a powerful synergism between radon and smoking on the development of lung cancer. Therefore the miners may have given an unrealistically high estimate of the risk of lung cancer due to radon exposure to the population at large. Further studies are under way to investigate this point.

Some have claimed that the EPA's "action level" of 4 pCi L^{-1} for radon is unrealistically low, especially since the EPA estimates that as many as 10% of all United States homes may have radon concentrations above the action

---

[23] Radioactive decay over geological time is believed to be the origin of helium in natural gas, the major terrestrial source of helium. Helium is currently relatively cheap, since it is a byproduct of natural gas recovery. There is concern that once natural gas reserves become depleted, helium will become very scarce.

[24] R. Berger, "The carcinogenicity of radon," *Environ Sci Technol*, **1990**, 24, 30–31.

[25] Aspects of this controversy are discussed in a letter by M.J. Welch under the title, "Health effects of radon," *Chem Eng News*, March 13, **1989**, 3.

[26] *New Scientist*, September 29, **1988**, 29.

level. Those making such claims suggest that an action level in the range 10–20 pCi L^{-1} would be more appropriate. Ironically in Canada, where the action level is set at 20 pCi L^{-1}, others assert that many Canadian homes having radon concentrations between 4 and 20 pCi L^{-1} are unsafe because they are above the EPA action level[27].

What should the homeowner do if the home is found to contain radon above the action level? The remedies boil down to "better ventilation." Opening the windows works, but invalidates the whole concept of the energy efficient home. More practical is a pipe inserted through the concrete basement floor to the soil below, from which the radon is escaping. A small pump can be used to void the contaminated air from beneath the basement directly to outdoors[28]. However, radon mitigation systems need to be monitored to check their continuing effectiveness[29]. Even if effectiveness could be guaranteed, the cost of bringing every U.S. home within the action level of 4 pCi L^{-1} has been estimated at $270,000 per cancer death avoided[30].

4.2.2 *Formaldehyde*[31]

Formaldehyde is a component of numerous resins (e.g., phenol-formaldehyde and urea-formaldehyde). Free formaldehyde gas can be released from the resins either if excess formaldehyde was used in their formulation or if the polymer hydrolyzes. Acid and humidity expedite the latter process. Resins containing formaldehyde are used as bonding agents in products such as plywood, particle board, and glass fiber insulation, and hence are widely used in home and office construction: insulation, wall sheathing, flooring, and cupboards: see Table 4.2 below. These materials may continue to liberate free formaldehyde for many years. Mobile homes often contain rather high concentrations (up to 10 ppm) of formaldehyde[32] (Problem 10). Urea-formaldehyde foam insulation (UFFI) was used in the late 1970's in homes; its use was discontinued because of the release of formaldehyde from

---

[27] This raises an interesting point. The EPA action level is a guideline. In the heat of discussions about public safety, it becomes construed as a "safety standard." These terms are not interchangeable: compare the analogous comments in Section 4.1 concerning the interpretation of TLV's.

[28] North American government publications on steps to reduce radon exposure include "A citizen's guide to radon" U.S. EPA, 1986 (OPA-86–004), and "Radon—you and your family," Health and Welfare Canada, **1989** (H49–39/1989). Construction techniques to minimize radon entry are discussed by R.S. Dumont and D.A. Figley, "Control of radon in houses," National Research Council of Canada, Publication CBD 247, February **1988**.

[29] R.J. Prill, W.J. Fisk, and B.H. Turk, "Evaluation of radon mitigation systems in 14 houses over a two-year period," *J Air Waste Management Assoc.*, **1990**, 40, 740–746.

[30] W.W. Nazaroff and K. Teichman, "Indoor radon," *Environ Sci Technol.*, **1990**, 24, 774–782.

[31] R.B. Gammage and K.C. Gupta, Chapter 7 in *Indoor air quality,* Eds P.J. Walsh, C.S. Dudney, and E.D. Copenhaver, CRC Press, Boca Raton, FL, 1984.

[32] K. Sexton, M.X. Petreas, and K.S. Liu, "Formaldehyde exposures inside mobile homes," *Environ Sci Technol,* **1989**, 23, 985–988.

the foam. In several countries government assistance has been provided to homeowners to remove UFFI from their homes.

Table 4.2: Formaldehyde emission rates from products found in homes

Products	Emission rate $\mu g/g/day$
Plywood and particle board	< 0.1–9
Panelling	0.8–2
Glass fiber insulation	0.3–2.3
Clothing and drapery	0–5

Formaldehyde gas has an irritant odour, detectable at concentrations above about 0.2 ppm. Respiratory impairment in animals has been described with formaldehyde concentrations down to 0.3 ppm (guinea pigs) and 0.5 ppm (mice). In studies of residents of mobile homes and "UFFI homes" with high levels of formaldehyde, symptoms included drowsiness, nausea and headaches as well as respiratory ailments. The gas is mutagenic in the Ames assay. Long term exposure of rodents to high (14 ppm) concentrations of formaldehyde produced nasal cancers in several animals, but there is insufficient evidence yet to classify formaldehyde as a proven human carcinogen.

Since formaldehyde is a reducing agent, devices have been constructed to remove it by chemical reaction with an oxidant, such as a canister containing $Al_2O_3/KMnO_4$ (one tradename for which is Purafil). As in any chemical process, the capacity of the device is limited by the amount of the active chemical ($KMnO_4$); its capacity for formaldehyde is further limited because other gases can be oxidized[6]. Increased ventilation is therefore the recommended way to reduce the concentration of formaldehyde. The suggestion has been made that a polymeric amine such as polyethylenimine could be used as a coating on furnace filters to remove formaldehyde by chemical reaction[33].

The release of formaldehyde from materials used in home and office construction is accelerated at reduced pressure. Consequently mechanical exhausting of the air from a building by means of extractor fans may be less effective in reducing the concentration of formaldehyde in the building than would be predicted from the increased number of air changes per hour. The reduced pressure so created may be sufficient to speed up the rate of release of the gas into the air of the building. Ventilation systems which operate at balanced pressure, by using fans to move air into as well as out of the building, can minimize this effect.

Several European countries have set standards for formaldehyde in residential indoor air; most are near 0.1 ppm maximum. This figure has been

[33] H.D. Gesser and S. Fu, "Removal of aldehydes and acidic pollutants from indoor air," *Environ Sci Technol,* **1990,** 24, 495–497.

adopted also by the American Society of Heating, Refrigeration, and Air Conditioning Engineers.

4.2.3 Combustion products

There are various potential combustion sources within buildings, especially homes. Some of the indoor pollutants thus formed are shown below.

Pollutant	Sources
CO_2	Space heaters, gas cooking ranges, respiration
CO	Space heaters, woodstoves, tobacco smoke, fumes from garages
NO_X	Gas cooking ranges, space heaters
particulate matter	Woodstoves, tobacco smoke

Carbon dioxide.

Carbon dioxide is not poisonous, although it is an asphyxiant in very high concentrations. Elevated atmospheric $p(CO_2)$ impairs the release of this gas from the lungs. In airtight office buildings particularly, $p(CO_2)$ has been measured as high as 2000 ppm (recall that the outdoor concentration of CO_2 is about 350 ppm). Complaints from office-workers of fatigue and inability to concentrate are frequent when $p(CO_2)$ exceeds about 800 ppm[34][35]. These symptoms are relieved by increasing the ventilation rate. Note however that the TLV for industrial workplaces is an order of magnitude higher.

Carbon monoxide.

Carbon monoxide is toxic at low concentrations because it reacts irreversibly with reduced hemoglobin, forming carboxyhemoglobin, which has CO bonded as a ligand to the vacant binding site on the Fe(II)-porphyrin. This prevents uptake of oxygen by the hemoglobin. At a concentration of 10 ppm, 2% of the average person's hemoglobin is inactivated in this way; at 100 ppm, the proportion is 15%. Mild carbon monoxide poisoning is characterized by headaches, fatigue, and lowered alertness. The following are additional comments on some of the sources given in the Table above.

Kerosene space heaters: These inevitably produce high indoor concentrations of CO_2. Older models emit considerable amounts of CO due to incomplete combustion of the kerosene fuel. Those manufactured since about 1970 incorporate a catalytic device to ensure complete combustion. However, they emit more NO_X than the older models because the fuel burns hotter. Kerosene space heaters are also potential sources of high indoor

[34] K. Kreiss and M.J. Hodgson, Chapter 6 in *Indoor air quality*, Eds P.J. Walsh, C.S. Dudney, and E.D. Copenhaver, CRC Press, Boca Raton, FL, 1984.

[35] Reference 7, p. 13.

concentrations of respirable particles, NO_X, and SO_2, especially if they are poorly maintained[36].

Woodstoves: The traditional "Franklin" style of stove is not air tight. Both combustion gases (notably CO in this context) and particulate matter can leak into the room around the joins. This was less of a problem in Franklin's day, since his home was not air tight! Fuel-efficient, air tight woodstoves do not suffer from leakage in this way, except when the stove door is opened for stoking.

Tobacco smoking: The smouldering cigarette burns at a lower temperature, and much less completely, than the one which is being actively smoked. It is estimated[37] that during the 12 minutes the average cigarette is alight it is only smoked actively for 8–10 three second periods. The rest of the time it is smouldering. A single cigarette may release more than 100 mg of carbon monoxide into the room.

Fumes from garages: In homes with attached garages, leaving the car in the garage with the motor running almost ensures the infiltration of carbon monoxide into the living area. A different hazard exists in apartment and office buildings with underground garages, where car engines operate as a matter of course; a poor design of the ventilation system may lead to exhaust air from the parking garage being drawn into the ventilation intake, and contaminating the working or living areas above.

Nitrogen oxides

Relatively little NO_X is produced by combustion sources such as woodstoves and cigarettes, because the temperature of combustion is too low. For example, about 60 μg of NO_X is produced when one cigarette is smoked. Significant amounts of this pollutant are formed by gas cooking ranges and kerosene space heaters, where the temperature of the flame is high. Concentrations above the standard for NO_X (50 ppb in the United States; in Canada, the "desirable" level is < 32 ppb, and "acceptable" is < 110 ppb) can be recorded in kitchens equipped with gas cooking stoves, but not in similar homes where electric ranges are used[38].

In one office building, complaints about the quality of the air decreased when the lighting was changed from sunlight-simulating fluorescent lamps to the ordinary "cool-white" type. Although it was suggested that the sunlight-simulating lamps might have been producing photochemical smog

[36] B.P. Leaderer, P.M. Boone, and S.K. Hammond, "Total particulate, sulfate and acidic aerosol emissions from kerosene space heaters," *Environ Sci Technol,* **1990,** 24, 908–912.

[37] S.A. Glantz, Chapter 9 in *Indoor air quality,* Eds P.J. Walsh, C.S. Dudney, and E.D. Copenhaver, CRC Press, Boca Raton, FL, 1984.

[38] T. Godish, *Air quality,* Lewis Publishers Inc, Chelsea, MI, 1986 Chapter 11.

indoors[39], there is insufficient evidence to say whether this speculation was correct.

Particulate matter

The particulate matter from tobacco smoke and from woodstoves contains relatively large amounts of polycyclic aromatic hydrocarbons, many of which are carcinogenic (Chapter 3). The concentration of the carcinogenic hydrocarbon benzo(a)pyrene (often abbreviated BaP) is sometimes used as a monitor for the total amount of polycyclic aromatic hydrocarbons (PAH), but since the composition of the PAH fraction varies markedly and BaP is only a minor component, this may be misleading. In areas where a substantial fraction of domestic heat is obtained by burning wood, outdoor concentrations of BaP may rise to 10–100 ng m^{-3}, compared with 0.7 ng m^{-3} as the mean concentration of this substance in United States cities[40] (Problem 11).

Calle and Zeighami[40] have collected risk assessment data from several sources, all involving exposure to PAH's, in an attempt to estimate the health risk due to residential wood burning. In order to compare different studies, they have presented their data in terms of the "% excess risk of lung cancer for each 1 ng BaP per m^3" if exposure occurs for a 70-year lifetime. Considering workers in different industries and residents of urban areas in the United States and the United Kingdon, the excess risk was 0.2–5% per lifetime ng of BaP per m^3, with the single exception of cigarette smokers, for whom the corresponding figure was ca. 60%, regardless of the number of cigarettes smoked per day. Depending upon the assumptions, they made estimates of an excess lung cancer mortality that ranged from almost zero to a high of 25% additional deaths due to lung cancer as a result of residential wood burning.

The risk of lung cancer and cardiovascular disease among smokers is now so well known as to require no further elaboration here. A single cigarette can release close to 5×10^{12} particles, containing 5 μg of total PAH, into the air. Typical indoor levels of particulates are 20–60 μg m^{-3} in the absence of smoking, but concentrations of 100–700 μg m^{-3} are common where smoking is permitted. Passive, or second hand smoking is the term used to describe the inhalation of tobacco smoke by non smokers. Controversy exists over whether passive smoking carries an increased risk of lung cancer. Significantly elevated rates of death from lung cancer have been found among the non–smoking wives of Japanese and Greek smokers, but not among the non-smoking wives of American smokers. The Japanese and Greek data seem to implicate the husbands' smoking as the cause of the wives' disease. Why

[39] K. Kreiss and M.J. Hodgson, "Building-associated epidemics," Chapter 6 in *Indoor air quality,* Eds P.J. Walsh, C.S. Dudney, and E.D. Copenhaver, CRC Press, Boca Raton, FL, 1984.

[40] E.E. Calle and E.A. Zeighami, Chapter 3 in *Indoor air quality,* Eds P.J. Walsh, C.S. Dudney, and E.D. Copenhaver, CRC Press, Boca Raton, FL, 1984.

is this relationship not seen also in the United States? Glantz[41] points out that American women are much more mobile and spend less time in the home than those in the more traditional Japanese and Greek societies; as a result they have more opportunity for exposure to passive smoking outside the home, and spend less time in the smoke-filled home environment. Consequently the differences in the environments between smoking and non-smoking American wives of smokers tend to be evened out.

Further reading

1. P.J. Walsh, C.S. Dudney, and E.D. Copenhaver (Eds), *Indoor air quality,* CRC Press Inc., Boca Raton, Florida, 1984.
2. W.J. Fisk, F.J. Offermann, R.K. Spencer, B. Petersen, D.T. Grimsrud, and R. Sextro, *Indoor air quality control techniques,* Noyes Data Corp., Park Ridge, N.J., 1987.
3. T. Godish, *Air quality,* Lewis Publishers Inc., Chelsea, MI, 1986, Chapter 11.
4. R.J. Milko, *Indoor air quality,* Background paper for parliamentarians, Library of Parliament, Ottawa, 1985.

4.3 Problems

1. Express the TLV, given the table in mg m^{-3}, in the units ppm and mol L^{-1} for chlorine gas, CFC-11, and lead chromate.
2. Make assumptions about the physical characteristics of the particles in order to estimate the number of particles per cm^3 in air containing the TLV of coal dust.
3. Decide whether the TLV is exceeded in each of the following cases.
   (a) SO_2 (1.6 ppm) and chlorine (0.9 ppm)
   (b) benzene (1.5 ppm) and toluene (85 ppm)
   (c) heptane (180 ppm, TLV = 1600 mg m^{-3}), octane (100 ppm, TLV = 1450 mg m^{-3}), and nonane (25 ppm, TLV = 1050 mg m^{-3}).
   (d) CFC-11 (300 ppm), diethyl ether (120 ppm) and methyl isocyanate (0.035 ppm)
4. (a) Derive the composite TLV relationship given as Equation 1, stating any assumptions you make.

[41] S.A. Glantz, "Health effects of ambient tobacco smoke," Chapter 9 in "Indoor air quality," Eds P.J. Walsh, C.S. Dudney, and E.D. Copenhaver, CRC Press, Boca Raton, FL, 1984.

(b) Establish, where appropriate, a composite TLV if the following liquids should vaporize. The compositions of the liquids are given by weight.

(i) CFC-11 (60%) and CFC-12 (40%)
(ii) CFC-11 (10%) and CFC-12 (90%)
(iii) Heptane (20%), octane (30%), and nonane (50%): use TLV data from Problem 2c
(iv) Nitrobenzene (20%) and acetone (80%)
(v) Acetone (48%), diethyl ether (48%), and methyl isocyanate (2%)

5. (a) Obtain curves for the effect of the number of air changes per hour (range 0.1–5.0 ach) on the steady state concentration (in ppm) of CO_2 in an office building under these assumptions. Take the outdoor $p(CO_2)$ as 330 ppm.

(i) Volume = 20,000 m^3; rate of CO_2 emission = 5.0 kg h^{-1}
(ii) Volume = 20,000 m^3; rate of CO_2 emission = 10.0 kg h^{-1}

(b) An office building of volume 15,000 m^3 has a natural infiltration rate of 0.15 ach and the ventilation system can effect a further 1.3 ach. When the ventilation system recirculates 75% of the building air, $p(CO_2)$ in the building is 840 ppm. What proportion of fresh air should be used if $p(CO_2)$ is not to exceed 550 ppm?

(c) For the conditions in part (b), how long will it take for $p(CO_2)$ to drop from 840 ppm to 600 ppm?

6. (a) Derive a kinetic model for the build-up of formaldehyde in a two-compartment home. Label the compartments B for basement and L for living area, and obtain relationships from which the steady state concentration of formaldehyde in each compartment can be calculated.

(b) Calculate the steady state concentration of formaldehyde with the following assumptions. Volume of basement, 100 m^3; volume of living area 300 m^3; rate of transfer of air from basement to living area, 0.1 ach; rate of transfer of air from living area to basement ≈ 0; rate of infiltration to basement 0.1 ach; rate of infiltration to living area 0.4 ach; emission of formaldehyde in basement 8 mg h^{-1}; emission of formaldehyde in living area 6 mg h^{-1}.

7. A one compartment home of volume 330 m^3 has an infiltration rate of 0.25 ach with doors and windows closed. During an episode of photochemical smog, the outdoor concentration of PAN is 85 ppb. If the family remains indoors, and the initial concentration of PAN inside is 18 ppb, how long will it be before the PAN concentration inside rises to 45 ppb?

8. Let us assume for this problem that further decay of the radon daughters to Pb-210 occurs almost instantaneously after the decay of a radon

atom. Express the EPA action level of 4 pCi L^{-1} in Bq m^{-3}, and as a "TLV" in both ppm and mg m^{-3}.

9. **(a)** Calculate the amount of energy released when one radon atom and its daughters decay to Pb-210.

(b) Calculate the average concentration of radon in the air if a worker accumulates one working level month of radon exposure.

10. A mobile home has a volume of 100 m^3 and a ventilation rate of 0.28 ach. If the concentration of formaldehyde measured in the home is 11 ppm, what is the rate of emission of formaldehyde from the materials in the home?

11. In cigarette smoke, the ratio of BaP to the total particulate matter is about 1.7×10^4 mg in 4.5×10^{12} particles. Assuming wood smoke to be similar, and stating any other necessary assumptions, calculate:

(a) The concentration of particles when the outdoor air analyzes for 15 μg L^{-1} of BaP;

(b) The concentration of BaP in the air if woodsmoke is considered as a "nuisance dust" and it is present at the TLV of 10 mg m^{-3}.

Chapter 5

Natural Waters

Introduction

Two thirds of the Earth's surface is covered with water. Most of this is ocean, as seen in the figures below.

Distribution of the Earth's surface water

Oceans	9.5×10^{19} mol (> 99%)
Lakes and Rivers	1.7×10^{15} mol
Atmosphere	7.2×10^{14} mol

As well, there is water deep underground, called "ground water." In this chapter, we examine the chemistry of these different natural waters, as well as some of the interactions between the water and either the atmosphere or the underlying rock.

5.1 Dissolved gases in natural waters

When any gas equilibrates with a solvent, the amount of gas which dissolves is proportional to the partial pressure of the gas. This statement, which is known as **Henry's Law**, can be written mathematically as follows:

$$[\mathrm{X, solvent}] = \mathrm{constant} \times p(\mathrm{X,g})$$

or, in its usual form:

$$\mathrm{K_H} = [\mathrm{X,solv}]/p(\mathrm{X,g})$$

The proportionality constant is an equilibrium constant, known as the Henry's law constant. The usual units for K_H are mol L^{-1} atm^{-1}.

An everyday example of Henry's law in action is seen when a soft drink is opened. The drink is manufactured by dissolving CO_2 in the drink at a pressure of about 2 atm. The unfilled space in the bottle or can contains CO_2 at this pressure. When the drink is opened it comes in contact with the air where $p(CO_2) = 3 \times 10^{-4}$ atm, so $CO_2(aq)$ comes out of solution to restore equilibrium. This causes the pleasant effervescence of the drink, which eventually goes flat.

Gases become less soluble if their solutions are heated. This is true for all gases and all solvents, although in this chapter our discussion will be restricted to water as the solvent. From this variation of K_H with temperature, we can deduce the underlying thermodynamic principle governing the dissolution of gases namely, that change from the gaseous state to the solution is a process for which ΔH° and ΔS° are both negative. Enthalpic stabilization accompanies dissolution, but the dissolved state is more ordered than the gas. A rise in temperature thus favours the gaseous state ($-T\Delta S^\circ$ for dissolution becomes more positive).

5.1.1 Solubility of methane in water

For the case of water as solvent, thermodynamics permits an understanding of the low solubility of non-polar gases such as methane (Problems 1 and 2). A common misconception is that methane dissolves poorly because water hydrogen-bonds to itself and methane does not, and this disrupts the hydrogen bonding in water. This is an enthalpic argument. In fact, as illustrated in Problem 3, ΔH° for the process $CH_4(g) \longrightarrow CH_4(aq)$ is negative. The low solubility is due to the very negative ΔS°, which has two causes: the intrinsically greater order of a condensed phase compared with the gas phase, and the property of water in ordering itself around the non-polar solute molecule, a phenomenon which has been likened to forming a miniature iceberg around the solute. This greatly reduces the entropy of the water (remember that we must consider the whole system, not just the methane). This property of water (reduced solvent entropy in the presence of non-polar solutes) will be encountered again in other contexts.

The following table gives the Henry's law constants for some common gases at 25°C.

Gas	K_H, mol L^{-1} atm^{-1}	Gas	K_H, mol L^{-1} atm^{-1}
H_2	7.8×10^{-4}	CO	9×10^{-4}
N_2	6.5×10^{-4}	O_2	1.3×10^{-3}
CO_2	3.4×10^{-2}	O_3	1.3×10^{-2}

In this chapter we shall be particularly interested in O_2 and CO_2.

5.1.2 Solubility of oxygen in water

The atmosphere contains 0.21 atm O_2, so from the quoted value of K_H at 25°C it is possible to write[1]:

$$[O_2,aq] = K_H \times p(O_2,g) = (1.3 \times 10^{-3} \text{ mol L}^{-1} \text{ atm}^{-1})(0.21\text{atm})$$
$$= 2.7 \times 10^{-4} \text{ mol L}^{-1}$$

Converting to mg L^{-1}, we have

[1]This calculation ignores a small correction, namely that $p(\text{total}) = 1$ atm includes the contribution from $p(H_2O,g)$ which is about 0.03 atm at 25°C. Consequently, our calculation overestimates $[O_2,aq]$ by about 3%.

$$
\begin{aligned}
[\mathrm{O_2,aq}] &= 2.7 \times 10^{-4}\ \mathrm{mol\ L^{-1}} \times 32\ \mathrm{g\ mol^{-1}} \times (1000\ \mathrm{mg\ g^{-1}}) \\
&= 8.7\ \mathrm{mg\ L^{-1}}
\end{aligned}
$$

This value applies to oxygen which is in equilibrium with the atmosphere. From the definition of ppm for aqueous solutions[2], this is the same as 8.7 ppm.

Dissolved oxygen is essential for the survival of aquatic life. Most fish species, for example, require 5–6 ppm of dissolved oxygen. Without sufficient oxygen, the fish suffocate. Fish kills are common if for any reason the oxygen supply is depleted. Possible causes include:

- thermal pollution
- decomposition of biomass e.g., algal blooms
- oxidizable substances in the water (sewage, factory effluents, agricultural run-off)

Thermal pollution results when water is used for cooling (e.g., in electricity generating plants), causing it to be returned to a river or lake at a higher temperature. Because the solubility of oxygen is lower at higher temperature, the warm water is less oxygenated. Thermal pollution is more often a problem in summer, when water temperatures are high to begin with.

All the other situations arise when the water contains oxidizable substances, and these substances are oxidized by microorganisms. Algal blooms occur when a water body is supplied with excessive amounts of nutrients. Living algae produce oxygen by photosynthesis, but when they die oxygen is required to oxidize their biomass back to carbon dioxide and water. Untreated or partially treated sewage, factory effluents, especially from food processing (meat packing plants, vegetable and fruit canneries), and animal feedlots (manure seepage) are common sources of oxidizable organic compounds in waterways: see Chapter 8.

The concentration of oxygen in the water is of crucial importance to aquatic life. We next consider a number of different descriptors of the oxygen status of a water body and how these quantities are measured.

Dissolved oxygen. The actual concentration of oxygen in the water. There are three common methods for this analysis.

---

[2] A value of 1 ppm for liquids and solids is defined by mass i.e., 1 gram in 10^6 grams, 1 mg per kg, etc. Since 1 kg of water occupies a volume of exactly 1 L, for the special case of water as solvent, 1 ppm is the same as 1 mg $\mathrm{L^{-1}}$.

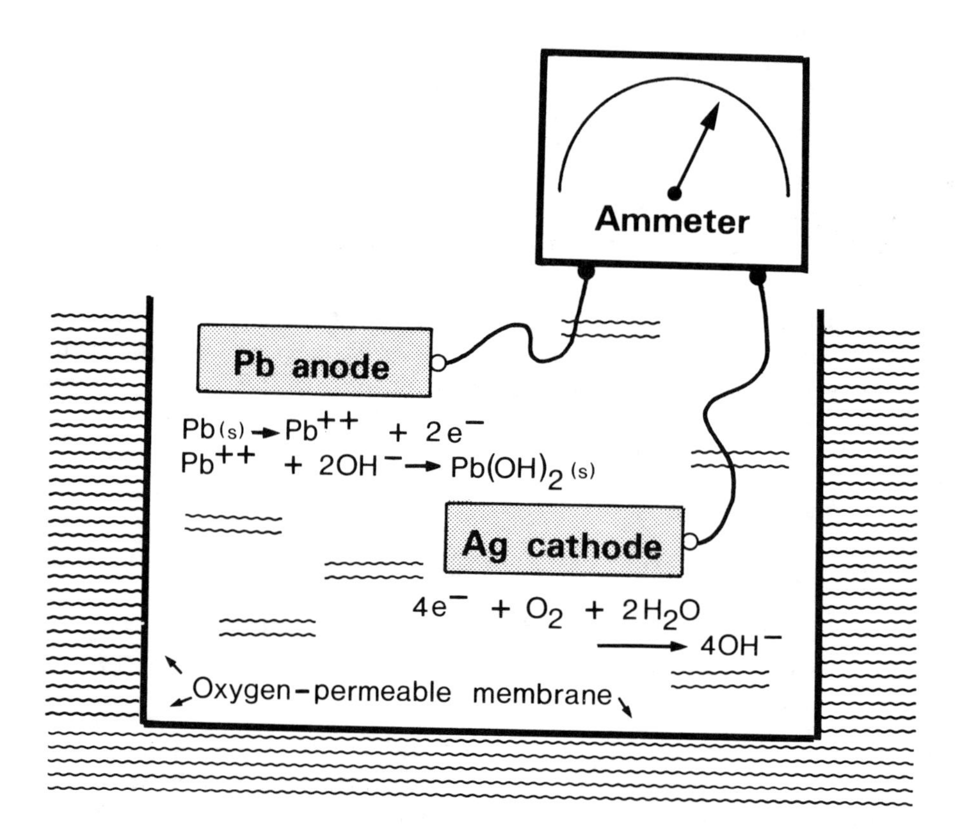

Figure 5.1: Schematic of electrochemical oxygen sensor.

1. Titration (Winkler's method). The relevant reactions follow (all substances are (aq) unless noted otherwise).

$$(1) \qquad Mn^{2+} + 2OH^- + \tfrac{1}{2}O_2 \longrightarrow MnO_2(s) + H_2O$$

$$(2) \qquad MnO_2(s) + 4H^+ + 2I^- \longrightarrow I_2 + Mn^{2+} + 2H_2O$$

$$(3) \qquad I_2 + 2Na_2S_2O_3 \longrightarrow Na_2S_4O_6 + 2NaI$$

The water sample is treated with manganese sulfate in alkaline solution; the precipitated manganese dioxide is used to oxidize I^- to I_2, which is then estimated by titration against standardized sodium thiosulfate solution (Problem 4).

2. Spectrophotometry. Several dyes are oxidized by aqueous solutions of O_2 to derivatives having a different colour. Methylene blue and indigo carmine are examples. For example, the pale yellow leuco form of indigo carmine is readily oxidized by molecular oxygen at pH < 10 to the dark blue indigo form, which can be quantitated spectrophotometrically.

3. Electrochemically (Makareth oxygen electrode). In the electrochemical oxygen sensor, oxygen diffuses into the cell through a thin, disposable polyethylene membrane. The electrode reactions are:

cathode (made of silver):

(4) $$O_2(g) + 2H_2O(l) + 4e^- \longrightarrow 4OH^-$$

anode (made of lead):

(5) $$4OH^- + 2Pb(s) \longrightarrow 2Pb(OH)_2(s) + 4e^-$$

overall:

(6) $$O_2(g) + 2H_2O(l) + 2Pb(s) \longrightarrow 2Pb(OH)_2(s)$$

From the Nernst equation, we can see that the potential across this cell depends only upon $p(O_2)$:

$$E_{cell} = E^\circ - \left(\frac{RT}{n\mathcal{F}}\right) \ln\left(\frac{1}{p(O_2)}\right)$$

In actual operation of the oxygen electrode, a constant potential is applied across the cell, and the current flow—directly proportional to $[O_2,aq]$—is measured. Use of a small battery as the power source makes the device lightweight and portable, and thus ideally suited to measurements in the field.

The other terms all refer to the amounts of oxidizable substances in the water, and hence relate indirectly to the oxygen status through the amount of oxygen that would be needed to oxidize them.

Total organic carbon (TOC). Measured by oxidizing all the organic matter to CO_2, and then analyzing the CO_2 formed gas chromatographically. TOC is usually reported in ppm of carbon. A TOC analyzer is normally a self-contained unit comprising a small furnace to vaporize and oxidize the sample, and a dedicated gas chromatograph in which the small amount of CO_2 can be separated chromatographically from the vast excess of water that is present.

Chemical oxygen demand (COD). Measured by reacting the water sample with a fixed amount of $Na_2Cr_2O_7/H_2SO_4$ under stated conditions of time and temperature, and then titrating the **unreacted** $Na_2Cr_2O_7$ against a standardized Fe^{2+} solution. Each mole of $Cr_2O_7^{2-}$ consumed is equivalent, in acidic solution, to 1.5 moles of O_2. In other words, one mole of $Cr_2O_7^{2-}$ can oxidize as much organic material as 1.5 mol of O_2 (Problem 5).

(7) $$Cr_2O_7^{2-} + 8H^+ \longrightarrow 2Cr^{3+} + 4H_2O + 3(O)$$

The 3(O) represents the available oxidant. For Fe^{2+} as the reducing agent, the stoichiometry is as follows:

(8) $$Cr_2O_7^{2-} + 14H^+ + 6Fe^{2+} \longrightarrow 2Cr^{3+} + 6Fe^{3+} + 7H_2O$$

Biochemical oxygen demand (BOD). Measured by incubating the water sample with aerobic microorganisms under stated conditions of time and temperature (usually 5 days, 25°C). The dissolved oxygen is measured at the beginning and end of the experiment, and the difference is the BOD

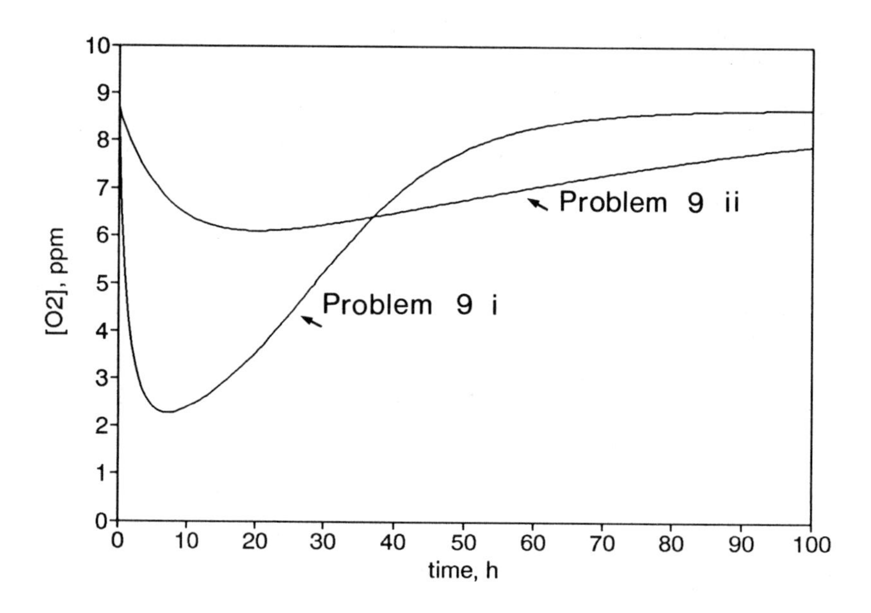

Figure 5.2: Effect on [O_2,aq] of a ten-fold change in the rate constant for oxidation of an oxidizable substance. The figure was constructed using the parameters of Problem 9.

(Problems 6 and 7). An important consideration in setting up the experimental protocol is that the oxidizable material is the limiting reactant. If it is not, then all the oxygen will have been consumed before the test ends, and the analyst will not know how much more would have been consumed if only there had been more oxygen available. To avoid this problem, samples having very high BOD will need to be diluted so that oxygen will not be the limiting reactant.

Each of these last three measures of the oxygen status of the water is arbitrary. Not all organics oxidize with equal ease. For example, carbohydrates, which are polyalcohols, oxidize rapidly, whereas alkanes have no functional group and oxidize very slowly. Slowly oxidizable compounds cause less oxygen depletion and hence are less of a threat to aquatic life, because O_2(aq) can be replenished from the air while oxidation is proceeding; this means that the concentration of O_2(aq) is reduced to a smaller extent but for a longer time (Figure 5.2, and Problem 9). Furthermore, a fixed dose of oxidizable material will have a more harmful effect on a natural water when the temperature is high, because oxidation will be faster, and replenishment of the oxygen from the air will be less likely to keep pace with depletion.

If the flow of a stream is taken to be constant with distance, Figure 5.2 can also be interpreted with the x-axis representing distance. In this format, Figure 5.2 portrays the depletion of oxygen as a function of distance downstream from a point source discharge (e.g., a sewage outfall).

Total organic carbon is an arbitrary measure of oxygen status because all carbon compounds are included, but they oxidize at different rates. Chemical oxygen demand goes some way to compensating for this, because acidic dichromate readily oxidizes functionalities such as alcohols and alkenes, but is very slow to attack alkanes, carboxylic acids, and aromatic rings. Even so, this reagent cannot really mimic the environment. Biochemical oxygen demand would seem at first to be the ideal approach, since oxidation is accomplished by microorganisms; however, both the choice of the time and temperature for the test, and the selection of the microorganisms themselves, are arbitrary parameters. All these measures thus have their uses, in terms of convenience of analysis, but none can reflect accurately what happens in a real environmental system.

5.1.3 Solubility of carbon dioxide in water

The solubility behaviour of CO_2 in water is inherently more complex than that of oxygen because CO_2 reacts with water.

$$CO_2(g) \rightleftharpoons CO_2(aq) \xrightleftharpoons{H_2O} H_2CO_3(aq) \xrightleftharpoons{-H^+} HCO_3^-(aq) \xrightleftharpoons{-H^+} CO_3^{2-}(aq)$$

For our purposes we will consider the total of $CO_2(aq)$ and undissociated $H_2CO_3(aq)$ together, and we will also regard $CO_2(aq)$ and $H_2CO_3(aq)$ as interchangeable[3]. On this basis, the value of K_H for the dissolution of CO_2 in water at 25°C is 3.4×10^{-2} mol L^{-1} atm^{-1}, and the concentration of dissolved CO_2 in equilibrium with the air ($p(CO_2,g) = 3.0 \times 10^{-4}$ atm) is 1.0×10^{-5} mol L^{-1}, or 0.44 ppm of CO_2.

When CO_2 dissolves in pure water, a little of the H_2CO_3 (but almost none of the HCO_3^-) dissociates, making the water slightly acidic. This is conveniently illustrated by posing the question, "What are the pH and the total carbonate concentration of a sample of water in equilibrium with the air?"

To solve this problem, we would first identify the relevant chemical equation, namely, the acid dissociation of H_2CO_3.

$$H_2CO_3(aq) \rightleftharpoons H^+(aq) + HCO_3^-(aq) \tag{9}$$

Proceeding as in first year chemistry, we would identify "initial" and "equilibrium" concentrations of these chemical species. In this example, we do

[3]The actual equilibria are:

$$CO_2(g) \xrightleftharpoons{K_H} CO_2(aq) \xrightleftharpoons{K_1} H_2CO_3(aq) \xrightleftharpoons{K_2} H^+(aq) + HCO_3^-(aq)$$

where $K_1 = 1.6\times10^{-3}$ and $K_2 = 3.2\times10^{-4}$ mol L^{-1}. The value of K_a normally quoted (and which we shall use throughout this book) is $K_1K_2 \approx 4 \times 10^{-7}$ mol L^{-1} at 25°C. Since $CO_2(aq)$ and $H_2CO_3(aq)$ equilibrate rapidly, there is no advantage to separating out these equilibria. The only point of chemical interest is that $H_2CO_3(aq)$ is actually a much stronger acid, $K_2 = 3.2 \times 10^{-4}$ mol L^{-1}, than the composite acidity constant K_a would indicate.

not need to write out the "initial" concentrations, because the concentration of H_2CO_3 remains constant, with the value 1.0×10^{-5} mol L^{-1} (see above), because it is in equilibrium with the atmospheric reservoir of $CO_2(g)$ at all times. Therefore we proceed as follows:

$$\begin{array}{lccc} & \underbrace{H_2CO_3(aq)} \longrightarrow & \underbrace{H^+(aq)} + & \underbrace{HCO_3^-(aq)} \\ \text{equilib conc.} = & 1.0 \times 10^{-5} & x & x \end{array}$$

Restating the previous point, the "x mol L^{-1}" which dissociates does not deplete the H_2CO_3 (i.e., we do not write $(1.0 \times 10^{-5} - x)$ because at equilibrium, the H_2CO_3 is replenished from the atmosphere.

$$\begin{aligned} \text{For } H_2CO_3, \quad K_a &= \frac{[H^+][HCO_3^-]}{[H_2CO_3]} \\ &= 4.2 \times 10^{-7} \text{ mol L}^{-1} \text{ at } 25°\text{C}. \\ \text{Hence:} \quad x &= 2.1 \times 10^{-6}, \text{and pH} = 5.67. \\ \text{Total carbonate} &= [H_2CO_3] + [HCO_3^-] = 1.2 \times 10^{-5} \text{ mol L}^{-1}. \end{aligned}$$

This gives the important result that pure water in equilibrium with the air is not at pH 7; it is slightly acidic due to the presence of dissolved CO_2. By the same token, even unpolluted rainwater is at pH ≈ 5.6 rather than 7. This topic is discussed further in Chapter 6.

As the pH rises, the total amount of dissolved carbonate ($= [CO_2] + [H_2CO_3] + [HCO_3^-] + [CO_3^{2-}]$) in equilibrium with the atmosphere increases. The atmosphere acts as an inexhaustible reservoir of $CO_2(g)$, maintaining its pressure at 3.0×10^{-4} atm. This is shown in Figure 5.3.

5.2 Dissolved solids in natural waters

The amount of solid dissolved in natural water varies widely. The values in Table 5.1 are typical of river and ocean water, although as we shall see, river water is quite variable in its mineral content. Ground water is at least as high in dissolved solids as lake and river water; sometimes it is much higher, with total dissolved solids exceeding 1000 ppm.

With the exceptions of Ca^{2+} and HCO_3^-, there is a parallel between the average concentrations of ions in fresh and in ocean water. There is relatively more Ca^{2+} and HCO_3^- in river water because rivers dissolve ancient rocks containing $CaCO_3$, whereas the oceans precipitate $CaCO_3$ in the form of marine organisms' exoskeletons.

5.2.1 Alkalinity

The alkalinity of a water sample is a measure of the concentration of bases it contains. The most important of these, following on from the previous section, are usually HCO_3^- and CO_3^{2-}. The pH of a water sample is not a

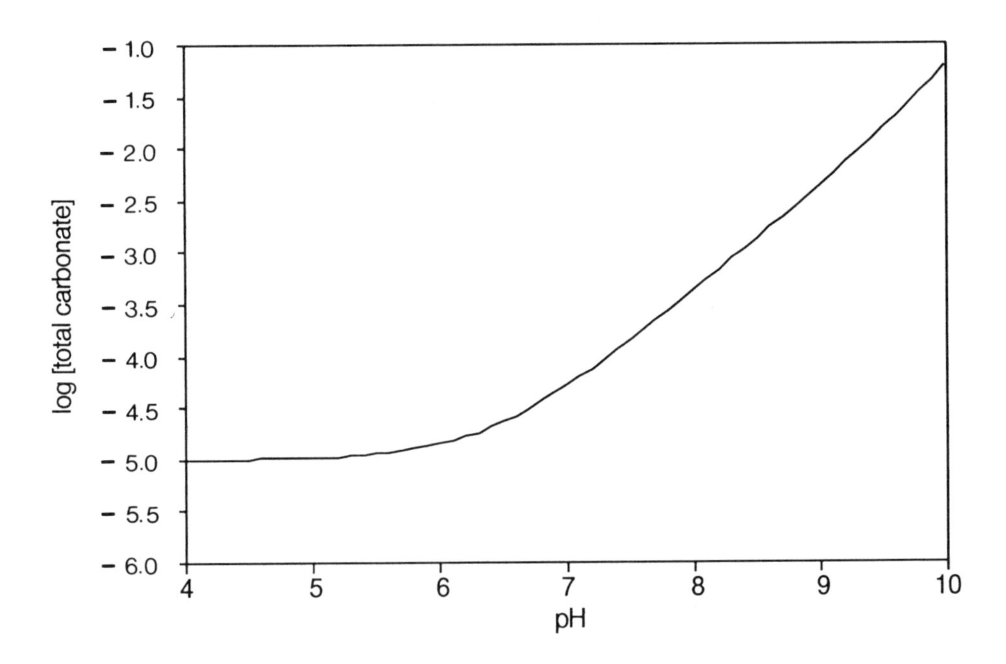

Figure 5.3: Calculated concentration of dissolved carbonate (H_2CO_3 + HCO_3^- + CO_3^{2-}) in equilibrium with 350 ppm of CO_2(g) as a function of pH.

good guide to its alkalinity because these weak carbonate bases are usually much more abundant than OH^-. Also, you cannot estimate the concentration of total carbonate at a given pH from Figure 5.3, because many natural waters are not in equilibrium with the CO_2 of the atmosphere (which was one of the premises under which Figure 5.3 was obtained).

The alkalinity of a solution is measured by titration against standard acid. Two terms are in common use, **total alkalinity** and **phenolphthalein alkalinity.** Remember that the sample is being titrated against acid, so the pH decreases as the titration proceeds. Total alkalinity is measured by using methyl orange as the indicator, when the colour change occurs at pH $\approx$ 4.3; the phenolphthalein alkalinity is obtained by using phenolphthalein as the indicator, and stopping the titration at pH $\approx$ 8.5. The relevant reactions follow:

phenolphthalein alkalinity:

(10) $$CO_3^{2-} + H^+ \longrightarrow HCO_3^-$$

total alkalinity:

(11) $$CO_3^{2-} + 2H^+ \longrightarrow H_2CO_3$$

(12) $$HCO_3^- + H^+ \longrightarrow H_2CO_3$$

Table 5.1: Typical concentrations of ions in river and sea water[4].

Ion	c(river), mol $L^{-1} \times 10^4$	c(ocean), mol L^{-1}
Cl^-	2.2	0.55
Na^+	2.7	0.46
Mg^{2+}	1.7	0.054
SO_4^{2-}	1.2	0.028
K^+	0.59	0.010
Ca^{2+}	3.8	0.010
HCO_3^-	9.5	0.0023

This is easier to understand in terms of a titration curve. For review, Figure 5.4a represents the titration of a weak base (ammonia) with hydrochloric acid. The equivalence point is marked. Figure 5.4b is a titration curve of a water sample against hydrochloric acid; initially it contains much more HCO_3^- than CO_3^{2-}. At pH 8.5, the phenolphthalein end-point, all the CO_3^{2-} has been converted to HCO_3^-. At pH 4.5, the methyl orange end-point, both the CO_3^{2-} and the HCO_3^- have been changed into H_2CO_3.

The carbonate composition of a natural water can be deduced in several ways[5]. The commonest are:

1. measure both "phenolphthalein alkalinity" (CO_3^{2-}) and "total alkalinity" ($HCO_3^- + 2CO_3^{2-}$);
2. measure **either** "phenolphthalein" **or** "total" alkalinity, and then obtain the ratio $[CO_3^{2-}]/[HCO_3^-]$ from the pH (Problem 8).

$$\text{For:}\quad HCO_3^-(aq) \rightleftharpoons H^+(aq) + CO_3^{2-}(aq),$$

$$K_a = \frac{[H^+][CO_3^{2-}]}{[HCO_3^-]} = 4.8 \times 10^{-11} \text{ at } 25°C.$$

The units of alkalinity are expressed either as mol H^+ per liter (i.e., in terms of how much acid is needed to neutralize all the bases) or as ppm of calcium carbonate. The latter involves determining how much $CaCO_3$ would be neutralized by the acid, and using this as a proxy for the actual mixture of CO_3^{2-}, HCO_3^-, and maybe other bases that might be present. The amount of $CaCO_3$ in the sample is usually reported in ppm (convert mol L^{-1}

---

[4] R.W. Raiswell, P. Brimblecombe, D.L. Dent, and P.S. Liss, *Environmental Chemistry,* Edward Arnold (Publishers), London, England, Chapter 3.

[5] In this discussion, the contribution of free hydroxide ion to the alkalinity has been ignored. This is a reasonable approximation for water samples having pH in the usual range of 5–8.5. Even at pH 9, the contribution of free $[OH^-]$ to the alkalinity is only 1×10^{-5} mol L^{-1}. The effect of free OH^- should always be included for very alkaline water samples, for example some industrial effluents.

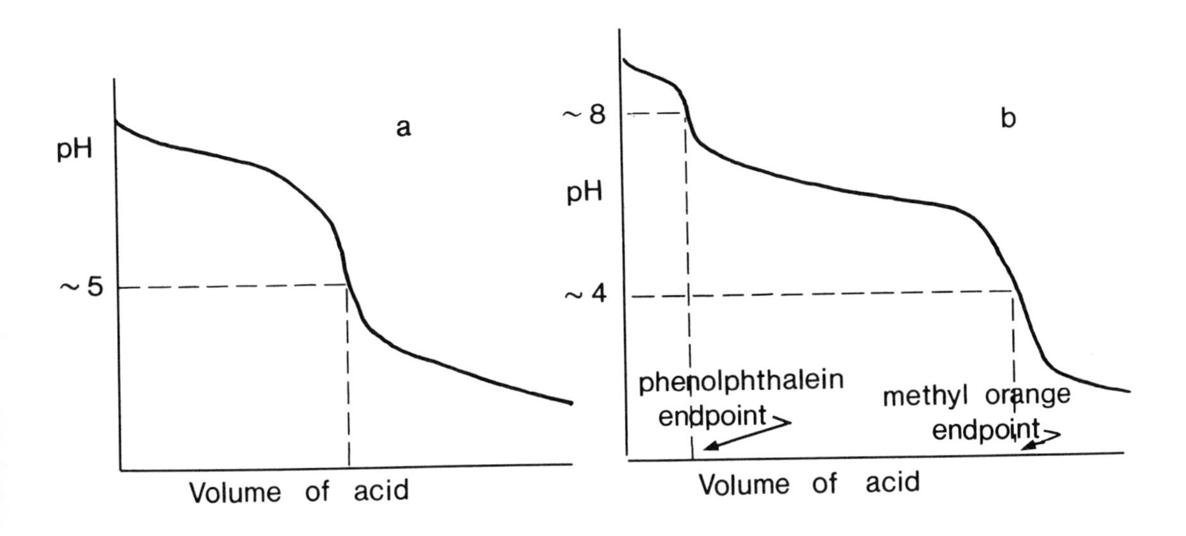

Figure 5.4: (a) Titration curve for 0.1 mol L^{-1} ammonia against 0.1 mol L^{-1} HCl, (b) Titration curve of a water sample against 0.01 mol L^{-1} HCl.

of $CaCO_3$ to mg L^{-1} (ppm), using the molar mass of $CaCO_3 = 1.00 \times 10^5$ mg mol^{-1}).

The speciation of total carbonate (Figure 5.5) shows that for the majority of natural waters (pH 6–8.5), HCO_3^- is the predominant species contributing to the alkalinity[6]. Very few natural waters contain much carbonate ion, notable as an exception being the "alkaline waters" of some parts of the United States Midwest.

It must be emphasized that **pH is not a good guide to alkalinity**. It is quite possible for a water sample at pH 7 to have total alkalinity in excess of 10^{-3} mol L^{-1} of H^+. This is because all bases, and not just OH^-, are titratable with acid and thus contribute to the total alkalinity.

[6]Speciation diagrams may not be familiar to all readers, but will be encountered several times in this book. The y-axis of Figure 5.5 represents the **fractions** of all the carbonate species ($H_2CO_3, HCO_3^-, CO_3^{2-}$) which are present at a given pH. At pH 8, for example, the fraction of HCO_3^- is close to unity, i.e., HCO_3^- is the predominating carbonate species. Note also that when $f(H_2CO_3) = f(HCO_3^-)$, the pH = pK_a of H_2CO_3 (about 6.3), and likewise the pK_a of HCO_3^- is deduced from the pH at which $f(HCO_3^-) = f(CO_3^{2-})$.

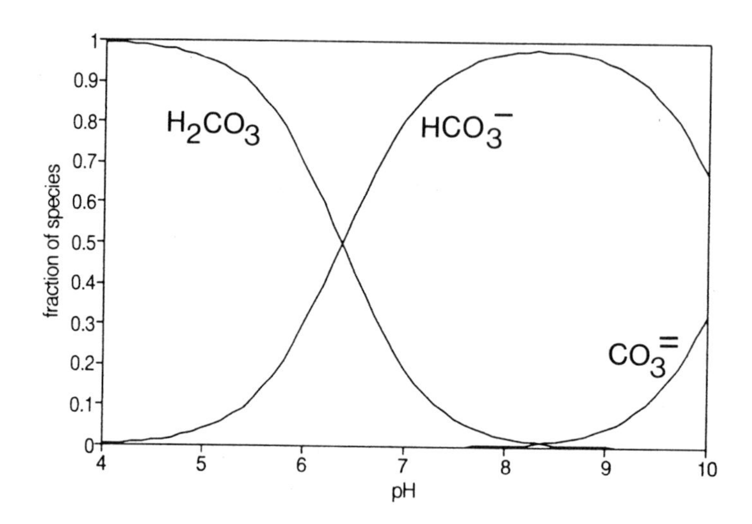

Figure 5.5: Speciation of H_2CO_3, HCO_3^-, and CO_3^{2-} with pH.

A second consideration is that the alkalinity can be higher than predicted by Figure 5.3, where it was assumed that the solution was in equilibrium with the CO_2 of the air. Subterranean waters usually come in contact with higher partial pressures of CO_2, up to ≈ 0.01 atm, thirty times greater than the atmospheric partial pressure. This is because microbial oxidation of soil organic matter increases $p(CO_2)$ at the expense of $p(O_2)$.

5.2.2 Hardness

The discussion of alkalinity has focussed on the anions that are present in the water sample. Their negative charges must, of course, be balanced by the positive charges of cations. The "hardness cations" are the "alkaline earths" of the periodic table, principally Ca^{2+} and Mg^{2+} (Problems 10–12). These are the cations which cause precipitation with soaps, see Chapter 8. Calcium is the ion whose concentration is usually analyzed, since limestone ($CaCO_3$) is a very common rock. The analysis is carried out by titrating the water sample against a standard solution of ethylenediaminetetraacetic acid (EDTA) at pH < 10. EDTA is supplied as its tetrasodium salt, $EDTA^{4-}$.

$$(13)\quad Ca^{2+} + EDTA^{4-} \rightleftharpoons (CaEDTA)^{2-};\quad K_{assoc.} = 5 \times 10^{10}\ L\ mol^{-1}$$

All these substances are colourless, so an indicator, Eriochrome Black T, is used to signal the end-point. Free Eriochrome Black T is dark blue in solution. Also added to the solution at the start of the titration is Mg^{2+}, which forms a red complex with Eriochrome Black T (K(association) $= 1 \times$

10^7 L mol^{-1}). Mg^{2+} also associates with $EDTA^{4-}$ (K = 5×10^8 L mol^{-1}), but two orders of magnitude less strongly than Ca^{2+}. Consider what happens during the titration with EDTA.

At the beginning of the titration, no EDTA has been added. The solution is red, due to the colour of the Mg^{2+}/Eriochrome complex. $EDTA^{4-}$ is now added; it complexes preferentially with Ca^{2+}, so the Mg^{2+} remains complexed to the Eriochrome indicator and the solution remains red. When the last of the Ca^{2+} has been complexed by $EDTA^{4-}$, free $EDTA^{4-}$ is present in solution. Since the association constant for $Mg^{2+}/EDTA^{4-}$ is larger than that for Mg^{2+}/Eriochrome, the Mg^{2+} detaches from the indicator and associates with the EDTA. The colour changes from red (Mg–Eriochrome) to blue (free Eriochrome).

The source of hardness, indeed of all dissolved solids, in fresh water is the dissolution of the underlying rock. Limestone ($CaCO_3$) yields Ca^{2+}, HCO_3^-, and CO_3^{2-}; gypsum ($CaSO_4$) affords Ca^{2+} and SO_4^{2-}. Water that has percolated through limestone has both hardness and alkalinity; water in contact with gypsum is hard but of low alkalinity, because the sulfate anion is so feebly basic ($K_b = 8 \times 10^{-13}$ mol L^{-1}).

5.2.3 Quantitative assessment of water hardness

In this section, we consider the reasons that water samples, for example those used for domestic water, often contain the high levels of dissolved materials. Waters in contact with $CaSO_4$ are easier to understand.

Calcium sulfate is slightly soluble in water:

$$CaSO_4(s) \rightleftharpoons Ca^{2+}(aq) + SO_4^{2-}(aq) \qquad (14)$$

$$K_{sp} = 3 \times 10^{-5}\ (\text{mol L}^{-1})^2 \text{ at } 25°\text{C}$$

From this, the maximum (equilibrium) solubility[7] is calculated to be $\approx 5 \times 10^{-3}$ mol L^{-1}.

The presence of calcium sulfate in water is sometimes called "permanent hardness." This is because, unlike hardness due to carbonates, it cannot be removed upon boiling. In many ways, it is the lesser problem, because it is less prone to leave scale in hot water boilers and hot water pipes.

Calcium carbonate is intrinsically much less soluble in water than calcium sulfate, $K_{sp} = 6 \times 10^{-9}$ (mol L^{-1})^2 at 25°C. However, because the carbonate anion is basic, $CaCO_3$ becomes increasingly soluble as the pH drops. The reaction may be summarized as Equation 15.

$$CaCO_3(s) + H_2CO_3(aq) \rightleftharpoons Ca(HCO_3)_2(aq) \qquad (15)$$

[7] Actually, simple K_{sp} calculations are notoriously poor for giving the correct solubility, because they omit important factors such as the need to use activities rather than concentrations, and the presence of ion pairs (in the present example, $CaSO_4^\circ$). For further discussion, see section 2.6 of this chapter and also S.O. Russo and G.I.H. Hanania, "Ion association, solubilities, and reduction potentials in aqueous solution," *J Chem Educ*, **1989**, 66, 148–153.

This reaction is responsible for the formation, over thousands of years, of caves and gorges in limestone areas, as CO_2-laden rainwater very slowly dissolves the rock. The process is slow because the equilibrium constant is small.

We can estimate the equilibrium constant for this reaction at 25°C.

(16)	$CaCO_3(s) \rightleftharpoons Ca^{2+}(aq) + CO_3^{2-}(aq)$	$K_1 = K_{sp}$ for $CaCO_3$
(17)	$H^+(aq) + CO_3^{2-}(aq) \rightleftharpoons HCO_3^-(aq)$	$K_2 = 1/K_a$ for HCO_3^-
(18)	$H_2CO_3(aq) \rightleftharpoons H^+(aq) + HCO_3^-(aq)$	$K_3 = K_a$ for H_2CO_3
(19)	$CaCO_3(s) + H_2CO_3(aq) \rightleftharpoons Ca(HCO_3)_2(aq)$	$K_4 = K_1 \times K_2 \times K_3$

where: $K_4 = (6.0\times10^{-9})(1/4.8\times10^{-11})(4.2\times10^{-7}) = 5.3\times10^{-5}$ $(\text{mol L}^{-1})^2$.

Taking these values for the equilibrium constants and $[H_2CO_3] = 1 \times 10^{-5}$ mol L^{-1}, we obtain 5×10^{-4} mol L^{-1} = 20 ppm as the equilibrium concentration of Ca^{2+}. This is more than an order of magnitude less than the hardness actually encountered in hard water areas. What is wrong?

Two things. First, ground water is closer to 5°C than 25°C; the equilibrium constants are temperature-dependent, particularly K_H for the dissolution of $CO_2(g)$ in water (6.3×10^{-2} at 5°, rather than 3.4×10^{-3} mol L^{-1} atm^{-1} at 25°). Second, $p(CO_2,g)$ underground is higher (up to ≈ 0.01 atm) than the atmospheric partial pressure. The concentration of $H_2CO_3(aq)$ is therefore higher than 1×10^{-5} mol L^{-1}. Recalculation gives a value for $[Ca^{2+}]$ that is closer to that found in hard water, namely 2×10^{-3} mol L^{-1}, or 80 ppm.

In some regions the underlying rock is "dolomitic limestone," and even higher levels of hardness are seen. Dolomitic limestone, which takes its name from the Dolomite mountains in Italy, is an approximately 1:1 composite of calcium and magnesium carbonates. It is more soluble than regular calcium carbonate:

$$(20)\quad \tfrac{1}{2}CaCO_3\cdot MgCO_3(s) \rightleftharpoons \tfrac{1}{2}Ca^{2+}(aq) + \tfrac{1}{2}Mg^{2+}(aq) + CO_3^{2-}(aq)$$

For this reaction $K_{sp} = 5.1 \times 10^{-7}$ (at 12°C). Water in equilibrium with dolomite is calculated (again on the basis of K_{sp} only) to contain 190 ppm of calcium and the equivalent quantity of magnesium (Problem 10).

Hardness due to carbonates is sometimes called "temporary hardness" because it can be removed by boiling. Upon heating, the reverse of Equation 15 occurs, mainly because of the reduced solubility of CO_2 in water at elevated temperatures (Problems 13–15).

$$(21)\qquad Ca(HCO_3)_2(aq) \rightleftharpoons CaCO_3(s) + H_2O(l) + CO_2(g)$$

Precipitation of scale (calcium carbonate) in hot water systems is a problem in hard water areas and is the main incentive for softening water. In industrial boilers for raising steam, prior softening of the water is absolutely mandatory, given the large volumes of water that must be processed and hence the large mass of scale that would otherwise be formed.

5.2.4 Soft water

From the foregoing discussion, it may be guessed that soft water contains only low concentrations of the hardness ions calcium and magnesium. Depending on its source, it may contain alkali metal cations, particularly sodium, or it may contain very little dissolved solid at all. Soft water is encountered commonly in regions where the underlying rock is granite, which is very insoluble. New England, much of eastern Canada, Scotland, and Scandinavia are all areas where soft water is common. Low cation concentrations are parallelled by low anion concentrations, so soft waters are usually low in alkalinity also. Rivers, lakes, and ground water in soft water areas typically have pH 6–7, whereas in hard water areas pH 7–8.5 is more common. Low alkalinity means that the water has little capacity to neutralize added acid, and acid rain is a much greater threat to aquatic life in these soft water areas than it is in limestone areas where added acidity can be neutralized (see Chapter 6).

At this point we will digress to discuss artificial methods of softening water. Softening is practised to protect water installations from scaly deposits, and because calcium and magnesium salts give scummy, or curdy, precipitates with soaps. Precipitation of the soap renders it ineffective for cleaning, as well as leaving an unsightly mess in the bath-tub. Water softening is the process of removing the hardness ions.

1. Lime softening. The raw water is treated with the stoichiometric amount of lime, $Ca(OH)_2$.

$$Ca(OH)_2 + Ca(HCO_3)_2 \longrightarrow 2CaCO_3(s) + 2H_2O \tag{22}$$

What is really happening here is an acid-base reaction between HCO_3^- and OH^-, followed by precipitation of the CO_3^{2-} thus formed as $CaCO_3$.

Because lime softening involves addition of one calcium compound (lime) to precipitate another, the exact stoichiometry must be maintained. This method is only applicable industrially, where the composition of the raw water can be closely monitored. Its benefit is the low cost of lime as a softening agent.

2. Ion exchange. This method is suitable for home or industrial use. A cation exchanger is an insoluble inorganic or organic polymer, which carries multiple negative charges on its backbone. An organic polymer, for example, might have carboxylate or sulfonate substituents on its carbon backbone. These negative charges are balanced by cations, but the cations are free to move within the structure of the polymer. When the ion exchanger is ready for use the associated cations are normally Na^+. When water passes over a bed of the ion exchange resin, the cations in the water can exchange with the cations associated with the resin. Designating the anionic sites on the polymer as (A^-) we have:

$$Ca^{2+}(aq) + 2Na^+(A^-) \longrightarrow 2Na^+(aq) + Ca^{2+}(A^-)_2 \tag{23}$$

Calcium (or magnesium) from the water becomes associated with the ion exchanger, and sodium leaves the resin and enters the water.

When most of the sodium ions have been exchanged for calcium or magnesium, the ion exchanger loses its effectiveness. It can be regenerated by passing a concentrated salt solution through the resin, whereupon the reverse reaction occurs.

$$(24) \qquad Ca^{2+}(A^-)_2 + 2Na^+(aq) \longrightarrow Ca^{2+}(aq) + 2Na^+(A^-)$$

The calcium released is discharged to the drain, together with the excess NaCl.

Since the reaction by which the resin is regenerated is the reverse of the reaction by which the exchanger operates, the chemical requirements for the design of a successful water softener can be specified. The equilibrium constant for the exchange Reaction 23 must be large enough that a low concentration of calcium will successfully displace Na^+ from the resin sites. However, it must not be **too** large, otherwise it will not be possible to get the calcium ions off the resin again; that would give a "one time only" water softener! A moderately large equilibrium constant for Reaction 23 means that the reverse Reaction 24 can be made to proceed if the regenerating solution of NaCl is strong enough, and that is how the resin is regenerated in practice.

Common ion exchangers are sulfonated polystyrene, cross linked for rigidity with divinylbenzene, and zeolites, which are inorganic structures based on linked tetrahedral arrangements of SiO_4 and AlO_4 units. Each oxygen bridges two other atoms, Si–O–Si or Si–O–Al, etc. to form a rigid network, similar to quartz, SiO_2. Quartz itself has no charge, but one negative charge on the framework is created for every aluminum atom which replaces silicon. This charge must be balanced by one mobile cation in the channels in the zeolite.

3. Deionized water. This is prepared by the use of two ion exchangers in series. The cation exchanger is similar to that just described, except that H^+ is the counter ion in the active form of the resin.

$$(25) \qquad Ca^{2+}(aq) + 2H^+(A^-) \longrightarrow 2H^+(aq) + Ca^{2+}(A^-)_2$$

The resin is regenerated with HCl rather than NaCl:

$$(26) \qquad Ca^{2+}(A^-)_2 + 2H^+(aq) \longrightarrow Ca^{2+}(aq) + 2H^+(A^-)$$

The second exchanger is an anion exchanger. The polymer backbone this time carries trialkylammonium ($-NR_3^+$) substituents, and the counter ions are OH^-. Writing (C^+) as a cationic site on the resin, and Cl^- as an example of an anion to be exchanged, the reaction for anion exchange is:

$$(27) \qquad Cl^-(aq) + (C^+)OH^- \longrightarrow OH^-(aq) + (C^+)Cl^-$$

This resin is regenerated with NaOH solution.

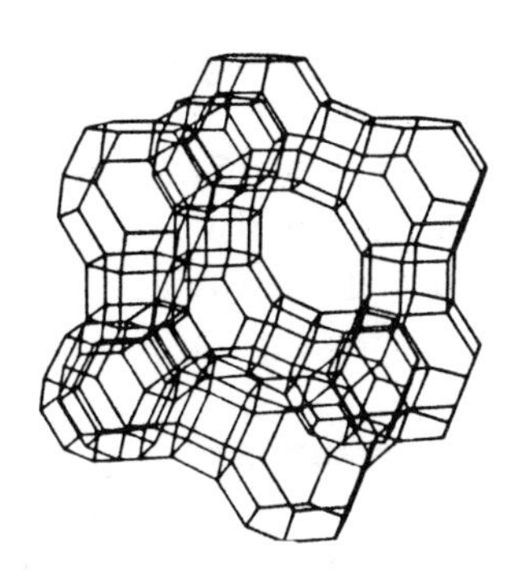
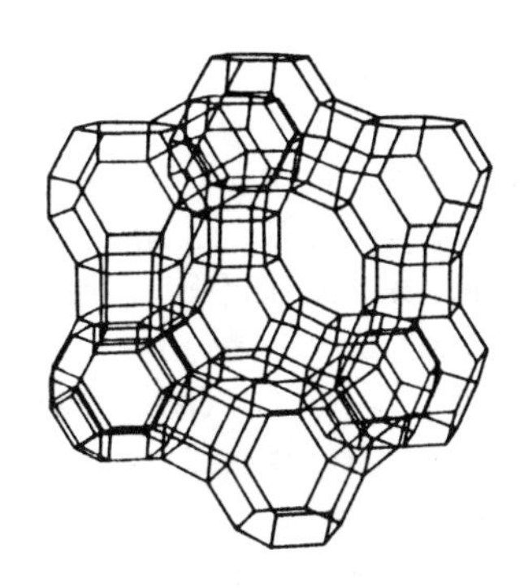

Faujasite, a zeolite *(stereo view)*

The cation exchanger replaced unwanted cations by $H^+(aq)$. Charge balance requires that the anion exchanger replaces unwanted anions by an equal number of OH^- ions. These react with the $H^+(aq)$ to form water, thus removing all the ions from the solution.

$$H^+(aq) + OH^-(aq) \longrightarrow H_2O(l) \tag{28}$$

4. Magnetic water softeners. Many such devices have been offered for sale over the years. Enthusiastic claims for these devices are often made, such as "softens the water while not taking out healthful minerals." Their purveyors have generally been held in the scientific community to be charlatans. Recent reports suggest that magnetic fields really can have some influence[8]; apparently scale is less likely to deposit in the pipes, perhaps because the crystalline form of the calcium carbonate is in some way altered. However, this does not constitute water softening; even if there is less deposition of scale in the pipe, the hardness ions are still present.

5.2.5 *Dissolved solids and irrigation*

As we have seen, all fresh waters contain some level of dissolved solids, though the actual amounts vary greatly. This raises an important issue when these waters are used for irrigation in order to increase agricultural production. Many parts of the world have a climate which is well suited to agriculture, but lack sufficient rainfall; these include parts of the southern

[8]See, for example, L. Pandolfo and R. Colale, "Magnetic fields and tap water," *Chim Ind (Milan)*, **1987**, 69, 88–89.

Canadian Prairies and many parts of the United States Mid- and Southwest. Irrigation has brought successful agriculture to areas that were formerly desert or near-desert; when you fly over Nevada or southern Alberta, it is easy to spot the green irrigated areas among the brown natural background. The irrigated areas of the United States Southwest can produce a crop of alfalfa hay every 30 days through the growing season. Besides agricultural irrigation, of course, an enormous amount of water is needed to supply the fast-growing cities of the United States Southwest (and to create green suburban lawns in the desert). A serious political issue between Canada and the United States is whether water from Canadian rivers, or from the Great Lakes, should be exported to feed the insatiable appetite of the southern United States for water.

Like other attempts to master nature technologically, large-scale irrigation comes with a price, in the form of increased salinity and aquifer depletion. Increased salinity arises because irrigation in a dry climate is inevitably accompanied by high rates of evaporation. The water evaporates, but the dissolved salts remain in the soil. Continued irrigation therefore leads to a buildup of salts in the soil, to the point at which plant growth becomes impossible because of the salinity of the soil. It is now thought that the decline of the ancient civilizations in the Tigris and Euphrates valleys may have been due to salinity problems reducing the fertility of the soil. In modern times, inorganic residues from fertilizers compound the problem.

Salinity was not recognized when irrigation was first introduced in North America, and some farms went through a complete cycle of desert—high productivity—reduced productivity—abandonment in as little as twenty years. Modern irrigation technology uses a kind of "back-flushing" approach: after irrigation for a certain time, a heavy application of water is made to the land in order to wash out the accumulated salts with the run-off, which is then returned to a convenient river. This approach extends the working life of the irrigated land but it also has costs, including:

1. the provision of two water channels to each farm; one for incoming water, and one for waste;
2. the cost to the individual farmer of the extra water for back-flushing;
3. the environmental degradation of the area which receives the waste water.

Irrigation water comes either from rivers or from underground aquifers. Both these sources raise political questions. Again, the examples are taken from the United States.

The Colorado is one of the rivers most extensively used for irrigation in the United States southwest[9]. The use of its waters for irrigation has substantially reduced its flow, to the point that its own ecology is threatened.

[9]United States Department of the Interior Report, "Colorado River water quality improvement program: saline water use and disposal opportunities," 1981.

In this instance, the problem of increased salinity is principally due to the use and re-use of the water for agricultural irrigation. Several tributaries of the Colorado have concentrations of dissolved solids in excess of 1 gram per liter. One suggestion is to divert some of the most contaminated "used" water to arid sites and allow it to evaporate, rather than returning it to the Colorado River.

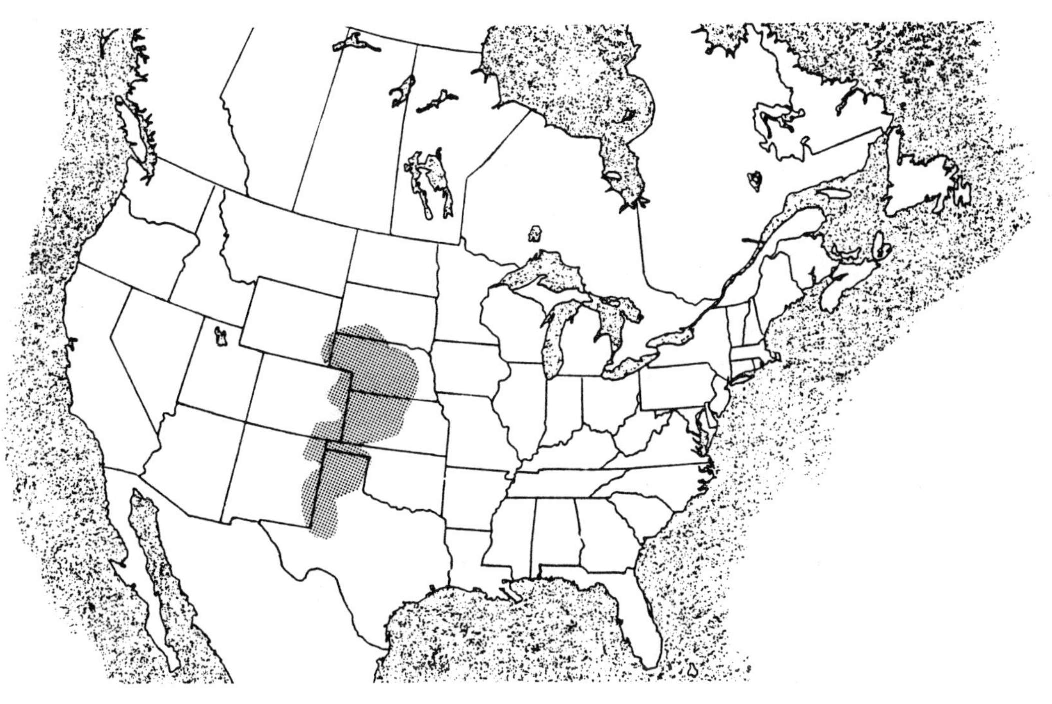

Ogallala Aquifer.

The closure of the San Luis Drain is a particularly well-publicized case, where public opinion forced the abandonment of an irrigation project in the San Joachim Valley in California, when the project was almost completed. The drainage waters were found to be extremely toxic to wildlife. This was not, as might have been expected, due to pesticide or other agricultural contaminants in the water. Instead, irrigation had led to the dissolution of selenium from the soil which is naturally high in selenium; the discharge water was found to cause death to wildlife through selenium poisoning[10].

Much of the area of the southern United States which we have been discussing is irrigated from the immense Ogallala Aquifer, which underlies parts of eight states. In some areas, the rates of water use from the aquifer

[10] J. Letey, C. Roberts, M. Penberth, and C. Vask, *An agricultural dilemma: drainage water and toxics disposal in/ the San Joachim Valley*, University of California, Publication 3319, 1986.

are one to three **orders of magnitude** greater than the rates of recharge. Put another way, up to 1000 years' worth of rainfall lies beneath the surface. By 1980, Texas alone had already consumed (mined) over 20% of its share of this water. Furthermore, in some parts of the aquifer the water contains high concentrations of dissolved salts even before it is used for irrigation. These issues suggest that intensive agricultural production on this land may not be sustainable in the long term.

Depletion of natural waters as result of irrigation is not confined to North America. The huge aquifer beneath the Sahara Desert is being consumed rapidly for agricultural irrigation in Libya; its rate of recharge is essentially zero, so that this amounts to mining a non-renewable resource. Likewise, the shores of the Dead Sea in Israel, and the Aral Sea in the U.S.S.R. have each retreated many kilometers in the last few decades. Agricultural irrigation has reduced the flow rate into these inland seas to the point that it is quite possible that these bodies of water may cease to exist in the foreseeable future. Depletion of the aquifer underlying Mexico City has occurred in order to provide drinking water for the city; so much water has been withdrawn that the land has subsided several meters. Tourist development on tropical islands can present this problem in a different form; excessive rates of removal of fresh water can cause infiltration of the fresh water aquifer by the sea. Readers interested in learning more about the management of underground waters should consult texts in hydrogeology.

5.2.6 Seawater[11] *(Problems 16–18)*

Seawater (Table 5.2) is a much more concentrated salt solution than almost any other natural water (Problem 18), with the exception of inland lakes in areas of high evaporation, such as the Dead Sea in Israel and the Great Salt Lake in Utah. Over the millennia, rivers carry dissolved minerals from the land, but the oceans lose only the water, and not the salts, by evaporation. As a result, the oceans are the major reservoir of "soluble" ions such as Na^+ and Cl^-. In the mid-nineteenth century, Usiglio evaporated large volumes of sea water; he found that about half the total volume could evaporate before any precipitate appeared, whereupon $CaCO_3$, $CaSO_4{\cdot}2H_2O$, and NaCl appeared, in that order, and in the ratio 1:5:250 by mass.

Over geological history, some seas have dried up and others formed. The legacy of these ancient seas is in deposits of **evaporites**. Evaporites typically contain gypsum ($CaSO_4$), sodium chloride (rock salt), or mixtures of sodium and potassium chlorides. Evaporite formation seems to require rather special conditions, where an arm of the sea is connected to the main body of the ocean by a narrow channel, so that evaporation occurs more quickly than the water in the arm can mix with the larger ocean. No evaporite formation appears to be taking place on Earth at the present time; the closest approach is in the Red Sea, where the Straits of Aden

[11] This discussion is based on that in Reference 4, Chapter 4.

Table 5.2: Ions in seawater.

Ion	Input from rivers 10^{10} mol /yr	Concentration mol L^{-1}	Residence time yr
Na^+	900	0.46	7×10^7
Mg^{2+}	550	0.054	1×10^7
K^+	190	0.010	7×10^6
Ca^{2+}	1220	0.010	1×10^6
Cl^-	720	0.55	1×10^8
SO_4^{2-}	380	0.028	1×10^7
HCO_3^-	3200	0.0023	1×10^5

(20 km across) provide a restricted access. As a result, the salinity of the Red Sea (4.1%) is slightly higher than the average of the oceans generally (3.75%).

A second pathway for removal of dissolved substances from sea water is the incorporation of calcium carbonate into the exoskeletons of marine animals. The exoskeletons of ancient marine organisms are the source of chalk and limestone minerals. An interesting paradox about sea water is that calcium carbonate does not spontaneously precipitate, but neither do sea shells on the beach dissolve. This suggests that the oceans are not far from equilibrium with respect to the system $CaCO_3(s)/Ca^{2+}(aq)/CO_3^{2-}(aq)$. Let's look at this in a little detail.

At 15°C, K_{sp} for $CaCO_3$ is 6.0×10^{-9} (mol $L^{-1})^2$. From the ionic concentrations given in the table earlier in this section, we can write:

$$Q_{sp} = c(Ca^{2+}, aq)\, c(CO_3^{2-}, aq) = (0.010)(2.7{\times}10^{-4}) = 2.7{\times}10^{-6}\ (\text{mol L}^{-1})^2$$

According to this calculation, the oceans are vastly supersaturated with respect to calcium carbonate, and we would expect calcium carbonate to precipitate spontaneously.

It turns out that there are two factors which have been overlooked in this simple calculation, namely **ionic strength** and **ion complexation.** We will consider them in turn.

1. Ionic strength. Strictly speaking, equilibrium constants should be written in terms of activities (which we can regard as "effective concentrations") rather than concentrations. For low concentrations, we can usually approximate by writing concentration where we really mean activity, and this is what is done in most places in this book. Since:

$$\text{activity} = \text{concentration} \times \text{activity coefficient}$$

the approximation is equivalent to saying that the activity coefficient is unity. The approximation fails in a solution of high ionic strength such as sea water. Activity coefficients appropriate for sea water are 0.26 for Ca^{2+} and 0.20 for CO_3^{2-}. Now the reaction quotient looks like this:

$$\begin{aligned} Q_{sp} &= a(Ca^{2+}, aq)\, a(CO_3^{2-}, aq) = (0.010 \times 0.26)(2.7 \times 10^{-4} \times 0.20) \\ &= 1.4 \times 10^{-7} \text{ (no units, activities are dimensionless)} \end{aligned}$$

This represents an improvement of nearly an order of magnitude, but we would still predict that the solution is supersaturated with respect to $CaCO_3$.

2. Complexation. The ionic strength/activity coefficient effect arises because at high ionic strength a particular ion (say, Ca^{2+}) is not completely free in solution; it will be surrounded by ions of opposite charge. This is a generalized effect in that the identities of these counter ions are not important. Besides this general effect the ion in question can associate with specific counter ions to form recognizable chemical species whose concentrations can be measured. For Ca^{2+}, these include $(CaSO_4)^\circ$ and $(CaHCO_3)^+$. These "tight" ion pairs are observable chemical species. Note that $(CaSO_4)^\circ$ for example is an aqueous entity, different from $CaSO_4(s)$. As an aside, the existence of these complexes is the reason that simple K_{sp} calculations frequently underestimate the true solubility of ionic compounds[8]. Calculations have shown that in seawater, Ca^{2+} and CO_3^{2-} are speciated as shown in the following table.

Species	K_{assoc}, L mol^{-1}	% of total
Calcium, total concentration: 0.010 mol L^{-1}		
Ca^{2+}(aq)	—	91
$(CaSO_4)^\circ$	2×10^2	8
$(CaHCO_3)^+$	2×10^1	1
Carbonate, total concentration: 0.00030 mol L^{-1}		
CO_3^{2-}(aq)	—	10
$(MgCO_3)^\circ$	3×10^3	64
$(CaCO_3)^\circ$	3×10^3	7
$(NaCO_3)^-$	2×10^1	19
Hydrogencarbonate, total concentration: 0.0023 mol L^{-1}		
HCO_3^-(aq)	—	75
$CaHCO_3^+$	2×10^1	3
$MgHCO_3^+$	1×10^1	12
$(NaHCO_3)^\circ$	2×10^0	10

We can now put the finishing touches to our Q_{sp} calculation.

$$\begin{aligned} Q_{sp} &= a(Ca^{2+}, aq)\, a(CO_3^{2-}, aq) \\ &= (0.010 \times 0.26 \times 0.91)\,(2.7 \times 10^{-4} \times 0.20 \times 0.10) \\ &= 1.3 \times 10^{-8} \end{aligned}$$

We are now within a factor of two to the value of K_{sp}, and have shown $CaCO_3$ in seawater to be **close** to saturation, as found experimentally.

5.2.7 *Other issues affecting the quality of natural waters*

So far, we have touched on oxygen depletion (thermally and through increased BOD), and agricultural irrigation (water depletion and increased salinity) as issues which affect the quality of natural waters. Other serious environmental questions are raised by the careless disposal of industrial and sewage wastes. High levels (high in this connection means parts per billion) of organic compounds are found in the Great Lakes and in the inland Aral Sea in the U.S.S.R., and in rivers such as the Danube, Rhine, and Mississippi. These substances profoundly affect the suitability of these waters as sources for drinking water (Chapter 7), as well as threatening wildlife. Acidification (Chapter 6) likewise compromises the continued existence of many natural species. Non-polar organic compounds accumulate in the food chain by the process known as biomagnification (Chapter 9); this effect can concentrate harmful substances from apparently innocuous levels in the water to concentrations toxic to animals at the top of the food chain. The release of mercury from industrial processes has caused contamination and disease in both natural and human populations (Chapter 10).

Further reading

1. R.W. Raiswell, P. Brimblecombe, D.L. Dent, and P.S. Liss, *Environmental Chemistry,* Edward Arnold (Publishers), London, England, 1980, Chapters 3 and 4.

2. V.L. Snoeyink and D. Jenkins, *Water Chemistry,* Wiley, New York, 1980, Chapter 4.

5.3 Problems

1. (a) Use the Henry's law constants from the text to calculate the ratio $[N_2,aq]/[O_2,aq]$ when water is equilibrated with air at 25°C.

   (b) Use the data below to calculate $\Delta H°$ and $\Delta S°$ for the reaction $N_2(g) \longrightarrow N_2(aq)$, and hence explain why the solubility of $N_2(g)$ in water decreases with temperature. $\Delta H_f^\circ(N_2, aq) = -14.7$ kJ mol^{-1}; $S°(N_2, g) = 0.19$ kJ mol^{-1} K^{-1}; $S°(N_2, aq) = 0.08$ kJ mol^{-1} K^{-1}.

2. The table below gives the solubility of $N_2(g)$ in water at various temperatures. The data all relate to $p(N_2) = 1.00$ atm, and are expressed as cm^3 of N_2 (corrected to STP) per liter of water.

t°C	$V(N_2)$	t °C	$V(N_2)$
0	23.3	30	12.8
5	20.6	35	11.8
10	18.3	40	11.0
15	16.5	50	9.6
20	15.1	60	8.2
25	13.9		

Calculate $\Delta H°$ and $\Delta S°$ for the dissolution of N_2 in water.

3. The solubility in weight percent of methane in water is given below, (p(methane) = 760 torr).

t, °C	Solubility	t, °C	Solubility
0	0.00396	30	0.00191
5	0.00341	40	0.00159
10	0.00296	50	0.00136
15	0.00260	60	0.00115
20	0.00232	70	0.00093
25	0.00209	80	0.00070

(a) Calculate $\Delta H°$ and $\Delta S°$ for the dissolution of CH_4 in water and verify the statement in the text that the process $CH_4(g) \longrightarrow CH_4(aq)$ is enthalpically favoured and entropically disfavoured.

(b) Look up the thermodynamic constants of $CH_4(g)$ to obtain ΔH_f° and S° for $CH_4(aq)$.

4. A sample of water is equilibrated with the atmosphere at 0°C and then analysed by the Winkler method (equations below).

$$Mn^{2+} + 2OH^- + \tfrac{1}{2}O_2 \longrightarrow MnO_2(s) + H_2O$$
$$MnO_2(s) + 4H^+ + 2I^- \longrightarrow Mn^{2+} + I_2 + 2H_2O$$
$$I_2 + 2Na_2S_2O_3 \longrightarrow Na_2S_4O_6 + 2NaI$$

A 50.00 cm^3 sample of oxygenated water is treated by the above reactions and the I_2 liberated is titrated against 0.01136 mol L^{-1} $Na_2S_2O_3$, of which 8.11 cm^3 are required to reduce all the I_2. Calculate the solubility of O_2 in water at 0°C in mol L^{-1}, and hence the Henry's Law constant for oxygen at 0°C.

5. The COD of a water sample is determined as follows: A test water sample (100.0 mL) and a control using pure water (100.0 mL) are separately heated with 25.00 mL of $Na_2Cr_2O_7$ in 50% H_2SO_4 for 2 hours. A 25.00 mL aliquot is withdrawn from each solution and titrated against a solution of ferrous ammonium sulfate: $[Fe^{2+}] = 4.024 \times 10^{-3}$ mol L^{-1}.

$$6Fe^{2+} + Cr_2O_7^{2-} + 14H^+ \longrightarrow 6Fe^{3+} + 2Cr3^+ + 7H_2O$$

The titers of the test water and the control water are 9.77 mL and 26.40 mL of Fe^{2+} solution respectively. Calculate the COD of the test water sample.

6. **(a)** A raw sewage sample has organic matter content of 720 mg L^{-1}. Assume for this problem that the organic matter can be treated as if it were glucose $C_6H_{12}O_6$. What is the O_2 requirement for the complete oxidation of 1.2×10^5 L of this sewage? Give your answer in mg of O_2.

(b) The 1.2×10^5 L of sewage in part (a) is accidentally discharged into a lake of capacity 3.5×10^6 m^3. Assuming uniform mixing, what is the additional BOD (in mg L^{-1}) that is placed on the waters of the lake as a result?

(c) Give two reasons why a large discharge of sewage would be more damaging to the aquatic life of a very warm lake than a very cold one.

7. **(a)** A waste sample has BOD 80 mg L^{-1} and is to be discharged into a lake whose dissolved oxygen content is 8.1 ppm. How many liters of waste can be added to each liter of lake water if the dissolved oxygen of the lake must be guaranteed not to fall below 6.3 ppm?

(b) In practice much more waste can be added to the lake without serious risk of the dissolved oxygen falling below 6.3 ppm. Explain.

8. Two identical 250.0 mL samples of freshly drawn well water have approximate pH 6.6–6.8. One sample is titrated against 0.0510 mol L^{-1} NaOH solution to a phenolphthalein end point (pH 8.3); 11.66 mL of titrant are needed. The other sample is titrated against 0.1000 mol L^{-1} HCl solution to a methyl orange end point (pH 4.3); 12.25 mL of titrant are needed. Calculate the concentrations, in mol L^{-1}, of H_2CO_3, HCO_3^-, CO_3^{2-} and Ca^{2+} in the well water. [Hint: consider carefully which reactions occur in the titrations. Take $K_a = 4.8 \times 10^{-11}$ mol L^{-1} for HCO_3^-, $K_a = 4.2 \times 10^{-7}$ mol L^{-1} for H_2CO_3, and assume that Ca^{2+} is the only cation present besides H^+.]

9. Calculate the concentration of $O_2(aq)$ with time following the introduction into the water body of an oxidizable substrate S. Use the following scheme.

$$O_2(g) \underset{k_2}{\overset{k_1}{\rightleftharpoons}} O_2(aq)$$

$$O_2(aq) + S \xrightarrow{k_3} \text{products}$$

Take initial $[O_2,aq] = 2.72\times10^{-4}$ mol L^{-1}, initial $[S] = 5.0\times10^{-4}$ mol L^{-1} and assume a 1:1 reaction as above, $K_H = 1.3 \times 10^{-3}$ mol L^{-1} atm^{-1}, $k_1 = 1.0 \times 10^{-4}$ mol L^{-1} atm^{-1} h^{-1} and k_3 = (i) 1000; (ii) 100 L mol^{-1} h^{-1}.

10. A water sample obtained from an area of dolomitic limestone has pH 7.2 and total alkalinity 2.3×10^{-3} mol L^{-1} of H^+.

(a) Calculate the concentrations of the major ions in the water. Take $K_a(H_2CO_3) = 4.2 \times 10^{-7}$ mol L^{-1}; $K_a(HCO_3^-) = 4.8 \times 10^{-11}$ mol L^{-1}.

(b) What would be meant if this water was described as being well buffered towards acid?

(c) A 100 mL sample of this water is titrated against 0.0105 mol L^{-1} EDTA (hardness determination). What volume of EDTA solution will be used? State any assumptions you need to make.

11. A water sample has pH 8.44 and a total Ca^{2+} concentration of 155 ppm. For this question, assume that the only ions present in the water are Ca^{2+}, HCO_3^-, and CO_3^{2-}.

(a) What are the concentrations of CO_3^{2-} and HCO_3^- in moles per liter?

(b) What volume of 5.02×10^{-2} mol L^{-1} HCl is needed to titrate 1.00 L of this water to pH 4.3?

(c) What is the total alkalinity of the water?

12. The hardness of a water sample is determined by titrating 100 mL of sample against 0.0100 mol L^{-1} EDTA solution. The Eriochrome Black T endpoint occurs at 11.20 mL EDTA solution. Calculate the hardness of the solution

(a) in mol L^{-1} of $CaCO_3$

(b) in ppm of $CaCO_3$.

13. **(a)** Use thermodynamic data to calculate $\Delta G°$ at 50°C for the reaction

$$CaCO_3(s) \rightleftharpoons Ca^{2+}(aq) + CO_3^{2-}(aq).$$

(b) Does $CaCO_3$ become more soluble or less soluble as the temperature rises? Is the low solubility of calcium carbonate primarily due to enthalpic or entropic factors?

14. A water supply contains 130 ppm of calcium in the form $Ca(HCO_3)_2$. The water is heated to 55°C in a domestic water heater.

(a) Calculate K at 55°C for the reaction below

$$Ca^{2+}(aq) + 2HCO_3^-(aq) \longrightarrow CaCO_3(s) + CO_2(g) + H_2O(l)$$

Thermodynamic data:	ΔH_f°(kJ mol^{-1})	S°(J mol^{-1}K^{-1})
$CaCO_3(s)$	−1206.9	92.9
$Ca^{2+}(aq)$	−542.8	−53.1
$HCO_3^-(aq)$	−692.0	91.2
$CO_2(g)$	−393.5	213.6
$H_2O(l)$	−285.8	69.9

(b) Estimate the mass of scale that is formed when 1000 L of the water supply mentioned above are heated to 55°C. Assume $p(CO_2) = 3.3 \times 10^{-4}$ atm.

15. (a) A water sample has a phenolphthalein alkalinity of 1.22×10^{-4} eq. L^{-1} = (mol H$^+$ L^{-1}) and pH 8.84. Calculate the concentrations of CO_3^{2-} and HCO_3^-, and express the Ca^{2+} concentration in ppm.

(b) The water sample in part (a) is heated from 15°C to 75°C. What is its composition at equilibrium? Take $p(CO_2) = 3.0 \times 10^{-4}$ atm, and estimate K for the reaction

$$Ca(HCO_3)_2(aq) \rightleftharpoons CaCO_3(s) + CO_2(g) + H_2O(l)$$

16. In seawater the ions CO_3^{2-} and HCO_3^- are present in concentrations of 2.7×10^{-4} and 2.3×10^{-3} mol L^{-1}, respectively. K_a for HCO_3^- can be taken as 3.7×10^{-11} mol L^{-1}.

(a) Calculate the pH of seawater based on the stoichiometric concentrations of HCO_3^- and CO_3^{2-}.

(b) Now repeat this calculation including activity coefficients and complexation effects. Take the activity coefficient of HCO_3^- as 0.6.

17. The present partial pressure of CO_2 in the atmosphere is 3.0×10^{-4} atm. Consider the effect on the composition of the oceans if $p(CO_2)$ rose to 6.0×10^{-4} atm and the atmospheric and ocean temperatures rose from 15°C to 25°C. Consider the effect of temperature on the equilibrium $CO_2(g)/CaCO_3(s)$. Use these temperature dependent data and ignore complexation and activity coefficients.

	15 °C	25 °C
$K_{sp}(CaCO_3)$, (mol L^{-1})^2	6.0×10^{-9}	4.6×10^{-9}
$K_a(H_2CO_3)$, mol L^{-1}	3.8×10^{-7}	4.5×10^{-7}
$K_a(HCO_3^-)$, mol L^{-1}	3.7×10^{-11}	4.7×10^{-11}
$K_H(CO_2)$, mol L^{-1} atm^{-1}	0.046	0.034

18. Use the tabulated data in the text (Section 5.2.6) to calculate the total dissolved solids in seawater in ppm.

Chapter 6

Acid Rain

Introduction

Acid rain, and its companions acid snow and acid fog, represent an environmental issue which has been in the public consciousness for many years. It is a worldwide problem which causes great environmental mischief, with damage to crops, to forests, to environmentally sensitive lakes, and also to buildings and engineering structures made of stone (limestone) and metal (iron and steel).

In this chapter, we will examine the sources of acid rain, the chemistry involved, and the effects of this pollution on the environment. We will conclude by studying some of the measures that have been proposed and taken to alleviate this problem.

Acid rain is not a new phenomenon. The term "acid rain" was coined as long ago as 1872, in a book published in England by R.A. Smith. Air pollution caused by burning coal had been recognized as a public nuisance in cities at least a century earlier, as the developing Industrial Revolution in Europe led to the virtual elimination of the forests in Britain and parts of continental Europe, and coal replaced wood as the principal energy source of the new industries. Smith was one of the first to document the acidic properties of this polluted air.

Acid rain is usually associated with heavy industry, whether in developed regions such as North America and Western Europe, or in more recently industrialized areas such as the Soviet Union, Eastern Europe, China and India. To date however most of the public outcry about this problem has been raised in the former regions.

Normal, unpolluted rainwater has a pH close to 5.6, in consequence of the raindrops being in equilibrium with the atmospheric concentration of carbon dioxide.

$$(1) \quad CO_2(g) + H_2O(l) \rightleftharpoons H_2CO_3(aq) \rightleftharpoons H^+(aq) + HCO_3^-(aq)$$

This reaction has already been described in Chapter 5, Natural Waters. Restating the point, even completely clean rain does not have a pH of 7.0, because it is not absolutely pure water; it contains equilibrium amounts of the atmospheric gases.

Acidic precipitation is generally defined as having pH lower than about 5.0; pH 4 to 4.5 is not uncommon, and isolated examples of rain and fog having pH lower than 2 have been recorded. To put this in context, vinegar and lemon juice have pH 3.0 and 2.2 respectively. Since even unpolluted rainwater is slightly acidic, it may be helpful to think of "acid rain" as rain that is more acidic than normal.

6.1 Sources of acid rain

In industrialized regions, the main causes of acid rain are **sulfur oxides** and **nitrogen oxides** in the atmosphere. These gases are present in trace amounts (ppb) even in natural, unpolluted air; they generally become a problem only when they occur in higher than normal amounts as a result of human activities. Even so, the absolute amounts of these gases are very small, even in polluted air, amounting to no more than a part per million or two. Acid rain results when these gases are oxidized in the atmosphere and return the ground dissolved in raindrops. SO_2 falls as H_2SO_3 and H_2SO_4:

$$(2) \qquad SO_2 + H_2O \longrightarrow H_2SO_3$$

$$(3) \qquad SO_2 \xrightarrow{\text{oxidize}} SO_3 \longrightarrow H_2SO_4$$

NO_X falls as HNO_3 (for more detail, see Chapter 3):

$$(4) \qquad NO \xrightarrow{\text{oxidize}} NO_2$$

$$(5) \qquad 2NO_2 + H_2O \longrightarrow HNO_2 + HNO_3$$

$$(6) \qquad NO_2 + OH \xrightarrow{M} HNO_3$$

In most acid rain areas, the sulfur oxides are the major contributor to the problem, but the nitrogen oxides predominate on the U.S. west coast where acid rain and photochemical smog (Chapter 3) are closely linked. In a few regions of the world, notably parts of Alaska and New Zealand, highly acidic rain falls naturally as a result of the emission of HCl and SO_2 from volcanoes.

Sulfur Oxides: Two important activities which lead to the release of sulfur oxides into the atmosphere are **coal burning** and the **roasting of metal sulfide ores**.

Coal typically contains 2–3% of sulfur by mass, and this sulfur is oxidized along with the carbon when the coal burns.

$$(7) \qquad S(\text{as organosulfur compounds or metal sulfides}) + O_2(g) \longrightarrow SO_2(g)$$

Not all coals are alike however. In North America, those from southern West Virginia, Kentucky, and the Canadian and American Rockies are low in sulfur while those from the American Midwest, northern Appalachia, and Nova Scotia have sulfur contents at the upper end of the range. Likewise, in Britain, low sulfur coals and lignite tend to occur mainly in south Wales and Scotland, while the coal in England is higher in sulfur (but more easily mined, and thus less expensive). The air pollution associated with the Industrial Revolution of the nineteenth century in Britain, Germany, and the eastern United States was mainly due to ready access to high-sulfur coals in these areas. The Peoples' Republic of China now faces the same predicament as it, too, struggles to industrialize using high-sulfur coal as the energy source for its heavy industries. Air pollution due to coal burning is especially serious in the Eastern European countries of East Germany, Poland and Czechoslovakia where the local deposits of coal are likewise high in sulfur.

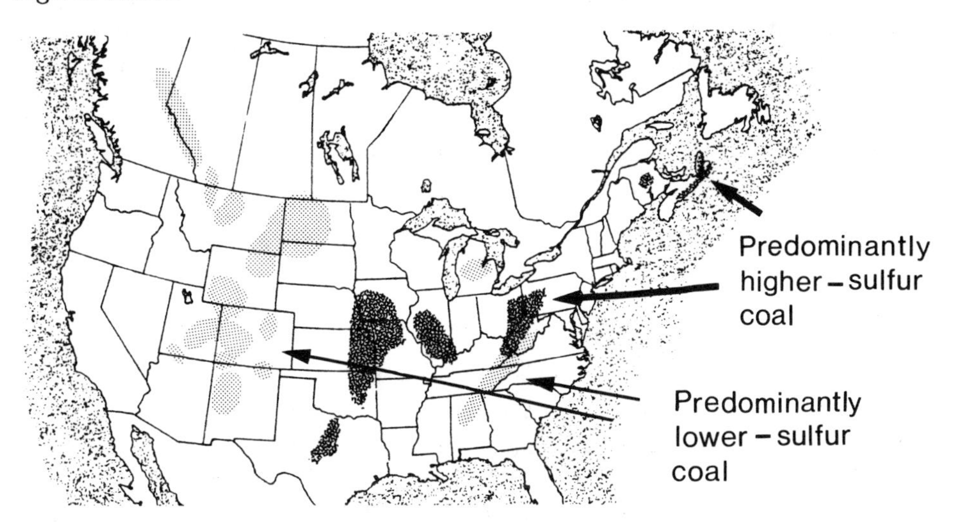

Industries such as the iron and steel industry and coal-burning thermal power plants use a great deal of coal. While 2–3% of sulfur may not sound like much, the sheer volume of coal consumed in these industries leads to prodigious emissions of sulfur dioxide (see Problem 3).

Until a generation ago, coal burning in domestic stoves and fireplaces was also a source of sulfur dioxide pollution in many countries. Restrictions on domestic coal burning, notably in Britain, have considerably reduced this problem, although the original impetus for these restrictions was the need to reduce air pollution by particulates (Chapter 3), which are also an unwanted byproduct of burning coal.

Many metals which are of commercial importance occur in nature as their sulfide ores. These metals include nickel as NiS, copper as Cu_2S, zinc as ZnS, lead as PbS, and mercury as HgS. The first step in recovering these metals from their ores consists of roasting the ore in air to give the metal oxide, which is subsequently reduced to the element, usually with coke.

(8) $$2MS\ (M{=}Ni,\ Zn,\ Pb) + 3O_2(g) \longrightarrow 2MO + 2SO_2(g)$$

(9) $$MO + C(s) \longrightarrow M + CO(g)$$

With Cu_2S and HgS, the metal is formed directly, e.g.:

$$(10) \qquad Cu_2S(s) + O_2(g) \longrightarrow 2Cu(s) + SO_2(g)$$

Since metal extraction is a very large-scale industry, the quantities of sulfur dioxide that are released through roasting are correspondingly very large. Even though some of the SO_2 is captured and converted into sulfuric acid at some of the largest smelting locations, thousands of tonnes of SO_2 may still be released daily (see Problem 13). The nickel smelter at Sudbury, Canada, remains the world's largest single point source emission of SO_2, despite pollution control measures which are discussed later in this chapter.

A minor source of SO_2 in the atmosphere is through the oxidation of hydrogen sulfide, which forms by microbial decay of organic matter and which is also released during the processing of "sour" natural gas. (The catalytic oxidation of H_2S to elemental sulfur is known as "sweetening" the gas.) The oxidation of H_2S in the atmosphere is believed to occur as follows[1].

$$(11) \qquad OH + H_2S \longrightarrow H_2O + SH$$
$$(12) \qquad SH + O_2 \longrightarrow SO + OH$$
$$(13) \qquad SH + OH \longrightarrow H_2O + S(\text{atomic sulfur})$$
$$(14) \qquad S + O_2 \longrightarrow SO + O$$
$$(15) \qquad SO + O_2 \longrightarrow SO_2 + O$$

Once again, we note the key role of the hydroxyl radical in bringing about the oxidation of an oxidizable atmospheric gas. Notice also that the reaction $S + O_2 \longrightarrow SO_2$ does not take place directly. The oxidation of atomic sulfur to SO_2 requires in two steps because removal of an oxygen atom from O_2 has a lower activation energy than insertion of the sulfur atom into the O–O multiple bond.

Nitrogen Oxides: As discussed in Chapter 3, combustion sources are the principal cause of acidic precipitation due to nitrogen oxides. This phenomenon is linked to that of photochemical smog, since the formation of HNO_3 is an important sink for removing the free radical species NO_2 from the atmosphere.

$$N_2 + O_2 \rightleftharpoons NO \xrightarrow{\text{oxidant}} NO_2 \xrightarrow{\text{OH, M}} HNO_3$$

Recall from Chapter 3 that the oxidant in the scheme above can be one of several species, including O_3, HO_2, and CH_3O_2.

Another route to HNO_3 is hydrogen abstraction from some suitable donor X–H by the nitrate free radical NO_3 (not to be confused with the nitrate anion NO_3^-).

$$NO_2 + O_3 \longrightarrow NO_3 \xrightarrow{\text{XH}} HNO_3$$

[1]R.P. Wayne, *Chemistry of Atmospheres,* Clarendon Press, Oxford, England, 1985, Chapter 5.

To date, discussions of acidic emissions from coal burning and metal smelting have focused mostly on SO_2. However, these processes also produce NO_X, because the nitrogen oxides are formed in small amounts whenever air is heated. Analyses of air polluted by acidic emissions show the amount of nitric acid to be typically one third of the total acid. The proportion of nitrate deposited appears to be growing steadily, as efforts to curb SO_2 emissions, especially through desulfurization of coal, achieve greater success. For example, Dillon *et al*[2] found that reductions in SO_2 emissions in eastern North America since 1976 were reflected in reduced deposition rates in central Ontario. No such reductions in nitrate emissions or deposition were found (Problem 3).

Schwartz[3] has discussed acid deposition with emphasis on the eastern United States. Extrapolating from the U.S. National Ambient Air Quality Standards for $SO_2(g)$ and $NO_2(g)$ (1.2 and 2.1 μmol m^{-3} respectively), he suggests that standards for (wet + dry) deposition rates of sulfate and nitrate might be set at 20–40 and 40–80 mmol m^{-2} yr^{-1} respectively. However, the emission rates for the whole northeastern U.S. (estimated at 130 and 120 mmol m^{-2} yr^{-1}) are well in excess of these proposed deposition rates. The point, of course, is that "what goes up must come down." Locally, deposition will be even higher than the average, since deposition is inevitably non-uniform.

6.2 Chemistry of acid rain

Unpolluted rainwater, as we have noted already, has pH close to 5.6 as a result of equilibration of raindrops with the ca. 350 ppm of CO_2 in the troposphere. This yields the weak acid H_2CO_3 for which $K_a = 4.2 \times 10^{-7}$ mol L^{-1} at 25°C.

In absolute terms, the concentrations of NO_2 and/or SO_2 are small—no more than 1 ppm or so—even in highly polluted air. As shown in the previous section, these gases are ultimately precipitated in rain as HNO_3, H_2SO_3 and H_2SO_4. H_2SO_4 and HNO_3 are strong acids, while H_2SO_3 has $K_a = 1.7 \times 10^{-2}$ mol L^{-1} at 25°C. Throughout this section $SO_2(aq)$ and $H_2SO_3(aq)$ will be taken as interchangeable, *cf.* $CO_2(aq)$ and $H_2CO_3(aq)$. Because HNO_3, SO_2, and SO_3 are all more soluble in water than CO_2, low concentrations of these acidic gases have a greater effect on the pH of rainwater than much

[2]P.J. Dillon, M. Lusis, R. Reid, and D. Yap, "Ten-year trends in sulphate, nitrate, and hydrogen deposition in central Ontario," *J Atmos Chem,* **1988,** 5, 901–905.

[3]S.E. Schwartz, "Acid deposition: unravelling a regional phenomenon," *Science,* **1989,** 243, 753–763.

higher concentrations of CO_2. This is shown quantitatively below.

For CO_2:

$$\begin{array}{rcll} CO_2(g) + H_2O(l) & \rightleftharpoons & H_2CO_3(aq) & K_H = 3.4 \times 10^{-2}\ \text{mol L}^{-1}\ \text{atm}^{-1} \\ H_2CO_3(aq) & \rightleftharpoons & H^+(aq) + HCO_3^-(aq) & K_a = 4.2 \times 10^{-7}\ \text{mol L}^{-1} \\ \hline CO_2(g) + H_2O(l) & \rightleftharpoons & H^+(aq) + HCO_3^-(aq) & K_c = 1.4 \times 10^{-8}\ \text{mol}^2\ \text{L}^{-2}\ \text{atm}^{-1} \end{array}$$

For SO_2:

$$\begin{array}{rcll} SO_2(g) + H_2O(l) & \rightleftharpoons & H_2SO_3(aq) & K_H = 1.2\ \text{mol L}^{-1}\ \text{atm}^{-1} \\ H_2SO_3(aq) & \rightleftharpoons & H^+(aq) + HSO_3^-(aq) & K_a = 1.7 \times 10^{-2}\ \text{mol L}^{-1} \\ \hline SO_2(g) + H_2O(l) & \rightleftharpoons & H^+(aq) + HSO_3^-(aq) & K_c = 2.1 \times 10^{-2}\ \text{mol}^2\ \text{L}^{-2}\ \text{atm}^{-1} \end{array}$$

Summarizing, the equilibrium constant for the overall reaction is larger in the case of SO_2 than of CO_2 because:

1. SO_2 is more soluble than CO_2;
2. H_2SO_3 is a stronger acid than H_2CO_3.

Consequently, a small concentration of $SO_2(g)$ has a greater influence on the pH of rain than a much larger concentration of $CO_2(g)$: 0.12 ppm of $SO_2(g)$ in equilibrium with rainwater will produce a pH of 4.30 in the water (Problem 1; contrast Problem 2), compared with the pH 5.6 produced by 350 ppm of $CO_2(g)$.

The chemistry of rain acidified by the sulfur oxides is complicated because the sulfur may be deposited in many different forms. It may either precipitate as $H_2SO_3(aq)$ as shown above, or it may first be oxidized to $SO_3(g)$ and precipitate as $H_2SO_4(aq)$. Deposition may occur either in the aqueous form (wet deposition) or in association with particulate matter (dry deposition) in which case much of the sulfur will deposit in the form of sulfite or sulfate ions rather than the free acids.

6.3 Oxidation mechanisms

6.3.1 *Oxidation of SO_2*

The situation is very complex[4], because oxidation can occur by three quite separate routes: homogeneously in the gas phase, homogeneously in the aqueous phase of raindrops, and heterogeneously on the surfaces of particles. The prevailing atmospheric conditions, especially the humidity and the concentration and composition of particulate matter, will determine the relative importance of these processes.

[4] B.J. Finlayson-Pitts and J.N. Pitts, Jr, *Atmospheric Chemistry*, Wiley, New York, **1986**, Chapter 11. See also the book by J.G. Calvert, *SO_2, NO, and NO_2 oxidation mechanisms*, Butterworth, Stoneham, Mass., 1984.

The reaction $SO_2(g) + \frac{1}{2}O_2(g) \longrightarrow SO_3(g)$ has $\Delta G^\circ_{298} = -71$ kJ mol^{-1}. Close to an emission source where $p(SO_2)$ is high and no SO_3 has yet formed, ΔG will be negative (spontaneous reaction as written). However, in dry air at 300 K the oxidation is imperceptibly slow and, as is well known in the laboratory, sulfur dioxide may be handled and stored without taking any precautions to exclude air. (Above about 400°C, and especially in the presence of catalysts, oxidation is rapid, as in the manufacture of sulfuric acid: see below.)

Homogeneous gas phase oxidation

The most important of the homogeneous mechanisms for the tropospheric oxidation of SO_2 involves the hydroxyl radical (Problem 4).

$$(16) \quad SO_2 + OH \xrightarrow{M} HSO_3 \qquad k_2 = 9 \times 10^{-13}\ \text{cm}^3\ \text{molec}^{-1}\ \text{s}^{-1}$$

In polluted air (i.e., $p(SO_2) \approx 1$ ppm), this reaction alone would generate a half life for SO_2 of 20–30 h.

The HSO_3 formed as this first intermediate is a free radical species and should not be confused with the anion HSO_3^-, which is produced in the acid dissociation of H_2SO_3. HSO_3 is subsequently oxidized to SO_3 by reaction with molecular oxygen.

$$(17) \qquad HSO_3 + O_2 \longrightarrow SO_3 + HO_2$$

In laboratory experiments with simulated polluted air, the HO_2 is reduced back to OH by NO, thus providing an explanation for the near constancy of [OH] in these experiments.

$$(18) \qquad HO_2 + NO \longrightarrow OH + NO_2$$

Other routes for homogeneous gas-phase oxidation of SO_2 are less important in practice (Problem 6). For example, oxidation by atomic oxygen is slow on account of the low concentration of atomic oxygen in the troposphere ($< 10^5$ molec cm^{-3}).

$$(19) \qquad SO_2 + O \xrightarrow{M} SO_3 \begin{cases} k = 7.7 \times 10^{-34}\ \text{cm}^6\ \text{molec}^{-2}\ \text{s}^{-1} \\ \Delta G^\circ_{298} = -303\ \text{kJ mol}^{-1} \end{cases}$$

From Reaction 19 a half life for SO_2 of about 3×10^4 h is computed, far longer than the observed half life of SO_2 in the troposphere. Oxidations by oxidants such as HO_2 and O_3 are slow because the rate constants are small[5]. For example:

$$(20) \quad HO_2 + SO_2 \longrightarrow SO_3 + OH \qquad k = \text{ca. } 10^{-18}\ \text{cm}^3\ \text{molec}^{-1}\ \text{s}^{-1}$$

[5] H. Sakugawa, I.R. Kaplan, W. Tsai, and Y. Cohen, "Atmospheric hydrogen peroxide," *Environ Sci Technol,* **1990,** 24, 1452–1462. Concentrations of $H_2O_2(g)$ are generally < 10 ppb, and are highest in the afternoon and lowest at night, consistent with HO_2 as a precursor.

Homogeneous aqueous phase oxidation

It is now known that SO_2 can be oxidized inside raindrops by hydrogen peroxide (Problem 5).

$$SO_2(aq) + H_2O_2(aq) \longrightarrow H_2SO_4(aq) \quad (21)$$

Hydrogen peroxide is a minor atmospheric constituent which can be formed by several reactions, the most important of which is the disproportionation of HO_2 radicals.

$$2HO_2 \longrightarrow H_2O_2 + O_2 \quad (22)$$

Disproportionation may occur in the gas phase, followed by dissolution of the H_2O_2 in the water droplet ($K_H = 10^5$ mol L^{-1} atm^{-1}); alternatively, the HO_2 radicals may first enter the aqueous phase ($K_H = 2 \times 10^3$ mol L^{-1} atm^{-1}), and then react together.

There are several aspects of the oxidation of $SO_2(aq)$ by H_2O_2 in rain or cloud drops that are not yet well understood[4]. These include the following.

1. What is the concentration of $H_2O_2(aq)$? Values between 10^{-7} and 10^{-4} mol L^{-1} have been put forward.
2. Is H_2O_2 the only oxidant to consider? Ozone is more abundant than H_2O_2 in the troposphere, and although it is less soluble in water ($K_H = 1.3 \times 10^{-2}$ mol L^{-1} atm^{-1}) it reacts rapidly with SO_2. In addition, polluted air usually contains NO_2 which, in the aqueous phase, is an oxidant for SO_2.

   $$SO_2 + NO_2 \longrightarrow SO_3 + NO \quad (23)$$

3. Could the dissolution of SO_2 in the raindrop i.e., $SO_2(g) \longrightarrow SO_2(aq)$ be rate limiting? This has been proposed to be the case, although the most recent reports suggest that the actual oxidation is the rate determining step, except perhaps for very large aqueous droplets and high oxidant concentrations.
4. How does speciation affect the reaction rate? The balance between $SO_2(aq)$, $HSO_3^-(aq)$, and $SO_3^{2-}(aq)$ depends on the pH. These three species react with H_2O_2 at different rates. Likewise the total concentration of S(IV) species in equilibrium with a fixed pressure of $SO_2(g)$ increases with pH (recall Figure 5.3, showing the corresponding behaviour with CO_2).
5. How does acid catalysis affect the oxidation of $SO_2(aq)$ by H_2O_2? The rate constant for this oxidation (in pure water) increases at low pH, but the solubility of S(IV) decreases at low pH, as just noted. The effects of rate constant and concentration thus tend to cancel.
6. Are there other catalysts which should be considered? Yes, the most important of them is $Fe^{3+}(aq)$. Iron, manganese, copper, and vanadium can all function as catalysts. These metals each have at least

two oxidation states interconvertible through one-electron transfers, and these allow oxidation of S(IV) to occur by way of radical chain reactions.

Heterogeneous oxidation on particles

The mechanism probably follows that involved in the industrial oxidation of SO_2, but is not known in detail.

$$SO_2 + \tfrac{1}{2}O_2 \xrightarrow{\text{catalyst}} SO_3 \longrightarrow H_2SO_4 \text{ (or } SO_4^{2-}\text{)}$$

Salts of vanadium, manganese, and iron are all effective catalysts for this oxidation, since all can undergo redox reactions. All are widely distributed environmentally; vanadium especially is found in particulate matter produced by burning coal, so that SO_2 and vanadium-bearing particles will frequently be found together.

Bahnemann *et al*[6] have suggested that H_2O_2 may also be formed on the surface of metal oxide or metal sulfide particles acting as electron-transferring semiconductors. The semiconductor(sc) provides the electrons to reduce O_2 photochemically.

$$(24) \quad 2\text{sc} \xrightarrow{2h\nu} 2\text{sc}^+ + 2\text{e}^-$$

$$(25) \quad 2\text{e}^- + 2\text{H}^+(\text{aq}) + \text{O}_2 \longrightarrow \text{H}_2\text{O}_2(\text{aq})$$

$$(26) \quad 2\text{sc}^+ + 2\text{H}_2\text{O} \longrightarrow \text{H}_2\text{O}_2(\text{aq}) + 2\text{sc} + 2\text{H}^+(\text{aq})$$

Summary

The reactions below appear to represent current thinking[3] on the most important oxidation processes for SO_2.

Gas phase:

$$(16) \quad SO_2 + OH \xrightarrow{M} HSO_3$$

$$(17) \quad HSO_3 + O_2 \longrightarrow SO_3 + HO_2$$

$$(29) \quad SO_3 + H_2O \xrightarrow{M} H_2SO_4$$

Aqueous phase:

$$(30) \quad HSO_3^- + H_2O_2 \xrightarrow{H^+} HSO_4^- + H_2O$$

$$(31) \quad HSO_3^- + O_3 \longrightarrow HSO_4^- + O_2$$

[6] D.W. Bahnemann, M.R. Hoffmann, A.P. Hong, and C. Kormann, "The Chemistry of Acid Rain," ACS Symposium Series No. 349, American Chemical Society, Washington, Chapter 10.

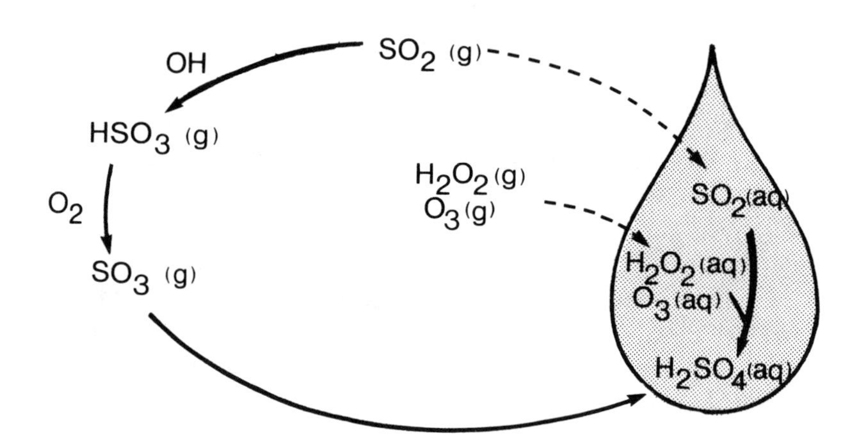

Summary of gas-phase and aqueous-phase oxidation of sulfur dioxide.

6.3.2 *Oxidation of NOx*

This subject is familiar from Chapter 3. Only a summary is provided here.

Gas phase:

$$(32) \quad NO + O_3 \longrightarrow NO_2 + O_2 \quad \text{(also with } HO_2, RO_2)$$
$$(33) \quad NO_2 + OH \xrightarrow{M} HNO_3$$
$$(34) \quad NO_2 + O_3 \longrightarrow NO_3 + O_2$$
$$(35) \quad NO_3 + RH \longrightarrow HNO_3 + R$$
$$(36) \quad NO_3 + NO_2 \longrightarrow N_2O_5$$

Aqueous phase:

$$(37) \quad N_2O_5 + H_2O \longrightarrow 2HNO_3$$

6.4 Oxidation and deposition of sulfur oxides

To preview the main result of this section, the half lives of SO_2 and SO_3 in the atmosphere are of the order of a few days. This means that acid precipitation can be expected over whatever distance is travelled by an air mass from a pollution source over the period of a week, or a little more. Assuming a wind speed of as little as 20 km h^{-1}, such an air mass will travel nearly 3500 km (2000 miles) over the course of a week.

While acid precipitation is more than a local pollution issue, it is not a global problem like that of the chlorofluorocarbons (Chapter 2). The latter substances remain for tens of years in the atmosphere, and therefore become evenly distributed around the globe. Acidic gases are deposited too rapidly for such universal mixing. By the same token, acidic emissions do not stay in the atmosphere long enough to migrate to the stratosphere; acid rain chemistry is exclusively a tropospheric phenomenon.

The rates of oxidation and deposition of SO_2 vary considerably with the conditions. These rates are usually both in the range 1 to 10% per hour i.e., pseudo first order rate constants are in the range 0.01 to 0.1 h^{-1}, but rates outside this range have been reported. In dry air, such as occurs over western Canada and the U.S. Southwest, rates of oxidation as low as 0.2% per hour have been recorded. Under these conditions, homogeneous gas phase oxidation by the OH radical is the predominant pathway. At the other extreme, rates up to 30% per hour have been reported under very humid conditions, when the oxidation occurs mainly in the aqueous phase. The rates of both extreme mechanisms depend on the solar intensity (winter $<$ summer) because sunlight is needed for the formation of the reactive oxidants (OH or H_2O_2). Figure 6.1 shows the proportion of SO_2 oxidation taking place in the aqueous phase as a function of the liquid water content of the atmosphere, based on the assumptions of Problem 5: namely that the only relevant mechanisms for oxidation are gas-phase attack by OH, and aqueous phase oxidation by H_2O_2.

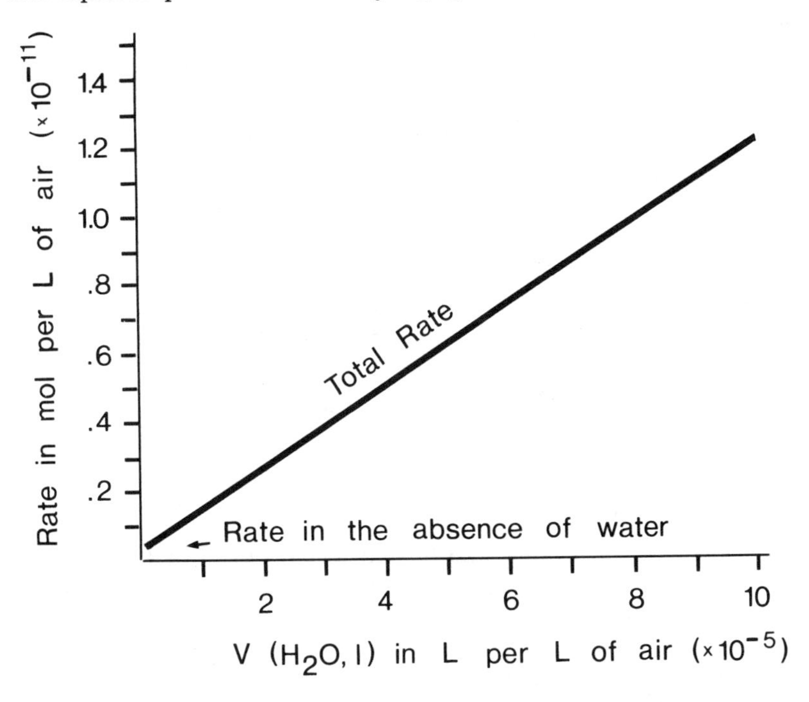

Figure 6.1: Rates of gas phase and aqueous phase oxidation of SO_2 as a function of the liquid water content of the atmosphere.

From the discussion so far, we can say that there are three processes which take place following the release of SO_2 to the atmosphere. These are oxidation of SO_2 to SO_3, deposition of H_2SO_3/SO_3^{2-}, and deposition of H_2SO_4/SO_4^{2-}.

$$SO_2 \xrightarrow{k_1} SO_3 \tag{38}$$

$$SO_2 \xrightarrow{k_2} \text{Deposition as } H_2SO_3 \text{ or } SO_3^{2-} \tag{39}$$

$$SO_3 \xrightarrow{k_3} \text{Deposition as } H_2SO_4 \text{ or } SO_4^{2-} \tag{40}$$

In this simplified model, k_1, k_2, and k_3 are all pseudo first order rate constants. Each reaction represents the sum of the several different mechanisms of oxidation or of deposition, and consequently the values of the rate constants will vary according to the prevailing conditions. Such a model would apply to the emission from a single point source under atmospheric conditions that remained constant over the whole lifetime of the emission plume. In real life there will be multiple emission sources and the weather conditions change continually.

Figure 6.2 shows the application of this model for values of the rate constants given in Problem 7. The concentration of SO_2 decays continuously with time, while that of SO_3 reaches a maximum and then falls again. At any time, the amounts of SO_2(g) and SO_3(g) will determine the ratio of sulfite:sulfate precipitated.

In Figure 6.2 the abscissa represents time. For the further assumption of constant wind speed, the abscissa could also depict the distance travelled by the plume from the emission source. We conclude that most of the precipitation should occur as sulfite close to the source, where little time for oxidation has elapsed. Farther from the source, the proportion of sulfur deposited as sulfate should increase and eventually predominate.

The complexity of real life compared with the simple model is shown in the recent analysis of acid deposition data from Ontario, Canada where many sites have been monitored for several years. These include sites in the vicinity of the large nickel smelter at Sudbury, Ontario, and cover a period in 1982/3 when this operation was closed down for several months by a strike. Two seemingly remarkable results of this study were[7]:

1. that most of the deposition close to Sudbury consists of material from upwind sources rather than from the local smelter;
2. that there was no appreciable change in the pattern or the total amount of deposition in Northern Ontario during the period of the strike, even though the Sudbury smelter is the largest single point source of acidic emissions in the world, accounting on its own for 1% of global SO_2 emissions.

[7] E.A. McBean, M. Kompter, J. Donald, S. Donald, and G. Farquhar, Technology Transfer Conference, Toronto, Ontario, 1987; Proceedings, paper A14.

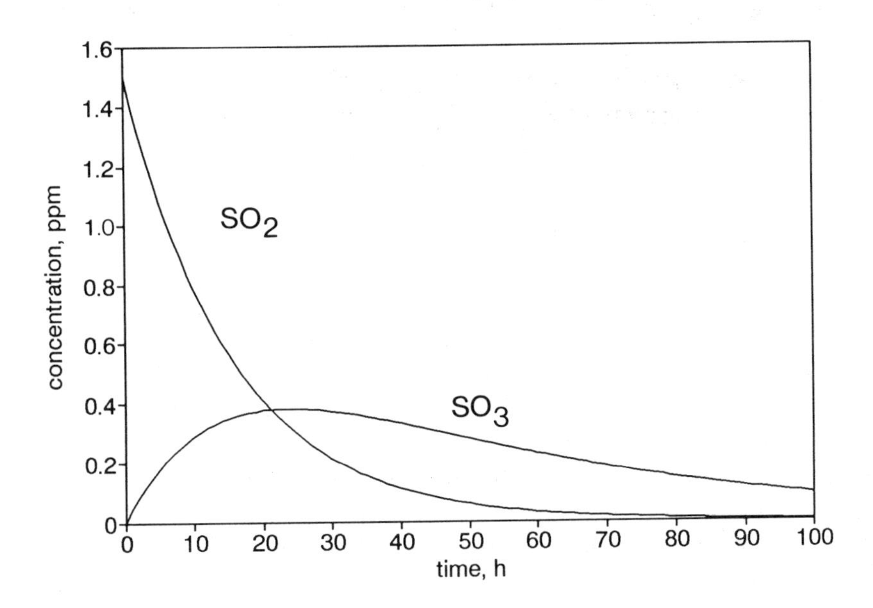

Figure 6.2: Concentrations of SO_2(g) and SO_3(g) as a function of the time elapsed since emission.

An explanation of these observations is that the 400 m "superstack" at Inco's Copper Cliff refinery injects its emissions above the tropospheric boundary layer, which separates the air near ground level from the "free" troposphere. This boundary lies as low as 300 m above ground level in the winter months. Pollution injected into the free troposphere will be carried away from the emission source with very little deposition nearby.

6.4.1 *Other aqueous phase tropospheric reactions*[8]

At this point, we digress to consider some chemical reactions which are not specific to the acid precipitation problem, but which also occur at least partly in the aqueous phase. This section relates to material covered in Chapter 3.

The realization that a major part of the tropospheric oxidation of SO_2 occurs in the aqueous phase, in fog and clouds, has prompted the examination of the role of aqueous chemistry in phenomena such as photochemical smog and ground level ozone. The table below gives some relevant Henry's Law constants.

[8] J. Lelieveld and P.J. Crutzen, "Influences of cloud photochemical processes on tropospheric ozone," *Nature*, **1990,** 343, 227–233; see also S.E. Schwartz, "Chemistry with a silver lining," pp. 209–210 of the same issue.

Table 6.1: Henry's Law constants (mol L^{-1} atm^{-1} at 298 K) for partition of tropospheric gases into water[8].

gas	K_H	gas	K_H	gas	K_H
N_2	6.5×10^{-4}	O_2	1.3×10^{-3}	NO	1.9×10^{-3}
CO	9×10^{-4}	O_3	1.3×10^{-2}	CO_2	3.4×10^{-2}
NO_2	6.4×10^{-3}	SO_2	1.2	NO_3	15.0
HO_2	$2.0 \times 10^{+3}$	CH_2O	$6.3 \times 10^{+3}$	H_2O_2	$7.4 \times 10^{+4}$

Atmospheric gases vary widely in their intrinsic solubilities in water. Thus O_2, N_2, O_3, NO, NO_2, CO, and CO_2 have limited solubilities, SO_2 and NO_3 have moderately high solubilities, and species such as HO_2, H_2O_2, HNO_3, and formaldehyde are scavenged extremely efficiently from the gas phase into aqueous droplets. This can disrupt the gas phase balance between other groups of gas-phase substances. For example, the efficient scavenging of HO_2 into aqueous droplets removes this powerful oxidant from the interstitial gas, thereby reducing the rates of oxidations, such as Equation 41 in the gas phase.

$$\text{(41)} \qquad HO_2 + NO \longrightarrow OH + NO_2$$

Furthermore, the HO_2 free radical functions as a weak acid in water (compare H_2O and H_2O_2).

$$\text{(42)} \quad HO_2(aq) \longrightarrow H^+(aq) + O_2^-(aq) \qquad K_a = 3.5 \times 10^{-5} \text{ mol L}^{-1}$$

This is significant because $O_2^-(aq)$ is a reducing agent (Reactions 43–44), unlike $HO_2(g)$, which acts as an oxidant, as noted above. Modelling studies have indicated that Reaction 43 has a significant impact on the tropospheric concentration of ozone.

$$\text{(43)} \quad O_2^-(aq) + O_3(aq) \xrightarrow{H_2O} 2O_2(aq) + OH(aq) + OH^-(aq)$$

$$\text{(44)} \quad O_2^-(aq) + CH_3O_2(aq) \xrightarrow{H_2O} O_2(aq) + CH_3OOH(aq) + OH^-(aq)$$

A related reaction is that between O_2^- and the HO_2 radical.

$$\text{(45)} \quad O_2^-(aq) + HO_2(aq) \longrightarrow O_2(aq) + HO_2^-(aq)$$

Since HO_2^- is the conjugate base of H_2O_2, this provides a route to H_2O_2 additional to those discussed in Section 6.3.

6.5 Effects of acidic emissions

6.5.1 *Effects on vegetation*

The effects on plants which must be considered are those of the gaseous pollutants themselves, and that of lowered pH.

Sulfur dioxide is very strongly phytotoxic (toxic to plants). Plant growth is inhibited at concentrations of SO_2 well below 0.1 ppm. Concentrations between 0.1 and 1 ppm can cause observable injury to plants and trees even if the length of exposure is as short as a few hours. To put this in context, SO_2 levels in large metropolitan areas such as New York City quite frequently reach 1 ppm. Combinations of gaseous pollutants often act synergistically; this is observed for the combinations SO_2/NO_2 and SO_2/O_3[9][10]. We note in passing that under conditions where coal is burned at the same time that the weather conditions favour photochemical smog, the atmosphere may simultaneously be polluted with sulfur dioxide, a reducing agent, and ozone, an oxidant. Under these conditions, SO_2 will be rapidly oxidized by ozone, or its further reaction products such as OH and H_2O_2.

Nitrogen dioxide also appears to be phytotoxic. However, the effects of NO_X on plants are not as clear-cut as those of SO_2, which are generally agreed to be exclusively adverse. The nitrate ion of nitric acid is a plant nutrient, and so the action of acid precipitation as a plant fertilizer must be considered. Indeed, in the unpolluted environment, nitrogen fixation as a result of lightning is a significant source of the nitrate available to plants (*cf.* Chapter 1). However, the combination of SO_2 and NO_X pollution seems from laboratory experiments to be more toxic to plants than SO_2 alone[10].

Excessive acidity is also harmful to plants. Leaves may be damaged below pH 3.5; soil chemistry will be altered well above this pH, and the problem will be most serious for poorly buffered soils, many of which tend to be naturally acidic. Gardeners are well aware of the preferences of certain plants for soils which are specifically alkaline, or alternatively, acidic. Few plants tolerate acidic soils however, and among other effects, the germination of seeds and the growth of seedlings may be inhibited.

Much controversy has been focused on the issue of whether acid rain is responsible for damage to forests[11]. Forests in Scandinavia, the Black Forest in Germany, and the forests of eastern North America have all experienced reduced productivity in recent years. The Black Forest has been especially hard hit, with some commentators predicting its "probable destruction" over the next few decades. The observed symptoms are consistent with long term acidification, namely a yellowing of the leaf, called chlorosis. The needles of conifers first become yellow, and then reddish-brown. Continued exposure causes the needles to drop and the trees to die. Acid rain seems to be a likely cause, since the effects of highly acidic precipitation on forest were clearly documented downwind of Sudbury, Ontario during the 1960's. Nevertheless it is difficult to obtain universal acceptance for this view, and the lack of consensus makes it difficult politically to pass legislation to restrict the emissions.

---

[9] P.K. Misra and R. Bloxham, Technology Transfer Conference, Toronto, Ontario, 1987: Proceedings, paper A11.

[10] Reference 1, p. 204

[11] S.N. Linzon, *Sulfur in the Environment,* Ed. J.O. Nriagu, Wiley, New York, 1978, Part II, Chapter 4.

An interesting point is that it is the base of a cloud which experiences the greatest reduction in pH when the cloud encounters acidic gases in the lower troposphere[12]. Mists and fog are therefore likely to be especially acidic, and this poses an additional threat to forests, which are often swathed in mist in upland areas. Adams *et al.*[13] have discovered that not all raindrops are equally acidic; those droplets having the lowest pH have radii near 0.5 mm.

6.5.2 Effects on health[14]

Sulfur dioxide and nitrogen dioxide are both irritants to the respiratory tract. In the case of gaseous acidity due to NO_2, the health effects of NO_2 itself are usually neglected, since the O_3 and PAN which also tend to be present are much more severely toxic (see Chapter 3). Several authors have made the point that in seeking a healthy environment, our actions on acid rain should be directed towards setting standards for acidic gases that can be tolerated by plants, since our forests, farmlands, and wilderness areas are much more susceptible to this problem than we ourselves.

At atmospheric levels of no more than the 1 to 2 ppm usually associated with this pollutant, sulfur dioxide tends to be absorbed high in the respiratory tract, and so it does not reach the far more sensitive alveoli; this occurs when the concentration reaches 25 ppm or so. The latter levels may be encountered by workers in industries such as smelting, tanning, papermaking, and sulfuric acid manufacture. However, the irritant effects of the gas at these concentrations (wheezing, coughing, tearing) are severe enough that actual injury is rare. Even so, respiratory irritation may occur at the lower levels associated with air pollution, especially in elderly people, and especially if particulates are also present (i.e., "London" smog, Chapter 3).

Data on the long-term health of workers exposed to SO_2 are not clear-cut: some studies show long term respiratory effects, but others do not. There appears to be a definite synergism between SO_2 and arsenic in the predisposition of arsenic smelter workers towards respiratory cancer. Experimental studies in animals also implicate SO_2 as a promoter of carcinogenesis.

Acidic emissions comprise both acidity (H^+) and particulate matter, and it is difficult to separate out these effects. Spengler *et al*[15] have summarized the information available. Interpretation is complicated because many jurisdictions routinely monitor sulfate as a proxy for acidity, rather than H^+ itself. Mortality rates show significant associations with total particles and with [sulfate] levels on preceding days; hospital admissions for respiratory ailments associate strongly with these two parameters.

---

[12] V.A. Mohnen, "The challenge of acid rain," *Sci Am*, **1988**, 259, 30.

[13] S.J. Adams, S.G. Bradley, C.D. Stowe, and S.J. de Mora, *Nature,* **1986,** 321, 842.

[14] C.M. Shy, Reference 11, Chapter 3

[15] J.D. Spengler, M. Brauer, and P. Koutrakis, "Acid air and health," *Environ Sci Technol,* **1990,** 24, 946–956

6.5.3 *Effects on buildings, etc.*

Limestone ($CaCO_3$) has been a preferred building material for many centuries. Even under conditions of very clean air it is subject to slow attack, by the same chemical processes which carve out caves and gorges (Chapter 5).

$$CaCO_3(s) + H_2CO_3(aq) \longrightarrow Ca^{2+}(aq) + 2HCO_3^-(aq) \tag{46}$$

However this process occurs exceedingly slowly and is not generally considered to be a problem. We can see this by examining the equilibrium constant for this reaction. As was shown in Chapter 5, this equilibrium constant has the rather small value of 5.3×10^{-5} mol^2 L^{-2}.

Acidic precipitation greatly increases the rate of dissolution.

$$CaCO_3(s) + H^+(aq) \longrightarrow Ca^{2+}(aq) + HCO_3^-(aq) \tag{47}$$

For this reaction, the equilibrium constant is much larger, the large value of K_c resulting from the very strongly favoured reaction of H^+ with the basic anion CO_3^{2-}.

$$\begin{array}{rcl}
CaCO_3(s) & \rightleftharpoons & Ca^{2+}(aq) + CO_3^{2-}(aq) \\
 & & K = K_{sp} = 6.0 \times 10^{-9}\ \text{mol}^2\ \text{L}^{-2} \\
H^+(aq) + CO_3^{2-}(aq) & \rightleftharpoons & HCO_3^-(aq) \\
 & & K = 1/K_a = 2.1 \times 10^{10}\ \text{L mol}^{-1} \\
\hline
CaCO_3(s) + H^+(aq) & \rightleftharpoons & Ca^{2+}(aq) + HCO_3^-(aq) \\
 & & K_c = 1.3 \times 10^{2}\ \text{mol L}^{-1} \\
\hline
\end{array}$$

The damage done to historical monuments is seen clearly in Figure 6.3 where the loss of detail in the face of General Abner Doubleday is very evident. Of particular cultural significance is the threat to the medieval stained glass of the great churches and cathedrals of Europe[16]. The following quotation illustrates the point. "For almost a millennium, in the case of the earliest stained glass windows, the glass escaped major damage....If stained glass windows are kept in situ in their present state of preservation, their total ruin can be predicted within our generation....The stained glass windows of Cologne Cathedral....now look like sheets of chalky plaster. Continuous etching by air pollutants has corroded the exterior surface of the glass, reducing its thickness year by year and giving the decomposed surface a so-called weathering crust....Each rain washes the crust away."

Iron and steel structures are well known for their susceptibility to corrosion, and the protection of such structures with paint costs billions of dollars annually. The chemistry of corrosion under atmospheric conditions is extremely complex, leading eventually to hydrated Fe_2O_3. Hydrogen ions catalyze the reaction, explaining why acid precipitation causes increased rates of corrosion.

[16] G. Frenzel, "The restoration of medieval stained glass," *Sci Am*, **1985**, 252, 126.

Figure 6.3: Etching of stonework by acid precipitation. The preserved memorial to General John Reynolds (right) provides a stark contrast to the corroding features of the statue of General Abner Doubleday (left). Gettysburg National Military Park, USA.

Steel protected with zinc i.e., galvanized steel is also subject to accelerated corrosion under acidic conditions. Normally, zinc offers good atmospheric protection; extrapolation of 5-year exposure tests indicates that protection may last for hundreds of years under optimum conditions such as those found in deserts. In heavily industrialized areas of Britain, galvanized steel may last as little as 5–20 years[17].

6.5.4 Effects on natural waters

These have probably been the most highly publicized effects of acid rain in the news media. Acidification is mainly a problem in areas where the underlying rocks provide poor buffering capacity. Rocks such as granite offer little protection by way of buffering, whereas chalk and limestone are able to react with the acid and hence to neutralize it (Chapter 5). The overall reaction with limestone can be considered as:

$$(48) \qquad 2H^+(aq) + CaCO_3(s) \longrightarrow Ca^{2+}(aq) + CO_2(g) + H_2O(l)$$

[17] J.O. Nriagu, in Reference 11, Chapter 1.

In reality, the $H^+(aq)$ reacts with $HCO_3^-(aq)$ that is contributing to the alkalinity of the water, and $CaCO_3(s)$ goes into solution to restore equilibrium.

As a result of the solid $CaCO_3$ going into solution, the pH of the lake is not changed significantly by the addition of the acidic rainwater. Lakes and streams in limestone areas are therefore fairly insensitive to acidic precipitation (Problem 15). In this sense the lake is said to be buffered, although the lake is not a "buffer solution" in the conventional sense of containing comparable amounts of a weak acid and its conjugate weak base.

The regions where acidic precipitation has been most recognized as a problem are areas of granitic rocks, namely Northern and Eastern Canada, the Northeastern United States (see Figure 6.9) and Scandinavia. In these locations not only are the forests under assault but the lakes are becoming progressively acidified. In these poorly buffered lakes a "normal" pH would probably be in the range pH 6.5 to 7. Today, many lakes in New York State, in Northern Ontario, and Scandinavia record pH levels of 5.0 and lower. Recent data for the Great Lakes basins suggest annual wet deposition rates of sulfur and nitrogen of almost 400,000 and 200,000 tonnes respectively[18].

Acidified lakes do not support the variety of life that can be found in their non-acidified counterparts[19]. Data provided by the National Research Council of Canada[20] indicate how progressive acidification results in the loss of aquatic organisms.

pH	Aquatic organism lost
6.0	Death of snails and crustaceans
5.5	Death of salmon, rainbow trout, and whitefish
5.0	Death of perch and pike
4.5	Death of eel and brook trout

Below pH ca. 4 the lake becomes a suitable habitat for white moss, which prefers an acidic environment. This plant forms a "felt mat," which may grow to 0.5 m or more thick, on the lake bottom. The mat prevents the exchange of nutrients between the water and the bottom sediments and also prevents the sediments from exerting any buffering action. The result is a lake whose waters are crystal clear, but whose waters support very few forms of aquatic life.

As we see from this information, the loss of game fish can be expected to be severe for lakes whose pH has already dropped to pH 5 or lower. By 1976, about half the lakes in the Adirondack Mountains of New York State had no fish in them, whereas forty years earlier almost all these lakes supported a

[18]E.C. Voldner and M. Alvo, "On the estimation of sulfur and nitrogen wet deposition to the Great Lakes," *Environ Sci Technol*, **1989**, 23, 1223–1232.

[19]D.W. Schindler, "Effects of acid rain on freshwater ecosystems," *Science*, **1988**, 239, 149–157.

[20]B. Henry, "Watch on the rain," *Science Dimension*, 1984/2, National Research Council of Canada, 7.

population of sport fish[1]. This observation correlates with comparisons of the alkalinity of Adirondack lakes today vs. sixty years ago[21]: of 274 lakes for which data were available, 80 % had suffered loss of alkalinity, the median loss being 50 μmol H^+ L^{-1}. The loss of the fish has serious consequences for regions like upstate New York and Northern Ontario, where tourism is a mainstay of the economy. Since one of the principal attractions for tourists is sport fishing, the loss of fish is soon followed by the loss of the tourists.

A special problem for aquatic life is that spawning, which generally takes place in the early spring, coincides with what is often the worst "pulse" of acidity of the year: the influx into the lake of the winter's accumulation of acid snow during the annual spring run-off. Figure 6.4 illustrates this effect in an Adirondack lake[22]. The result is decreased rates of hatching and reduced viability of the newly hatched fry.

Experiments on rejuvenating acidified lakes have been conducted by adding limestone to neutralize the excess acid[23]. The limestone is sprayed as a powder using aircraft. The principal costs are the cost of the limestone, delivery to the site and application, fish restocking, and ongoing monitoring. Maintaining a single 40 acre (20 ha) lake for 10 years costs $15,000 to $45,000 in 1986 U.S. dollars.

Another adverse effect of acidity upon a body of water is the accompanying increase in the aqueous concentrations of metal ions. The identities of these metals will depend on the composition of the underlying bedrock, but they may include toxic metal ions such as Cd^{2+}, Pb^{2+}, and Hg^{2+}. In each case, the metals are solubilized because of the reaction of H^+ with the basic anion with which the metal is associated. Such a reaction is similar in every respect to the dissolution of $CaCO_3$ in lakes which overlie limestone. Consider, for example, the case of PbS, for which $K_{sp} = 1 \times 10^{-28}$ mol^2 L^{-2}. In pure water, we calculate $[Pb^{2+}] = 1 \times 10^{-14}$ mol L^{-1}. At pH 4.1 we calculate (Problem 8) the value $[Pb^{2+}] = 2.5 \times 10^{-5}$ mol L^{-1} which is more than 10^9 times larger. Under some conditions, the acidification of water can compromise its suitability for drinking (see Chapter 7).

One metallic element which has been the subject of attention lately is aluminum. This element is highly toxic to fish, and is the major ion released upon neutralization of acid[24]. The aqueous chemistry of aluminum is quite

[21] C.E. Asbury, F.A. Vertucci, M.D. Mattson, and G.E. Likens, "Acidification of Adirondack lakes," *Environ Sci Technol,* **1989,** 23, 362–365. An earlier study by Kramer *et al,* for the U.S. National Research Council, carried the conclusion that the mean loss of alkalinity was 0–44 μequiv L^{-1}. For conflicting views and additional references, see correspondence in *Environ Sci Technol,* **1990,** 24, 384–390.

[22] J. Shurkin, "Lake sensitivity to acid rain," *Electric Power Research Institute Journal,* June **1985,** 16.

[23] "Treating acidic lakes and streams with limestone," *Electric Power Research Institute, (Technical Brief),* No. RP 2337 (1986); For recent arguments against the practice of liming, see S. Woodin and V. Skiba, "Liming fails the acid test," *New Scientist,* March 10, **1990,** 50–54.

[24] L.O. Hedin, G.E. Likens, K.M. Postek, and C.T. Driscoll, "A field experiment to test whether organic acids buffer acid deposition," *Nature,* **1990,** 345, 798–800.

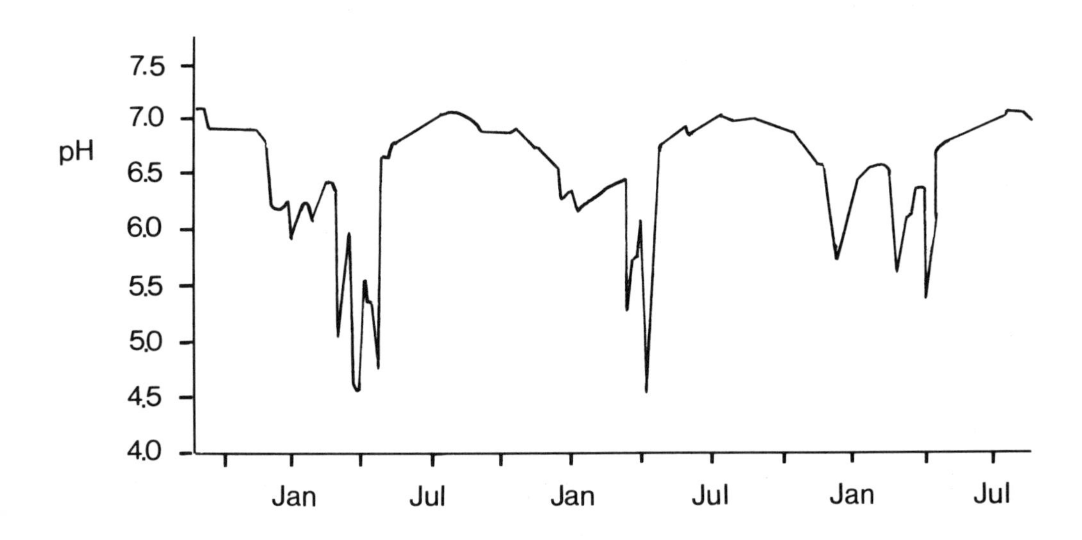

Figure 6.4: pH variation of an Adirondack lake during spring run-off; reproduced from Reference 21.

complex[25]; depending upon the pH, solutions of aluminum can contain Al^{3+}, $AlOH^{2+}$, $Al(OH)_2^+$ and numerous polynuclear hydroxo species such as $Al_2(OH)_2^{4+}$, and $Al_3(OH)_4^{5+}$. At higher pH, the $Al(OH)_4^-$ anion becomes important, since Al_2O_3 and $Al(OH)_3$ are amphoteric i.e., they dissolve in both acid and base. For our purposes, we will not consider the polynuclear complexes, as these form only at higher concentrations of aluminum. We will also take the solid phases $Al(OH)_3$ and hydrated Al_2O_3 as interchangeable. Figure 6.5 shows the speciation of aluminum with pH according to this model.

The hydrated Al^{3+} and $AlOH^{2+}$ ions are weak acids having pK_a about 5.5 and 5.6 respectively[26]. Most natural waters have pH 5–8, so they will contain significant amounts of Al^{3+}, $AlOH^{2+}$, $Al(OH)_2^+$, and $Al(OH)_4^-$ when acidity mobilizes aluminum from the bedrock.

$$(49)\quad \tfrac{1}{2}Al_2O_3.nH_2O(s) + 1H^+(aq) \rightleftharpoons Al(OH)_2^+(aq) + \tfrac{1}{2}H_2O(l)$$

$$(50)\quad \tfrac{1}{2}Al_2O_3.nH_2O(s) + 2H^+(aq) \rightleftharpoons AlOH^{2+}(aq) + H_2O(l)$$

$$(51)\quad \tfrac{1}{2}Al_2O_3.nH_2O(s) + 3H^+(aq) \rightleftharpoons Al^{3+}(aq) + 1\tfrac{1}{2}H_2O(l)$$

$$(52)\quad \tfrac{1}{2}Al_2O_3.nH_2O(s) + OH^-(aq) \rightleftharpoons Al(OH)_4^-(aq)$$

[25] R.B. Martin, "The chemistry of aluminum as related to biology and medicine," *Clin Chem,* **1986** 32, 1797.

[26] Based on thermodynamic data from H.M. May, P.A. Helmke, and M.L. Jackson, *Geochim Cosmochim Acta,* **1979,** 43, 861.

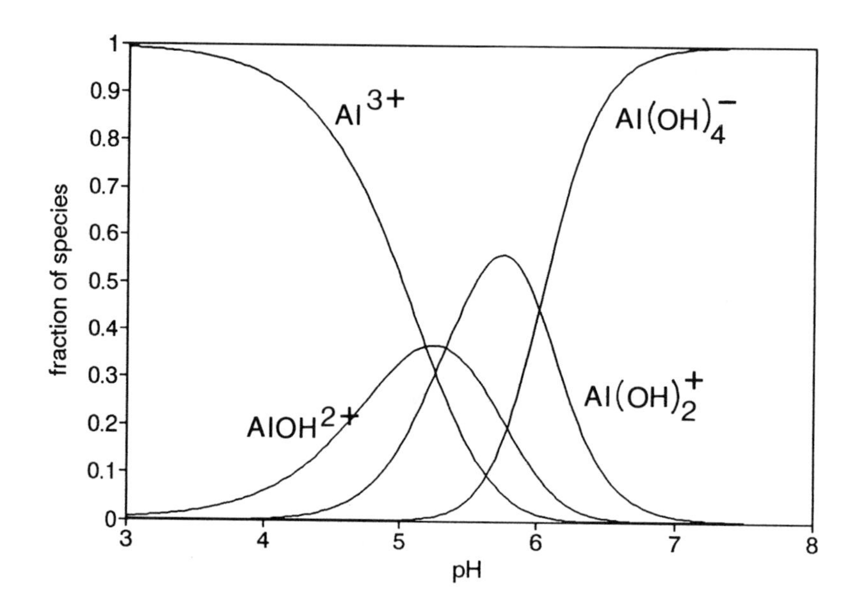

Figure 6.5: Speciation of aluminum over the pH range 3–9.

The equilibrium constants can be written as follows:

$$\begin{aligned} K_c(49) &= [Al(OH)_2^+]/[H^+] \\ K_c(50) &= [AlOH^{2+}]/[H^+]^2 \\ K_c(51) &= [Al^{3+}]/[H^+]^3 \\ K_c(52) &= [Al(OH)_4^-]/[OH^-] = [Al(OH)_4^-][H^+]/K_W \end{aligned}$$

Taking logarithms of each side of these equations, rearranging, and replacing $-\log_{10}[H^+]$ by pH, we obtain the relationships below.

$$\begin{aligned} -\log_{10}[Al(OH)_2^+] &= pK_c(49) + 1\ pH \\ -\log_{10}[AlOH^{2+}] &= pK_c(50) + 2\ pH \\ -\log_{10}[Al^{3+}] &= pK_c(51) + 3\ pH \\ -\log_{10}[Al(OH)_4^-] &= pK_c(52) - 1\ pH + pK_W \end{aligned}$$

Therefore if the logarithm of the dissolved aluminum concentration is plotted against pH, the slope of the curve will depend upon the pH: in other words, it will depend upon which of the equilibria 49 through 52 is most important at that particular pH. The equilibrium solubility of $Al(OH)_3$ (or $Al_2O_3.nH_2O$) over the pH interval 4–7 shows a minimum near pH 6, the result of the interplay between H^+ and OH^-in solubilizing the solid phase: see Figure 6.6, and also Problem 10.

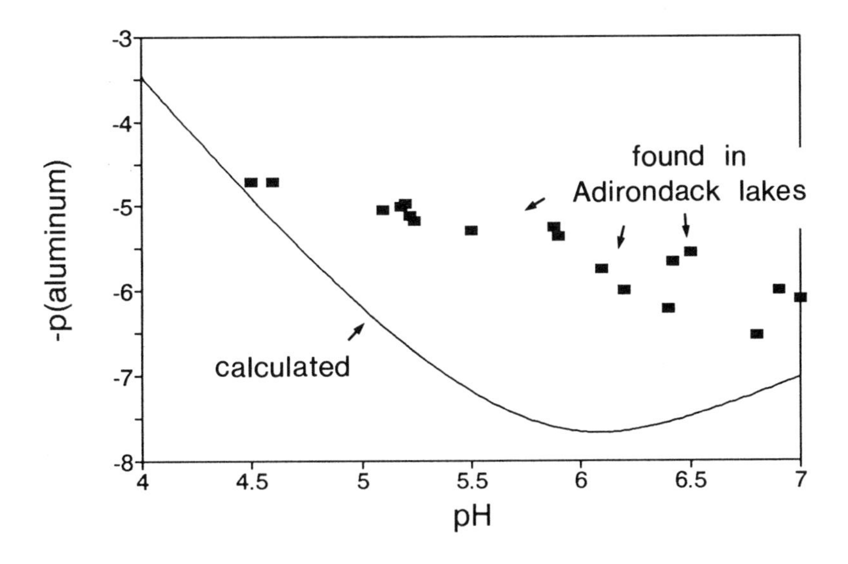

Figure 6.6: Calculated solubility of $Al(OH)_3(s)$ as a function of pH.

Driscoll and Newton have reported the total dissolved aluminum concentration in 19 lakes in the Adirondack region of New York State[27]. There is a definite trend towards higher aluminum concentrations in the more acidic lakes. The concentration of dissolved aluminum increases as $[H^+]^{1.6}$. However, the experimental solubilities are much greater than those calculated at equilibrium. Clearly, the calculated solubility of hydrated $Al_2O_3(s)$ gives only a minimum estimate of the total concentration of dissolved aluminum.

One possible explanation is that the lakewater samples include aluminum present in colloidal particles. Another is that where the water also contains fluoride ion, complexation with F^- increases the total concentration of dissolved aluminum, affording additional species such as AlF^{2+} and AlF_2^+ even at ppm concentrations of F^-[28] (Problem 11).

$$(53)\quad Al(H_2O)_6^{3+} + F^-(aq) \longrightarrow Al(H_2O)_5F^{2+} + H_2O(l)$$
$$\text{or:}\quad Al^{3+}(aq) + F^-(aq) \longrightarrow AlF^{2+}(aq) \quad K_C = 2.5 \times 10^6 \text{ L mol}^{-1}$$

But why is aluminum so toxic towards fish? The explanation currently favoured is that in a lake of, say, pH 5 we can expect an aluminum concentration of roughly 10^{-5} mol L^{-1}. The fish's blood is close to pH 7.4;

[27] C.T. Driscoll and R.M. Newton, "Chemical characteristics of Adirondack lakes," *Environ Sci Technol,* **1985,** 19, 1018.

[28] B.J. Plankey, H.H. Patterson, and C.S. Cronan, "Kinetics of aluminum fluoride complexation in acidic waters," *Environ Sci Technol,* **1986,** 20, 160.

its gill membrane is very thin in order to allow the efficient diffusion of oxygen from the water, and so a steep proton gradient is set up across the membrane. Because the experimental concentration of dissolved aluminum increases as a power of the H^+ concentration between 1 and 2, taking the example of the Adirondack lakes, a rise in pH of, say, two units corresponds to a decrease in the concentration of dissolved aluminum of about three orders of magnitude. As a result, a gelatinous precipitate of $Al(OH)_3$ forms on the fish's gills, leading to death by suffocation. Gotfryd[29] stresses the concept that aluminum and low pH act synergistically in affecting the biota of lakes and streams, suggesting that dissolution of aluminum from rock is a reason why moderate acidity (pH 5–6) appears to be much more harmful to aquatic life in the wild than in pure water in the laboratory. There may also be a synergy with fluoride, which has long been known to be toxic as an enzyme inhibitor; recent work suggests that the effects of fluoride as an enzyme inhibitor may actually be due to AlF_4^- and related species, rather than to F^- itself[30] (Problem 11).

An extreme example of natural acidity

In 1983, scientists at the University of Toronto reported a case of extreme natural acidification of a group of small lakes in the Canadian Arctic[31]. The "Smoking Hills" are situated on the Arctic Ocean, far from any habitation. They consist of cliffs, strata of which comprise combustible shales which have been burning continuously, probably for hundreds of years, and were likely ignited originally by lightning. Sulfurous smoke from the fires gives the region its name, and over the years it has acidified the small ponds in the immediate area to the extent that those ponds closest to the burning shales have pH 2 and lower. Even though the rock underlying this particular region is largely limestone, the assault of acidity over so many years has long since overcome the capacity of these ponds to resist a change in pH.

Table 6.2 presents a selection of data on the metal content of some of these ponds, as obtained by atomic absorption spectroscopy. As anticipated from the preceding discussion, the metal concentrations increase sharply as the pH drops (Problem 12).

Figure 6.7 is a titration curve for the addition of standard base to the filtered water of a pond having pH 2.0. The flat regions of the curve, which suggest a buffered solution, represent the neutralization of the acidic cation $Fe^{3+}(aq)$ at about pH 3.5, (whereupon hydrated $Fe_2O_3(s)$ precipitates), and the neutralization of $Al^{3+}(aq)$ to $AlOH^{2+}(aq)$ close to pH 5.

---

[29] A. Gotfryd, "Aluminum and acid: a sinister synergy," *Canadian Research,* July 1989, 10–11.

[30] M. Chabre, "Aluminofluoride and beryllofluoride complexes: new phosphate analogs in enzymology," *Trends in Biochemical Sciences,* **1990,** 15, 6–10.

[31] M. Havas and T.C. Hutchinson, "The Smoking Hills: natural acidification of an aquatic ecosystem," *Nature,* **1983,** 301, 23.

Figure 6.7: Smoking Hills, NWT, Canada.

Research on the vegetation in the vicinity of these ponds has shown that those plants which are most resistant to acid deposition release base to neutralize acid deposited on the leaves[32].

6.6 Acid rain abatement

This discussion will be restricted to the abatement of sulfur dioxide emissions. Because acidity due to the nitrogen oxides is associated mainly with automobile emissions, the limitation of NO_X emissions is discussed in Chapter 3 as part of the general topic of automobile emission controls. Also, in the case of the mixed SO_2 and NO_X emissions from power generating plants it has proved easier to reduce the former than the latter, for example by scrubbing with limestone slurry. Maximum thermal efficiency demands that combustion of the fossil fuel in these plants take place at the highest temperature possible, thereby maximizing NO_X formation also. As a result, the proportion of nitrate in the acidic depositions over eastern North America has been increasing over the past decade, as already noted. Urea has been suggested as a possible reagent for removing NO_X from stack gases[33].

$$(54) \qquad NH_2\text{-}CO\text{-}NH_2 + NO_X \xrightarrow{600-1100°C} N_2 + CO_2 + H_2O$$

[32] M. Ross, "Hardy herb thrives in acid rain," *Canadian Research*, November 1988, 10.
[33] *Chem Eng News*, April 18, 1988, 22

Table 6.2: Metal content of some ponds in the Smoking Hills (Canada).

Distance[a]	40	190	670	4400
pH	1.8	2.8	3.6	8.1
Acidity[b]	0.12	6.4×10^{-3}	6×10^{-4}	–
Alkalinity[b]	–	–	–	1×10^{-3}
SO_4^{2-}, ppm	16,000	380	160	110
Be, ppm	230	40	–	–
Mg, ppm	500	22	13	42
Al, ppm	590	18	3.8	–
Ca, ppm	370	45	49	61
Mn, ppm	64	2.8	3.5	–
Fe, ppm	2600	32	1.1	0.04
Ni, ppm	22	0.2	0.06	

[a] distance in meters from the nearest burning shale
[b] in mol L^{-1}

There are several possible strategies to prevent environmental damage by a pollutant. They include:

(a) minimizing the production of the pollutant;

(b) dilution of the pollutant so that it is no longer harmful;

(c) conversion of the pollutant to a harmless or, better, useable substance.

All these strategies have been explored in the case of sulfur dioxide. However, the two main SO_2–producing industries require different approaches. These, it will be remembered, are coal burning and the smelting of metal ores.

6.6.1 Coal burning power plants

Coal, as already mentioned, contains up to 2 to 3% sulfur by mass. Combustion of 1 tonne of a coal containing 2% of sulfur consumes about 80,000 moles of carbon, which will produce 80,000 moles of CO_2. Also present in the stack gases will be $4 \times 80,000 = 320,000$ moles of N_2, since combustion takes place in air, rather than in pure oxygen. The 600 moles of SO_2 that are also emitted represent less than 0.1% of the total volume of gases, too little for conversion of the SO_2 to sulfuric acid (as is done in the metal extraction industry, see below) to be economically attractive. Consequently, abatement measures have been directed towards dilution, or to preventing emission by chemical removal of SO_2 from the effluent gases.

Dilution consists in building tall stacks at the power plant, so as to disperse the effluent gases. While this spares the immediate locale from the

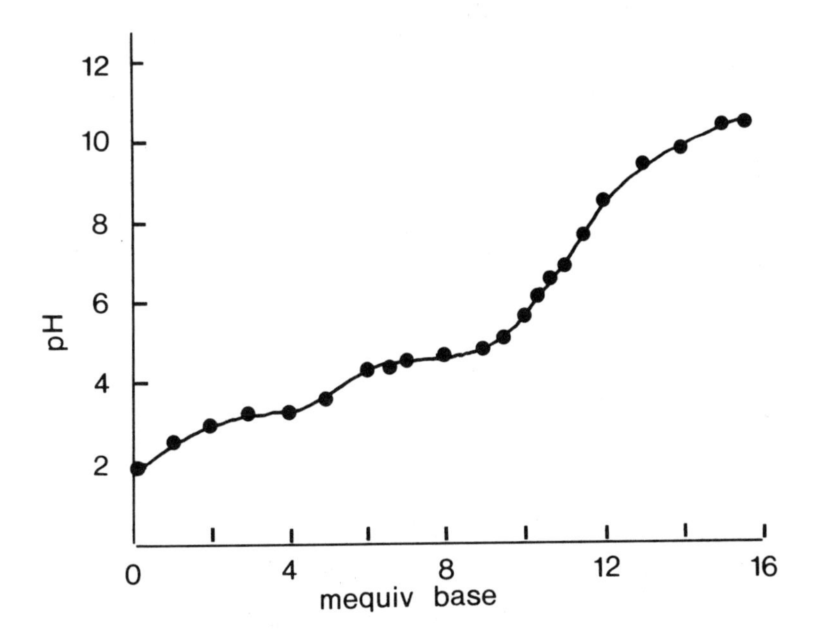

Figure 6.8: Titration curve for the addition of base to pond water of pH 2.0. Redrawn from Reference 31.

most serious effects of SO_2 pollution, the gas must ultimately be oxidized and deposited somewhere. At 2% of the mass of the coal, the SO_2 emitted amounts to 40 kg per tonne of coal; when one realizes that a modern power plant uses thousands of tonnes of coal a day, the magnitude of the problem is evident.

Most of the sulfur in coal occurs in the form of iron pyrites, FeS_2, which has a density of 4.5 g cm^{-3}. Coal itself has a lower density, 2.3 g cm^{-3}. If the coal is finely ground, most of the particles of coal will be physically distinct from the particles of pyrites. They may then be separated on the basis of density. The technique used is **oil flotation,** in which the particles are vigorously agitated with air and water, to which has been added oil and a small amount of surfactant. Oil flotation is generally useful for the separation of large quantities of solids having different densities; the composition of the water/oil/surfactant phase is adjusted so that the less dense particles are carried by surface tension at the air/liquid interface, and the denser particles are allowed to sink. The frothy air/oil/water mixture is then allowed to separate out in a second chamber, whereupon the lighter of the solid particles sink to the bottom. Applied to coal, this process is called **coal cleaning.** While cleaning adds to the cost of the coal, it is cheaper than removing SO_2 by the use of scrubbers once the coal has been burned.

Flotation is effective at removing inorganic sulfur, such as iron pyrites, from coal, but it leaves behind any sulfur that is organically bound. Organic

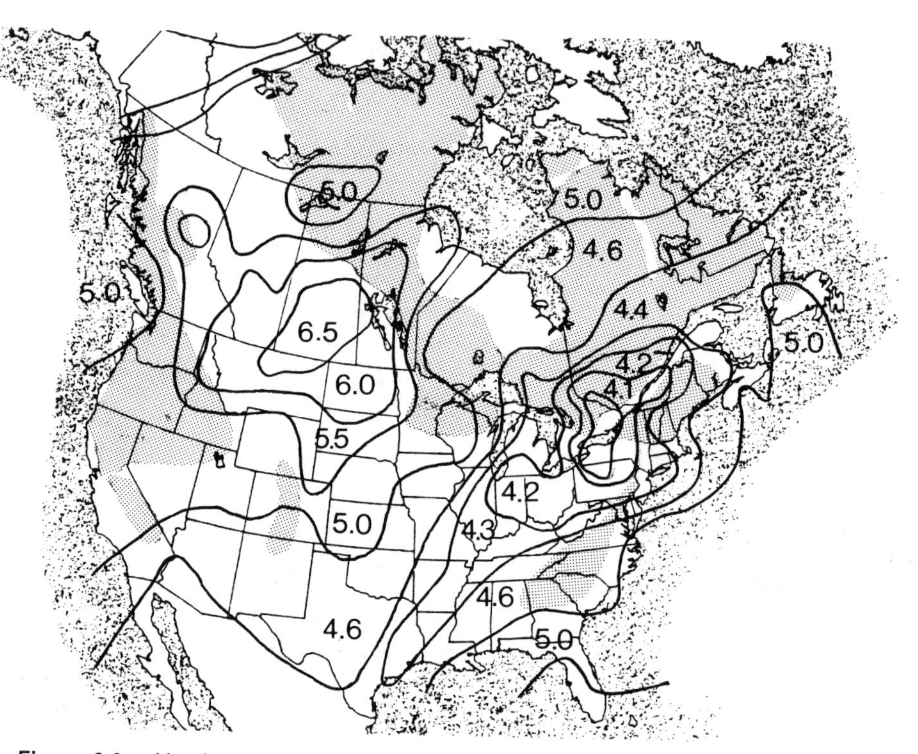

Figure 6.9: pH of precipitation. Shaded areas have granite as underlying rock.

sulfur occurs mostly as five membered thiophene rings in the coal; these replace the normal six membered benzene rings in the structure. Recent research[34] suggests that a promising approach may be to use genetically engineered microorganisms to extract the sulfur without breaking down the carbon skeleton of the coal.

Partial structure of coal.

Another method of preventing SO_2 emissions is to capture the SO_2 before it is released from the stack. Since SO_2 is an acidic gas, the method of choice is to trap it with base. Industrially, base does not mean the NaOH or KOH

[34] J. Haggin, *Chem Eng News,* August 29, 1988, 36.

which we might choose in the laboratory: they are too expensive. Lime, $Ca(OH)_2$, and limestone, $CaCO_3$, are the most commonly used. Most often, the stack gases are "scrubbed" with base by passing a fine spray of the lime or limestone slurry in water down the stack as the hot combustion gases are passing upwards. This process is often called Flue Gas Desulfurization.

$$SO_2(g) + Ca(OH)_2(aq) \longrightarrow CaSO_3(s) + H_2O(l) \quad (55)$$

$$SO_2(g) + CaCO_3(s) \longrightarrow CaSO_3(s) + CO_2(g) \quad (56)$$

Scrubbing equipment is expensive to install and maintain, and the cost of the large amount of lime or limestone is considerable. Not the least of the problems is the disposal of the thousands of tonnes per year of the $CaSO_3$ byproduct, which is obtained as an aqueous slurry, perhaps 98%+ water. This must be "dewatered" in huge holding tanks and eventually disposed of on the land: it represents a serious environmental problem itself. For all these reasons, electric power utilities tend not to install scrubbers unless they are required to do so through legislation.

Fluidized bed combustion with admixture of limestone is an alternative to scrubbing. Fluidized bed combustion involves burning finely ground coal on a screen through which air is passed from underneath, keeping the coal particles suspended until they burn. The small size of the particles increases the rate and efficiency of combustion. If pulverized limestone is added to the finely ground coal, the SO_2 formed through combustion reacts with the limestone (or the CaO produced from it at the high temperature) and is converted to a mixture of $CaSO_3$ and $CaSO_4$. Once again, the disposal of the byproduct is a difficulty, since it has no commercial value.

An area of current research is to oxidize SO_2 and NO_2 photochemically (using UV lamps rather than sunlight) and then to neutralize the acids formed with $Ca(OH)_2$, Na_2CO_3, or NH_3 to yield the corresponding sulfate and nitrate salts. The advantage of this method, if it were commercialized, would be the economic value of the ammonium salts thus formed as fertilizers, unlike the calcium salts discussed so far.

6.6.2 Metal extraction

Sulfides are the ores of interest in this context. As noted earlier in the chapter, SO_2 emissions result when the ores are roasted in air. Just as with a large modern power station, it is the size of a smelter that determines its environmental impact. For example, the Inco Ltd smelter at Sudbury, Ontario processes about 35,000 tonnes of ore per day having a typical analysis as follows:

Ni	1.3%	Cu	1.3%	Co	0.04%
Fe	20%	S	10%	CaO	4%
MgO	4%	SiO_2	40%	Al_2O_3	10%

The principal sulfides in the ore are FeS_2, $CuFeS_2$, NiS and CuS. The absolute amount of nickel in the ore is rather small, but the amount of sulfur by mass is 7 to 8 times greater than the amount of nickel. The ore is initially concentrated by oil flotation to remove the rock (SiO_2, Al_2O_3 etc), known as **gangue**. The various sulfides—nickel, copper, and iron—can also be separated by flotation on the basis of their different densities.

The production of nickel exemplifies many different aspects of chemistry, and illustrates the impetus for new technology that has come from environmental considerations. The traditional process at Sudbury has involved roasting the separated sulfide ores, giving a nickel stream and an iron ore recovery stream. Sulfur dioxide results from roasting the concentrated ore.

$$2NiS(s) + 3O_2(g) \longrightarrow 2NiO(s) + 2SO_2(g) \quad (57)$$

$$4FeS_2(s) + 11O_2(g) \longrightarrow 2Fe_2O_3(s) + 8SO_2(g) \quad (58)$$

The nickel oxide is reduced to the metal with coke, and the iron ore shipped off site for iron production. The poor economics of producing iron this way, coupled with environmental factors, have prompted a total overhaul of the process. Current efforts to reduce SO_2 emissions focus mainly on increasing the efficiency of obtaining the concentrate from the gangue in order to reduce the amount of sulfur which is roasted (Problem 13). The objective is to eliminate roasting this sulfur, and to return it to the ground. A second change, planned for 1994, is to roast the remaining CuS-NiS concentrate with pure oxygen, rather than with air. As seen in the next paragraphs, this change facilitates the subsequent recovery of SO_2, in order to meet the objective of reducing SO_2 emissions from the 1988 level of 685,000 tonnes to 265,000 tonnes by 1994.

In the Sudbury operation, sulfur dioxide is recovered as liquid SO_2 for resale and is also oxidized to sulfuric acid, most of which is used elsewhere in the production of fertilizers (Chapter 8). Over 1 million tonnes of sulfuric acid and 100,000 tonnes of liquid SO_2 are shipped from Sudbury every year. A slightly different approach is taken by Cominco, operator of a large lead and zinc smelter in Western Canada, which also produces sulfuric acid, but converts most of it directly into ammonium sulfate for use as fertilizer.

The chemical reactions involved in the conversion of SO_2 to H_2SO_4 are shown below.

$$SO_2(g) + \tfrac{1}{2}O_2 \xrightarrow[\text{catalyst, 450°C}]{V_2O_5} SO_3(g)$$

$$SO_3(g) \xrightarrow[H_2SO_4]{\text{dissolve in conc.}} \underbrace{H_2SO_4.SO_3(l)}_{\text{oleum}} \xrightarrow{H_2O} H_2SO_4$$

The exothermic conversion of SO_2 to SO_3, $\Delta H° = -99$ kJ mol^{-1}, is carried out at moderate temperature: a compromise between maximizing the equilibrium concentration of SO_3 (low temperature, Le Chatelier's principle) and attaining equilibrium rapidly (favoured by high temperature), see Problem 14. The substitution of oxygen for air in the roasting process will enhance the conversion of SO_2 to SO_3, by providing a "higher strength" SO_2 gas stream.

Environmental considerations dictate that the conversion of SO_3 to H_2SO_4 is carried out in two steps. The SO_3 stream is dissolved in concentrated sulfuric acid to give "oleum"; this reaction occurs rapidly and efficiently. Water is then added to produce the conventional "concentrated" sulfuric acid (93% H_2SO_4). The dissolution of SO_3 in water to give sulfuric acid directly is rather slow, and it is difficult to ensure that all the SO_3 is trapped by the water. Failure to do so would result in the formation of a corrosive fog of sulfuric acid droplets around the plant.

Figure 6.10: "Superstack" near Sudbury, Ontario. Constructed in 1972, the 400m stack towers above the town of Copper Cliff.

In principle, sulfuric acid manufacture is attractive in that it converts a polluting waste product (SO_2) into a useful commodity (H_2SO_4). In practice, the economics of the process are very unfavourable. First, the capital costs of the sulfuric acid manufacturing part of the operation are very expensive; more than $100 million for a plant this size. The greater the proportion of the sulfur dioxide that is to be trapped, the more expensive the facility. Second, the selling price of smelter grade sulfuric acid is very low, currently about $20–50 per tonne if bought in bulk. Thus there is the dilemma of a high manufacturing cost combined with a low selling price for the product.

Smelter grade sulfuric acid must compete with the purer "virgin" sulfuric acid, which is manufactured from elemental sulfur. Elemental sulfur is very cheaply available, both from vast subterranean deposits in Texas, and also as a byproduct of the "sweetening" of sour (that is, containing H_2S) natural gas in the Western Canadian gasfields. Even virgin sulfuric acid sells in bulk for only $60–100 per tonne[35].

The environmental impact of acidic emissions can also be reduced by dilution (dispersion of the stack gases after they leave the plant). The 400 m high "superstack" at Inco's Sudbury operation is probably the best example of the genre (see Figure 6.9). This stack has had an enormous impact upon the immediate locality of Sudbury, even though it does not change the total mass of SO_2 released. In the 1960's, the effect of SO_2 on the vegetation down-wind of the smelter was so severe that nothing green could survive. Today, because of the superstack, trees, shrubs, and grass have reappeared. The environmental price, of course, is that the effects of the plume of acidic gas emissions are felt at much greater distances from the smelter than previously, although at lower concentrations. However, an encouraging observation is that aquatic life seems able to recover rather quickly once acid deposition is reduced[36].

6.7 Political and economic considerations

As we saw in Section 6.4, the distance travelled by an acidic plume during the period in which SO_2 and SO_3 are being deposited means that acid rain is not just a local problem: SO_2-polluted air may, and does, cross national boundaries. The United States complains about the pollution from copper smelters in Mexico, Canada complains about pollution from power generating plants in the Ohio Valley and the U.S. Midwest, and Sweden complains about emissions in Britain and West Germany.

Several methods have been devised to pinpoint the source of acid rain deposited at a distant location[1]. Since the amounts of trace elements such as vanadium and manganese differ according to the source of a fuel, it may be possible to determine the origin of a particular air mass by analyzing the trace metals which are deposited along with acidity. This technique has implicated the Soviet Union as the source of much of the acidic deposition in the Arctic, and the U.S. Midwest as the origin of much of the acidic deposition in the northeastern U.S. and eastern Canada. In other work, the inert gas SF_6 has been deliberately introduced into the stack gases from power plants; the SF_6 travels with the plume without reaction, and thereby acts as a tracer. These experiments strongly implicate acidic emissions from Britain as the source of acidic precipitation in Sweden.

[35] *Chem Eng News,* April 28, 1986, 13.

[36] J.M. Gunn and W. Keller, "Biological recovery of an acid lake after reductions in industrial emissions of sulphur," *Nature,* **1990,** 345, 431–433.

The trans-boundary nature of acidic emissions causes continuing international (and, in the U.S., inter-state) friction. The costs of acidic precipitation, in terms of structural, ecological and environmental damage, run to many billions of dollars each year. In the U.S. Northeast alone the annual cost of acidic emissions has been estimated at 5 **billion** dollars, made up in these categories[37].

Forests	$ 1.75 × 10^9$
Agriculture	1.00
Corrosion of buildings, bridges etc.	2.00
Tourism and fishing	0.25
Total	$ 5.0 × 10^9$

Forster[38] has attempted to assess the economic impact of acidic precipitation in Canada, but points out the difficulty of attributing damage to acid rain with any certainty. For example, the yields per hectare of most crops in Ontario rose over the period 1957–1977, even though the acidity of precipitation increased during this time. Does this mean that acidity improves crop yields? No, because other factors also changed. Fertilizer applications were increased, and new crop varieties were introduced; in consequence, it is difficult to isolate the effects of acid precipitation. The inability of economists to document unequivocally the precise cost of acidic precipitation on any particular industry (farming, forestry, fishing, tourism...) makes it difficult for politicians to reach consensus on legislative programs against acid rain.

Unfortunately, the estimated costs of SO_2 reduction strategies are themselves very expensive. Switching from high to low sulfur coal and coal cleaning are generally cheaper options than scrubbing sulfur dioxide from the stack gases. In North America, much of the heavy industry has grown up in the east. Local, high-sulfur coals are less expensive to obtain than the low sulfur coals from the Rockies, some of which have sulfur contents as low as 0.3%. Several years ago, the estimated costs of reductions of SO_2 emissions by power utilities ranged from $1–2 billion for a 40% reduction to $2–4 billion for a 50% reduction to $5–6 billion for a 70% reduction[37]. Sizeable increases in electricity rates, perhaps up to 20% or more, would be needed to pay for these reductions, and these would be welcomed neither by industry nor the domestic consumer. It should be clear that pollution, in this case acid rain, is an expensive proposition: it is expensive in the damage it causes, and it is expensive to avoid or to clean up. One also suspects that while we, as members of the general public, regard pollution control and environmental clean up as a "Good Thing," our enthusiasm may wane when the costs come out of our pockets.

[37]T.D. Crocker and J.L. Regens, "Acid deposition control," *Environ Sci Technol,* **1985**, 19, 112.

[38]B.A. Forster, "Economic impact of acid precipitation: a Canadian perspective," in *Economic perspectives on acid deposition control,* Ed. T.D. Crocker, Butterworth, Boston, 1984, Chapter 7.

The introduction of tall stacks to disperse SO_2 emissions in an effort to improve air quality in the vicinity of power stations has improved air quality in the immediate district, because relatively little of the acidic emissions falls close to the source. Much of the damage takes place in remote locations, often in another country or state. There is therefore little incentive for the "exporter" of acid emissions to "clean up its act" since the damage occurs elsewhere. For example, it is very difficult to persuade the citizens of country "A" to accept higher charges for electricity when it is not A's crops or lakes that are at risk. The citizens of country "B," who are on the receiving end, may have quite a different opinion, however! In North America, the governments of Canada and many Northeastern U.S. States have been pressing for U.S. Federal legislation to limit acidic emissions. Likewise in Europe, acid rain is a very important political issue in Norway and Sweden, since these countries are the unwilling importers of this pollution.

To end this chapter on a more positive note, many nations are now taking steps actively to reduce acidic emissions. The 1990 U.S. Clean Air Act is a very significant piece of legislation in the North American context. It calls for an overall reduction of NO_X of 15% per year for each of six years. Part of this is to be achieved by a reduction of NO_X in auto emissions from 1.0 g/mile to 0.6 g/mile (0.6 to 0.4 g/km) beginning in 1994. SO_2 emissions are to be cut from 17 million tons to 7 million tons, with the following targets for coal fired power stations—1995: 2.5 lb SO_2 per million Btu ($\approx$ 1 kg per 10^9 J) 2000: 1.2 lb SO_2 per million Btu ($\approx$ 0.5 kg per 10^9 J). In Canada, the Ontario government has announced the target of a 50% reduction by 1994 of the acidic emissions from the province's major polluters, and has required regular reporting on the steps that are being taken to achieve these reductions. Many countries in the European Economic Community have agreed to cuts in emissions of 30% and more. Countries in the "30% club" include France (50% by 1990), West Germany (50% by 1993), and Sweden (60% by 1995).

Further reading

1. *Atmospheric Chemistry,* by B.J. Finlayson-Pitts and J.N. Pitts, Jr., Wiley, New York, 1986, Chapters 1 and 11.

2. A good journalistic account of acid rain (which includes references) has been given by Fred Pearce, *Acid Rain,* Penguin Books, Harmondsworth, England, 1987. Most of the examples are taken from the U.K.

3. *The Acid Rain Sourcebook,* Eds. T.C. Elliott and R.C. Schwieger, McGraw-Hill, New York, 1984. Most of the thrust of this book is towards the U.S.

6.8 Problems

1. (a) Calculate the pH of rainwater if SO_2 is the only acidic gas present and its concentration is 0.12 ppm. State what assumptions you make in this calculation.
   (b) Now extend this calculation by determining the additional $[H^+]$ that will be contributed by the equilibrium concentration of CO_2 (1.0×10^{-5} mol L^{-1}) that is also present.

2. Nitrogen dioxide reacts with water as follows:

$$2NO_2(g) + H_2O(l) \longrightarrow HNO_2(aq) + HNO_3(aq)$$

   Calculate the concentration of $[H^+]$ that would be contributed to rainwater due to this reaction by 100 ppb of $NO_2(g)$. ΔG_f° (all kJ mol^{-1}, 25°C): $NO_2(g)$ 51.3; $H_2O(l)$ −237.2; $HNO_2(aq)$ −50.6; $HNO_3(aq)$ −111.3; K_a for $HNO_2 = 4.5 \times 10^{-4}$ mol L^{-1}.

3. A power plant burns 10,000 tonnes per day of coal containing 2.35% of sulfur by mass. The stack gases contain SO_2 from the coal plus 150 ppm of NO_x (at the stack).
   (a) Taking an average molar mass of 38 g mol^{-1} for NO_x, calculate the total amount of acidic emissions from the plant per day.
   (b) How do the ratio, and the absolute amounts, of these emissions change if the plant switches to cleaned coal with a sulfur content of 0.30%?

4. The reaction

$$SO_2(g) + OH(g) \xrightarrow{M} HSO_3(g)$$

   has a reported rate constant of 9×10^{-13} cm^3 $molec^{-1}$ s^{-1}.
   (a) What is the value of the true third order rate constant for this reaction? State assumptions.
   (b) What is the half life for the reaction if $[OH] = 1 \times 10^7$ molec cm^{-3}?
   (c) What is the half life for the reaction under the same conditions in Mexico City, $P_{atm} = 640$ torr?

5. Suppose the only reactions important in oxidizing SO_2 are:

$$SO_2(g) + OH(g) \xrightarrow{M} HSO_3(g) \quad k_2 = 9 \times 10^{-13}\ cm^3\ molec^{-1}\ s^{-1}$$

$$SO_2(aq) + H_2O_2(aq) \longrightarrow H_2SO_4(aq) \quad k = 1 \times 10^3\ L\ mol^{-1}\ s^{-1}$$

   Make these assumptions: $[OH] = 5 \times 10^6$ molec cm^{-3}; temperature = 300 K. $K_H(SO_2) = 1.2$ mol L^{-1} atm^{-1}; $p(SO_2,g) = 1.0$ ppm; $K_H(H_2O_2) = 1 \times 10^5$ mol L^{-1} atm^{-1}; $p(H_2O_2,g) = 1$ ppb.

(a) Calculate the ratio of the rate of gas phase oxidation/rate of aqueous phase oxidation as a function of the amount of liquid water in the atmosphere over the range 0 to 0.1 g L^{-1}.

(b) How does the half-life of SO_2 vary as the water content changes?

6. The energy of a photon is given by Einstein's equation

$$E_{photon} = hc/\lambda$$

(a) Show that for λ in nm, $\Delta E(kJ\ mol^{-1}) = 1.2 \times 10^5/\lambda$

(b) The S–O bond energy in SO_2 is 550 kJ mol^{-1}. Make the best estimate you can of the wavelength of radiation that is capable of cleaving this bond and state any assumption you must make.

(c) The deposition rate of sulfur oxides from the troposphere is in the range 3% per hour. What is the residence time of the sulfur oxides in the troposphere under these conditions?

(d) Comment on whether photolysis is likely to be important in the tropospheric chemistry of SO_2.

7. A simplified scheme for oxidation and deposition of SO_2 is

$$SO_2 \xrightarrow{k_1} SO_3$$

$$SO_2 \xrightarrow{k_2} \text{deposition as sulfite}$$

$$SO_3 \xrightarrow{k_3} \text{deposition as sulfate}$$

The rate constants are all pseudo first order.

(a) Deduce the rate expressions for the loss of SO_2 with time and the production of SO_3 with time.

(b) Near a point source the SO_2 concentration in the atmosphere is 20 ppm. The windspeed is 8 km h^{-1} Calculate the rates of deposition of SO_3^{2-} and SO_4^{2-}both 8 km and 80 km downwind of the source under the conditions

$$\begin{aligned} \text{Rate of oxidation of } SO_2 &= 8.0\% \text{ per hour} \\ \text{Rate of deposition of } SO_2 \text{ as } SO_3^{2-} &= 2.5\% \text{ per hour} \\ \text{Rate of deposition of } SO_3 \text{ as } SO_4^{2-} &= 3.0\% \text{ per hour} \end{aligned}$$

8. At 25°C PbS has $K_{SP} = 8.0 \times 10^{-28}$ (mol $L^{-1})^2$.

(a) What is the concentration of Pb^{2+}(aq) in pure water that is in equilibrium with PbS(s)?

(b) What is the concentration of Pb^{2+}(aq) in a lake that has been acidified to pH 4.10? State any assumptions that you must make.

9. The aqueous chemistry of aluminum may be oversimplified as follows:

$$Al(OH)_3(s) \rightleftharpoons Al^{3+}(aq) + 3OH^-(aq) \qquad K_{sp} = 1.3 \times 10^{-33}\ (mol\ L^{-1})^4$$
$$Al^{3+}(aq) + H_2O(l) \rightleftharpoons AlOH^{2+}(aq) + H^+(aq) \quad K_a = 1.0 \times 10^{-5}\ (mol\ L^{-1})$$

(a) Calculate the concentration in ppm of aluminum of soluble Al = ($[Al^{3+}] + [AlOH^{2+}]$) at equilibrium with $Al(OH)_3$ at pH 5.28.

(b) Show by calculation what happens when water in a lake of pH 5.28, which is in equilibrium with aluminum-bearing rock (assume $Al(OH)_3$), comes in contact with a fish's gills at pH 7.4.

10. Use the thermodynamic data below for the system $Al_2O_3{\bullet}nH_2O(s)/Al^{3+}(aq)/AlOH^{2+}(aq)/Al(OH)_2^+(aq)/Al(OH)_4^-(aq)$ to calculate the solubility of monomeric aluminum over the pH range 4–7. ΔG_f°, kJ mol^{-1}: $Al(OH)_3(s)$, −1155; $Al^{3+}(aq)$, −489.4; $AlOH^{2+}(aq)$, −698.3; $Al(OH)_2^+(aq)$, −905.8; $Al(OH)_4^-(aq)$, −1312; $H_2O(l)$, −237.2; $OH^-(aq)$, −157.3.

11. Plankey *et al* have measured rate constants for the formation of the complex ion $AlF^{2+}(aq)$ from $Al^{3+}(aq)$. There are several paths, as shown below (all species are (aq)).

$$Al^{3+} + F^- \xrightarrow{k_1} AlF^{2+} \qquad k_1 = 32.6\ L\ mol^{-1}\ s^{-1}$$
$$AlOH^{2+} + F^- + H^+ \xrightarrow{k_2} AlF^{2+} + H_2O \qquad k_2 = 3.61 \times 10^3\ L^2\ mol^{-2}\ s^{-1}$$
$$Al^{3+} + HF \xrightarrow{k_3} AlF^{2+} + H^+ \qquad k_3 = 1.40\ L\ mol^{-1}\ s^{-1}$$
$$AlOH^{2+} + HF \xrightarrow{k_4} AlF^{2+} + H_2O \qquad k_4 = 1.1 \times 10^3\ L\ mol^{-1}\ s^{-1}$$

Given K_a for HF = 6.8×10^{-4} mol L^{-1} and K_a for $Al^{3+}(aq) = 1.0 \times 10^{-5}$ mol L^{-1} calculate the initial total rate of formation of AlF^{2+} in a water system where $c(Al^{3+}) = 2.0 \times 10^{-5}$ mol L^{-1}, $c(F^-) = 1.0 \times 10^{-6}$ mol L^{-1} and the pH has values of: (a) 3.00 (b) 4.00 (c) 5.00.

12. Assume that iron ore can be approximately represented by $Fe(OH)_3(s)$ for which $K_{sp} = 1.0 \times 10^{-38}\ (mol\ L^{-1})^4$. Calculate the equilibrium concentration of Fe^{3+} in two of the extremely acidic northern lakes which overlie iron ore and which have pH 1.8 and 3.6. Express your answer in ppm.

13. A nickel ore has the following partial composition by mass: Ni, 1.4%; Cu, 1.3%; Fe, 7.2%; S, 9.1%. A plant processes 35,000 tonnes of ore per day; 17% of the sulphur is converted into H_2SO_4 and 30% of the sulfur is released to the atmosphere. Calculate:

(a) the volume of SO_2 (in m^3 at STP) released to the atmosphere each day.

(b) the mass in tonnes of H_2SO_4 produced each day.

(c) the mass of SO_2 emitted for each tonne of nickel produced.

14. At 450°C the reaction

$$SO_2(g) + \tfrac{1}{2}O_2(g) \longrightarrow SO_3(g)$$

has $K_P = 24\ atm^{-1/2}$. SO_2 (initial pressure 2.0 atm) and air (initial pressure 20 atm) are passed over a catalyst at 450°C. Under these conditions 97% of the SO_2 is converted to SO_3. Did the reaction reach equilibrium?

15. **(a)** A 1.0 L sample of lake water is titrated against 1.05×10^{-3} mol L^{-1} HCl to a methyl orange endpoint. The volume of HCl required is 8.48 cm^3. What is the total alkalinity of the lake in mol H^+ per litre?

(b) Given $[CO_2,aq] = 1.0 \times 10^{-5}$ mol L^{-1} and pH = 6.33, what are the concentrations of HCO_3^- and CO_3^{2-} in the above lake?

(c) Is the above lake well buffered or poorly buffered? Could this lake be located close to your home? Explain.

(d) The lake contains 2.0×10^6 m^3 of water. During the spring runoff 4.2×10^4 m^3 of water having pH 4.15 (acidity assumed to be entirely present as $H^+(aq)$) are added to the lake in one day. What is the pH of the lake at the end of the day? (Remember: initial pH was 6.33 and assume $[CO_2, aq]$ is constant at 1.0×10^{-5} mol L^{-1}).

Chapter 7

Drinking Water

Introduction

No other public health or medical innovation comes close to having the importance of a safe clean supply of drinking water. Those of us who live in the developed parts of the world can turn a tap and obtain safe, fresh drinking water without a second thought. World-wide, the story is very different; millions of people—mostly women—must spend many hours every day carrying water, often of dubious quality, from a distant well to their homes. Between 15 and 20 million babies die every year as result of water-borne diarrheal diseases such as typhoid fever, ameboid dysentery, and cholera. Cholera, a disease endemic to Asia, appeared in Peru in 1991, and at the time of writing appears to be spreading through South America. But for municipal water purification, Western cities too would be rife with disease, as indeed they were as recently as the last century.

In this chapter we examine the chemistry of the treatment of water for drinking, with particular emphasis on the methods that are used for disinfection of water.

7.1 Sources of water

Potential drinking water sources are classified either as **ground water** or **surface water.** Ground water comes from deep in the ground, from underground aquifers, into which wells are bored to recover the water. These wells may be from tens to hundreds of meters deep. Water in such aquifers may be replaced only very slowly. In an extreme case, such as the Ogallala aquifer in the United States Great Plains, the "fossil water" as it is called has been estimated to be thousands of years old, and constitutes a resource which is virtually non-renewable, at least in our lifetime. Certain parts of this aquifer are being depleted, principally for agricultural irrigation, 1–2 orders of magnitude faster than the replacement rate by rainfall (see Chapter 5).

As the previous paragraph implies, communities which depend on underground aquifers may deplete their supplies if they draw on them too heavily.

Besides exhausting the water, this may have the undesirable side effect of causing the city to sink as the water is withdrawn. Exactly this problem has occurred in Venice, Italy and in London, England. Both cities are engaged in costly flood control measures in an attempt to keep out water at times of high tides, from the ocean and the River Thames, respectively. A different problem is developing in Mexico City: not only is the city sinking because of water removal, but aquifer depletion is so serious that the water supply may be exhausted before the middle of the next century.

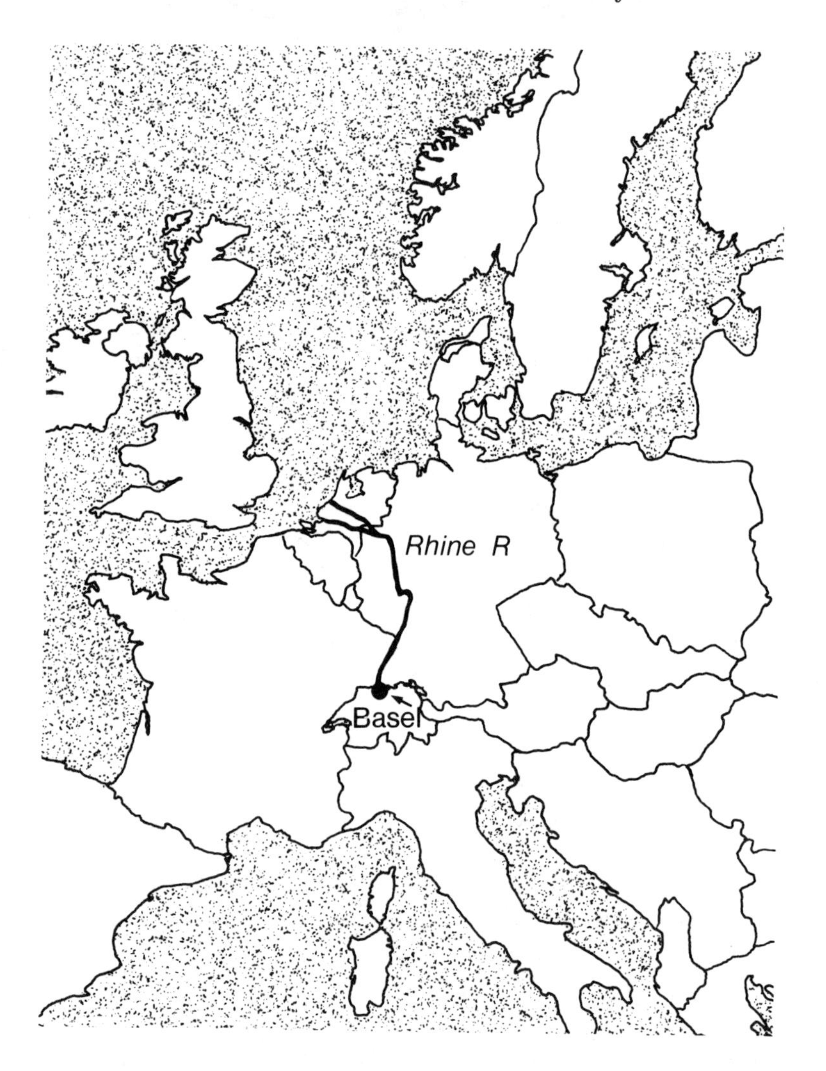

Figure 7.1: The Rhine River.

Surface water is drawn from a lake or a river; it almost always has a higher content of suspended materials than ground water, and consequently requires more processing to make it safe to drink. In addition, many major waterways, such as the Great Lakes and the Mississippi River in North America, and the Rhine and Danube Rivers in Europe are used for drinking and for other purposes by a large number of communities. Communities lying down-river draw water which has potentially (and more often actually) been contaminated by sewage outfall and industrial use upstream. For communities near the mouth of the Rhine, as little as 40% of the water withdrawn is "new" water i.e., has not previously been discharged by another city. The task of the municipal or the regional water engineer is to make this rather unpromising raw material fit to drink. A chemical spill following a warehouse fire in Switzerland in 1986 contaminated the whole of the Rhine with pesticides for several days[1], killing fish and other aquatic life for hundreds of kilometers downstream, and affecting the drinking water supplies of communities not only in Switzerland, but in Germany, France, and the Netherlands.

Ground water tends to be less contaminated than surface water because organic matter in the water has had time to be decomposed by soil bacteria. The ground itself acts as a filtering device so that less suspended matter is present. Indeed, filtration through sand was the first successful method of municipal water treatment. Its introduction in London, England in the middle 1800's led to an immediate decline in the incidence of water-borne disease. Some communities use "re-injection" to replenish their underground aquifer. This technique involves injecting surface water from a lake or river back underground in order to allow these natural purification processes to take place.

7.2 Outline of water treatment

There are four steps in a typical water treatment program.

1. Primary settling. The water is brought into a large holding basin to allow particulate matter to settle. Lime may be added at this stage if the pH of the water is below 6.5.

2. Aeration. The clarified water is agitated with air, which promotes the oxidation of easily oxidizable substances in the water. These reducing agents would otherwise consume the chlorine or other disinfecting material to be added later in the treatment process.

A problem substance in this context is Fe^{2+}, since certain rock formations contain iron in the reduced state, for example pyrites, FeS_2, in granitic areas, or $FeCO_3$ in carbonate areas. $FeCO_3$, in particular, is modestly soluble in water, $K_{sp} = 4.0 \times 10^{-5}$ mol L^{-1} at 25°C. In the presence of atmospheric oxygen, $Fe^{2+}(aq)$ is oxidized to $Fe^{3+}(aq)$, which precipitates as $Fe(OH)_3(s)$ at any pH higher than about 3.5. Such "iron staining" leaves an unsightly brown deposit in washbasins and toilets. Iron staining is quite common with water from domestic wells; besides the staining, which fortunately can be removed easily with acid, the dissolved iron gives the water an

[1] *Chem Eng News,* November 17, **1986,** 4. For a later perspective, see P.L. Layman, *Chem Eng News,* February 23, **1987,** 7.

unpleasant metallic taste. In a municipal treatment plant, aeration oxidizes any Fe^{2+} to Fe^{3+}[2] (Problem 6).

$$\frac{-d[Fe^{2+}]}{dt} = 8 \times 10^{13}[Fe^{2+}][OH^-]^2 \bullet p(O_2) \text{ L}^2 \text{ mol}^{-2} \text{ atm}^{-1} \text{ min}^{-1}$$

This reaction is also discussed in Chapter 8 under the heading "Acid mine drainage." Aeration is not practical for the individual homeowner. One possibility is the use of special filter cartridges which contain both an oxidant and a filter. They are replaced at intervals when the oxidant is all consumed or the clogging of the filter creates excessive back pressure.

3. Coagulation. Primary settling of the raw water is not sufficient to remove the finest particles, such as colloidal minerals, bacteria, pollen, and spores. Their removal is necessary to give the finished water a clear, sparkling appearance. The commonest filter aid used in water treatment is **filter alum** $Al_2(SO_4)_3.18H_2O$. In the pH range 6–8, $Al(OH)_3(s)$ is formed.

$$Al^{3+}(aq) + 3HCO_3^-(aq) \longrightarrow Al(OH)_3(s) + 3CO_2(aq)$$

At these pH values, $Al(OH)_3$ is close to its minimum solubility (Chapter 6) and at equilibrium very little aluminum is left dissolved in the water (see Problem 1). The acidic cation $Al^{3+}(aq)$ reacts with $HCO_3^-(aq)$ and hence reduces (slightly) the alkalinity of the water. Although the amount of alum added is controlled carefully at the water treatment plant so as not to leave significant amounts of aluminum in the water, this may not be true of the domestic user who buys alum in order to clarify a cloudy water supply: see Section 7.8, aluminum as a toxic metal in water. Two incidents during 1988 and 1989 in the United Kingdom showed that accidental contamination of finished water with excessive amounts of alum is nevertheless possible, even at a municipal water treatment plant.

Aluminum hydroxide is a very gelatinous precipitate, as is evident if a dilute solution of alum in a test tube is neutralized by NaOH: a milligram or so of precipitated $Al(OH)_3$ will fill the whole test tube! Because of this property, the precipitate settles very slowly, and as it does so, it carries down with it the fine particles in the water. Secondary settling is thus an integral part of the coagulation treatment.

Other useful coagulating agents include ferric sulfate and activated silica. Ferric hydroxide is gelatinous like aluminum hydroxide, and the chemistry is analogous. The coagulating effect of activated silica is probably due mostly to the formation of gelatinous substances such as alkali metal silicates.

4. Disinfection. Disinfection is the most essential part of water treatment. Filtration and coagulation afford a material that is pleasant to look at, but it is disinfection that makes the water safe to drink. Disinfection kills any bacteria and viruses which have escaped filtration but, at least

---

[2]The oxidation of $Fe^{2+}(aq)$ by aeration is rapid at the pH of a drinking water supply, say, $pH > 6$.

as important, it prevents recontamination during the time the water is in the distribution system. In the suburbs of a large city, the water may have been in the distribution system for five days or more before it is drunk. Five days is plenty of time for these "missed" microorganisms to multiply; furthermore, leaks and breaks in the water mains permit recontamination, especially at the extremities of the distribution system where the water pressure is low.

Recontamination is a serious problem in the urban slums found at the fringes of the rapidly growing cities of the Third World, e.g., in Africa, Asia, and South America[3]. These slums usually grow faster than the city can extend its distribution system to them. The populace must depend either on contaminated ground or surface waters, or on municipally-treated water of sub-optimum quality. Reasons include the following.

- Low pressure at the outer edges of the distribution system, and many breaks because of inadequate installation, allow contamination from the ground. The high pressure in a properly installed system means that the flow through any leak is always from inside to outside the pipe. Low pressure at the fringe of the system occurs when the city grows so rapidly that the demand on the system exceeds its capacity.
- Individual homes lack their own faucets. Water is drawn from a common, often dirty, communal outlet.
- The first two problems both exist when the city water supply is broken into (illegally) at a point close to the newly developing "barrio," because the latter has not yet been supplied by the city.

Chlorine is the most commonly used disinfecting agent. Other disinfectants are **ozone, chlorine dioxide,** and **ultraviolet radiation.** Chlorine is unique in that it is the only one of the group to possess residual disinfectant activity; in other words, it maintains its protection of the drinking water throughout the distribution system. All the other disinfectants mentioned must be followed with a low dose of chlorine in order to preserve the protection.

Not all four of the basic steps will be needed in every water treatment plant. Ground waters, in particular, require much less treatment than surface waters. In Guelph, Ontario, the ground water supply needs no settling, aeration, or coagulation: it is fit to drink as obtained from the ground. A little chlorine (0.16 ppm) is added to protect it in the distribution system. At the other end of the scale, water that is contaminated with organics, both anthropogenic and natural, may require extra treatment (see later).

Before discussing municipal water treatment in detail, we should remember that even in the developed world municipal water is by no means available everywhere. Millions of people in rural areas depend on individual wells, springs or streams, for their drinking water. Surface water and

---

[3] *Surveillance of drinking water quality,* World Health Organization, Geneva, 1976, p. 63.

shallow ground water are particularly vulnerable to pollution, and the users normally consume this water without any treatment or disinfection[4]. In developed countries, public health departments usually offer free analysis of individual water, especially for bacterial content, but the citizens of poorer countries are less fortunate in this regard.

7.3 Chemistry of different disinfectants

7.3.1 Chlorine

When chlorine dissolves in water the equilibria below are rapidly established.

$$
\begin{array}{rcll}
Cl_2(g) & \longrightarrow & Cl_2(aq) & K_H = 8.0 \times 10^{-3} \text{mol L}^{-1} \text{a} \\
Cl_2(aq) + H_2O(l) & \longrightarrow & H^+(aq) + Cl^-(aq) + HOCl(aq) & K_C = 4.5 \times 10^{-4} \text{mol}^2 \text{ L}^{-2} \\
HOCl(aq) & \longrightarrow & H^+(aq) + OCl^-(aq) & K_a = 3.0 \times 10^{-8} \text{mol L}^{-1}
\end{array}
$$

Hydrochloric acid, which is completely dissociated into $H^+(aq)$ and $Cl^-(aq)$, is a product of the second reaction, and hence chlorination is a process which reduces the total alkalinity of water. From these equilibrium constants we can calculate the speciation of "Cl^+" or "active chlorine," by which is meant Cl_2, HOCl, and OCl^-. All these species are oxidizing agents; Cl^- is not. This is shown in Figure 7.2, from which we see that free chlorine molecules will only be present in solution below about pH 1 (Problem 2). They are completely unimportant in drinking water.

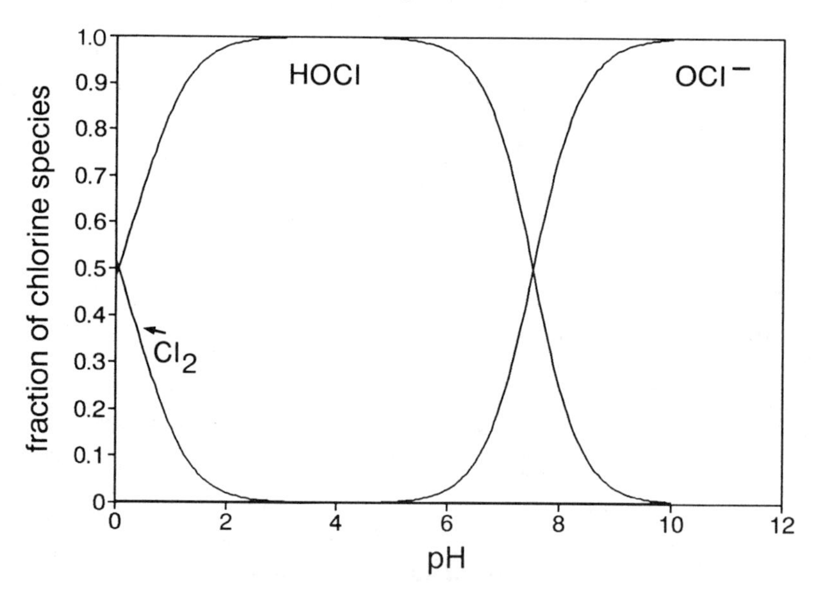

Figure 7.2: Speciation of active chlorine as a function of pH

[4] *Rural Water Supplies,* World Health Organization, Regional Office for Europe, Copenhagen, Denmark, 1983. This booklet describes processes for providing water treatment to small communities.

Of more importance in the context of drinking water is the speciation between hypochlorous acid and the hypochlorite ion. Hypochlorous acid has pK_a 7.5, so that at pH 7.5 the concentrations of HOCl and OCl^- in the solution will be equal. This is significant because HOCl is approximately 100 times more effective a disinfectant than OCl^-, so the amount of chlorine required to achieve a specified level of disinfection (or the time to achieve disinfection for a fixed dose of chlorine) depends on the pH of the water. Water at higher pH requires a larger dose of chlorine, as can be deduced from the data given in Problems 3 and 4, where 7.6 times as much chlorine must be used at pH 8.5 as at pH 7.0 to obtain the same extent of disinfection. The reason that HOCl is a stronger disinfectant than $ClO^-$is that HOCl, a neutral molecule, can penetrate the cell membranes of microorganisms much more easily than the ion ClO^-. HOCl is more destructive because ClO^- does not reach the interior of the cell.

Since some of the chlorine will have been used up in destroying microorganisms, as well as in some purely chemical reactions which will be considered shortly, the amount of chlorine in the finished water is less than the total amount that was used originally. The following terms are used by water engineers.

- **Chlorine dose**: the amount of chlorine originally used.
- **Chlorine residual**: the amount remaining at the time of analysis.
- **Chlorine demand**: the amount used up, i.e., the difference between the chlorine dose and the chlorine residual.
- **Free available chlorine**: the total amount of HOCl and ClO^- in solution.

Figure 7.3 shows the relationship between chlorine dose, chlorine demand, and chlorine residual (Problem 7). If the chlorine demand is zero, the graph is a straight line of unit slope, passing through the origin. When there is a chlorine demand, the residual stays at zero until the demand has been met, and then increases in direct proportion to the additional dose (Problem 8).

In practical terms, the chlorine is supplied as the bulk liquid under pressure (Cl_2 has a normal boiling point of $-35°$ C at 1 atm.). It is injected at a controlled rate in a large tank of water, such that the residence time of the finished water in the tank is about 20–60 minutes. A typical concentration of chlorine in the finished water is 1 ppm or less.

Drawbacks to the use of chlorine[5]

Although elemental chlorine is a cheap and effective disinfectant, there can be problems with its use. These include so-called "taste and odour" problems, and questions about the toxicity of chlorine and chlorinated byproducts.

[5]B. Hileman, "The chlorination question," *Environ Sci Technol,* **1982** 16, 15A.

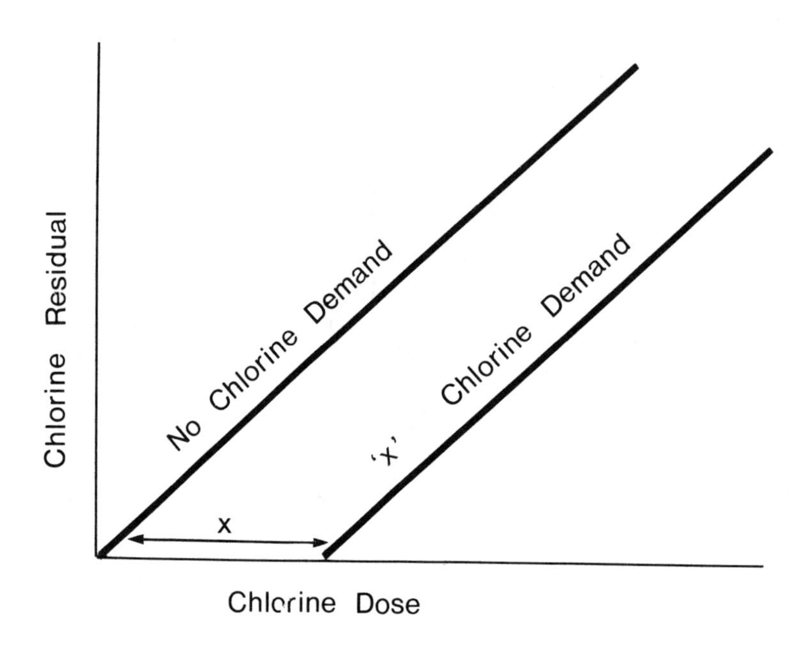

Figure 7.3: Chlorine dose, chlorine demand, and chlorine residual

Taste and odour

The commonest of these occur in industrialized areas where discharges from industry cause the water (almost always surface water) to be contaminated with phenol or its derivatives. Examples include the manufacture of certain herbicides or their precursors, the pulp and paper industry, the manufacture of phenol for use in plastics such as phenol-formaldehyde, and the manufacture and use of pentachlorophenol as a wood preservative. Phenols are the source of trouble because they are chlorinated very readily to **chlorinated phenols,** which have penetrating "antiseptic" odours. Odour thresholds for these compounds are in the ppb (μg/L) range; at the ppm level, they make the water completely unusable for drinking or cooking.

Cl substituents	Odour threshold, ppb[6]
none	> 1000
2–	2
4–	250
2, 4–	2
2, 6–	3
2, 4, 6–	> 1000

[6]Data from R.H. Burttschell, A.A. Rosen, F.M. Middleton, and M.B. Ettinger, "Chlorine derivatives of phenol causing taste and odor," *J Am Water Works Assoc.*, **1959** 51, 205.

Chlorination of phenol is an example of electrophilic aromatic substitution; phenol is so much more reactive than, say, benzene, that no Lewis acid catalyst is required, and the reaction takes place rapidly even in aqueous solution. The odour-causing 2−, 2,4−, and 2,6− chlorinated phenols are among the products because phenol is chlorinated preferentially in the ortho and para positions. In cases where phenols are present in the water supply, the options are to use a disinfectant other than chlorine, or to remove the contaminants by the use of activated charcoal. These will both be discussed later, but we may note here that activated charcoal treatment is expensive and few communities use it.

Another drawback to the use of chlorine has been much publicized in recent years, namely the formation of trihalomethanes such as $CHCl_3$. These substances do not arise by halogenation of dissolved methane, as might be supposed, since the $CHCl_3$ is not accompanied by other halogenated methanes such as CH_2Cl_2 and CCl_4, but other trihalomethanes such as $CHBrCl_2$ are formed. It is now recognized that trihalomethanes (THM's in the current vogue for acronyms) are formed by the action of hypochlorite upon the acetyl ($-C(=O)CH_3$) groups in humic acids, which are breakdown products of plant materials such as lignins. This reaction is analogous to the well known iodoform reaction in organic chemistry.

$$-C(=O)CH_3 + 3HOCl \longrightarrow -CO_2H + CHCl_3 + 2H_2O$$

Toxicity

Fears have been expressed about both chlorinated phenols and trihalomethanes as toxic contaminants in drinking water[5]. 2-Chlorophenol has been found to be fetotoxic in rats, while 2,4,6-trichlorophenol in large doses is carcinogenic in rats and mice. In practice, the latter substance has been of only minor concern in drinking water because the other chlorophenol congeners make the water so unpleasant to drink. Chloroform has been the subject of much more attention from the news media; it is a promoter, but not an initiator, of carcinogenesis in rodents. Consequently its presence in water raises the spectre of exposure to carcinogens through drinking water. In reality, these fears may have been overstated; the amounts typically present in the water are very small (a few ppb). Putting this into perspective, until very recently, many childrens' cough mixtures contained several **percent** of chloroform as a cough suppressant. Yet to put the other side of the argument, a study of Canadian chlorinated drinking water supplies by the Ames assay indicated that 1 in 4 samples had mutagenic activity. Further, epidemiological studies have shown elevated cancer levels in communities which chlorinate, compared with those which do not. However, the increases are very small and it is difficult to be sure that other confounding factors are not present.

There is no evidence that chlorine itself is carcinogenic. A seven-generation study of rats provided with drinking water supplemented with

100 ppm of chlorine showed no ill effects on breeding stock or progeny, even though the taste of chlorine is so strong at this concentration that the animals will drink the water only if no other fluids are available.

Before leaving the topic of chlorine as a disinfectant, note that the majority of municipalities still do use chlorine to disinfect drinking water. The benefits in terms of protection from water-borne diseases far outweigh the possible hazards from, for example, ten or twenty ppb of trihalomethanes. Disinfection with chlorine may (possibly) cause a miniscule risk of cancer in old age; no disinfection means a substantial chance of dying from typhoid fever or cholera as a child or young adult. Some commentators would have us switch from chlorine to something else, even though the toxicology of the alternatives and their potential byproducts has been less well studied than that of chlorine[7]. The next sections are devoted to some of these alternatives to chlorination.

7.3.2 Chlorine dioxide[8]

Chlorine dioxide ClO_2 is a very effective disinfectant, about twice as powerful as HOCl, and also about twice as expensive (Problem 11). Typical dose rates are 0.1–5 ppm in the water to be treated. Chlorine dioxide was first used as a municipal water disinfectant in Niagara Falls, New York in 1944. A 1977 survey showed that about one hundred municipalities in the United States were using chlorine dioxide, as well as thousands of European communities. A dozen or so Ontario municipalities currently use chlorine dioxide as an alternative disinfectant during periods when the water suffers from "taste and odour" problems.

A major drawback to the use of chlorine dioxide compared with chlorine is that chlorine dioxide cannot be stored; it has to be prepared and used on site, whereas chlorine can be delivered in tank cars. Chlorine dioxide is chemically unstable; it is an endothermic compound and is known to decompose explosively. It is generated by one of the two reactions below and used immediately.

$$(1) \quad 10NaClO_2 + 5H_2SO_4 \longrightarrow 8ClO_2 + 5Na_2SO_4 + 2HCl + 4H_2O$$

$$(2) \quad 2NaClO_2 + Cl_2 \xrightarrow{pH < 3.5} 2ClO_2 + 2NaCl$$

Reaction 1, the disproportionation of the ClO_2^- ion to ClO_2 and Cl^-, affords chlorine-free chlorine dioxide. Reaction 2 is less useful when ClO_2 is being used to combat taste and odour, since the product is inevitably contaminated with hypochlorous acid, thus defeating the objective of using a chlorine substitute. The precursor in either case is sodium chlorite $NaClO_2$, which is a powerful oxidizer, and has to be stored carefully.

[7] R.J. Bull, "Health effects of drinking water disinfectants and disinfectant byproducts," *Environ Sci Technol*, **1982**, 16, 554A.

[8] J. Katz, *Ozone and chlorine dioxide technology for disinfection of drinking water*, Noyes Data Corp., Park Ridge, N.J., 1980

Unlike chlorine which (as we saw in its reaction with phenol) behaves characteristically as a chlorinating agent, chlorine dioxide is an oxidizing agent. Trihalomethanes and chlorinated phenols are therefore not formed when chlorine dioxide is used. Its reactions with organic compounds generally lead to the introduction of an oxygenated functional group into the molecule. This is illustrated in its reaction with phenol. Chlorine dioxide, which owes much of its reactivity to being a (reasonably) stable free radical, is simultaneously reduced through a one-electron change to the chlorite ion.

$$ClO_2 + C_6H_5OH \longrightarrow ClO_2^- + [C_6H_5OH]^{\bullet +} \xrightarrow{H_2O} C_6H_4(OH)_2$$

Studies on the toxicology of the chlorite ion and of ClO_2 have shown that sodium chlorite causes hemolysis at 50 ppm. As a safety precaution, it is recommended[9] that finished water contains no more than 1 ppm of ClO_2 (which can therefore be reduced to no more than 1 ppm of ClO_2^-). Animal studies have shown the lowest levels of ClO_2 or $NaClO_2$ which cause observable toxic effects.

Toxicant	Observed effect	Lowest observed effect, ppm
$NaClO_2$	Red blood cell damage, –rats, 70 days –mice, 30 days	20–100
$NaClO_2$	Lower weaning rates –mice, 40 days	100
ClO_2	Red blood cell damage –mice, 30 days	>100
ClO_2	Reduced growth rates –rat pups, 40 days	100

Since chlorine dioxide leaves no residual activity in the water, it is necessary to treat the finished water after disinfection with a light dose of chlorine to maintain its protection in the distribution system. Note that any substances responsible for taste and odour in the water have already been oxidized by ClO_2, and hence are no longer present in the water by the time chlorine is added.

[9] L.W. Condie, "Toxicological problems associated with chlorine dioxide," *J Am Water Works Assoc*, June **1986**, 73.

7.3.3 Ozone[6]

The use of ozone to disinfect water dates to 1893 in the Netherlands and 1901 in Germany. Ozonation is now used in about 1000 communities in Europe. The first use in North America was in Indiana (1941). In Canada, some 18 Quebec municipalities employ ozonation to disinfect their drinking water, the largest being the city of Montreal.

Ozone is prepared by passing a high voltage electric discharge ($\approx 15,000$ V) through dry air, and then absorbing the ozone in water ($K_H = 1.3 \times 10^{-2}$ mol L^{-1} atm^{-1}). Under optimum conditions, up to about 6% of the air can be converted to ozone, although in practice, the ozonized air used to disinfect water contains about 1% ozone (Problem 10).

$$3O_2(g) + \text{energy} \longrightarrow 2O_3(g)$$

Like chlorine dioxide, ozone is a slightly more powerful disinfectant than chlorine, but a little more expensive; it cannot be stored or transported, but must be made on site. The chemistry of ozone is much like that of ClO_2, in that it is an oxidizing agent: for example, hydroxylation is the characteristic reaction with phenol. Ozone is of course free of any complications from formation of chlorinated derivatives. It decomposes rather rapidly in water, the kinetics of the reaction being pH dependent[10] (Problem 9).

$$\text{rate} = 2.2 \times 10^5 [OH^-]^{0.55} [O_3]^2 \text{ mol L}^{-1} \text{ s}^{-1}$$

Since ozone leaves no residual, post-treatment chlorination is necessary.

The equipment needed to generate ozone is expensive, and offers economy with large scale operation. For this reason, only large municipalities (e.g., the city of Montreal in Canada, and regional water authorities in Europe) tend to favour the use of ozone. Chlorine dioxide, on the other hand, may be generated using inexpensive equipment, which is why it finds application as a replacement for chlorine on occasions when taste and odour problems are experienced. By contrast, ozone will be used all the time or not at all.

7.3.4 Ultraviolet radiation

As noted in our study of atmospheric chemistry, ultraviolet radiation having wavelengths below 300 nm is very damaging to life including, of course, microorganisms. Low pressure mercury arc lamps are available having their output principally at 254 nm; they are known in fact as germicidal lamps. These lamps are very efficient, with up to 40% of their electrical input being converted to 254 nm radiation (Problem 12). The destructive effect of

[10] M.D. Gurol and P.C. Singer, "Kinetics of ozone decomposition: a dynamic approach," *Environ Sci Technol,* **1982,** 16, 377–383.

254 nm radiation is due to its absorption by DNA, which leads to photochemical reactions of the bases, particularly the pyrimidines[11].

Ultraviolet disinfection is used in some 2000 communities in Europe. Although a prototype plant was built in Marseilles, France as early as 1910, the technique was not used on a large scale until 1955, in Switzerland, by which time the necessary high-efficiency UV lamps had been developed. UV disinfection leaves no residual, and so as with ClO_2 and O_3, post-treatment with a little chlorine is required. Small scale UV units for individual use are now available for rural consumers whose well water has a high microbial count. At the other end of the scale, the world's largest UV disinfection system, in London, England, treats 14.5 million gallons of water daily.

In order for UV disinfection to be successful, the radiation must penetrate the water. Suspended matter (which scatters light), coloured material and dissolved organics (which compete to absorb the radiation) must all be absent or minimal if the UV method is to be used.

Angehrn[12] has compared UV disinfection with ozonation and chlorination. The strong points of the UV method are as follows.

- Short contact time: 1–10 s. Ozone and chlorine both require contact of 10–50 minutes, and this necessitates the construction of a large reaction tank. UV disinfection can be run on a "flow-through" basis.
- Low installation costs. Ozone generators are complex and expensive to install; chlorine metering equipment is less so. The reaction tanks mentioned above are also a high cost item. The low cost of installation has made UV disinfection attractive to some of the smaller water treatment facilities in Europe.
- Not influenced by pH or temperature. Chlorination and ozonation work best at lower pH, chlorine because more of it is in the HOCl rather than the OCl^- form; ozone because it decomposes more rapidly at high pH (Problem 9). Both the latter methods require longer contact at lower temperature.
- No toxic residues. As we have seen, this is now a matter of public concern, especially with chlorine. UV disinfection adds nothing to the water, provided that the water is initially free of organics which could be changed photochemically.

Table 7.1 is a cost comparison between the various disinfectants[13]. Note the cost effectiveness of UV for the smallest plants, its second place cost relative to chlorination for larger plants, and the extremely high cost of ozonation (associated with capital costs) for a small installation.

---

[11] G. Beddard, "Biological effects of ultraviolet radiation," in *Light, Chemical Change and Life,* Ed. J.D. Coyle, R.R. Hill, and D.R. Roberts, Open University Press, England, 1982, Chapter 5.2.

[12] M. Angehrn, "Ultraviolet disinfection of water," *Aqua,* **1984,** 2, 109.

[13] R.L. Wolfe, "Ultraviolet disinfection of potable water," *Environ Sci Technol,* **1990,** 24, 768–773.

Table 7.1: Cost of various disinfectants as a function of capacity of the water treatment plant.

Disinfectant	Plant capacity[a]	Cost[b]
Cl_2	0.05	26.0
	0.5	3.0
	1.0	1.7
ClO_2	0.05	52.5
	0.5	11.1
	1.0	8.6
O_3	0.05	72.5
	0.5	9.6
	1.0	6.6
UV	0.05	21.9
	0.5	7.2
	1.0	5.3

[a] in millions of US gallons per day [1 US gal = 3.8 L]
[b] in US cents per 1000 US gallons

7.3.5 Analysis of residuals

Iodometric titration is a simple and reliable method. The reactions are given below. Note that although ClO_2 functions as a one electron oxidant in its reactions with organics (see above), it is a five electron oxidant with I^-, with the chlorine changing from the +4 oxidation state in ClO_2 to the −1 oxidation state in the reaction product Cl^-.

$$HOCl + 2I^- + H^+ \longrightarrow I_2 + Cl^- + H_2O$$
$$ClO_2 + 5I^- + 4H^+ \longrightarrow \tfrac{5}{2}I_2 + Cl^- + 2H_2O$$
$$O_3 + 2I^- + 2H^+ \longrightarrow I_2 + O_2 + H_2O$$

In each case, the iodine liberated is titrated against standard sodium thiosulfate (Problem 5).

7.4 Organics in drinking water

Prospective drinking water supplies may become contaminated with organic compounds in many ways. Some of these are natural, as in the decay of biological materials; others are the result of human activities which cause contamination by natural substances (food processing and meat packing plants; manure from feedlots on farms) or by synthetic compounds (insecticides and herbicides used in agriculture; seepage from unsecured municipal waste dumps or industrial waste dumps; sewage outflows; water used for cooling purposes in industry; and even equilibration of organic air pollutants between lakes and the atmosphere). Several of these issues are taken up in more detail in other chapters.

The presence of organics in drinking water came to prominence in the 1960's with the development of gc/ms (gas chromatography combined with mass spectrometry), which made possible the detection and quantitation of organic compounds in water at the sub-ppm level. In a now-classic study carried out in the late 1960's, the lower Mississippi River, from which the city of New Orleans takes its drinking water, was found to be contaminated with literally hundreds of organic compounds, and these contaminants were carried through the water treatment process into the finished drinking water[14]. Public complaints about the poor taste of the water were instrumental in initiating this study. These were some of the substances detected:

- **Herbicides:** Alachlor, butachlor, atrazine, cyanazine, propazine, simazine.
- **Insecticides:** Chlordane, heptachlor, dieldrin, endrin, DDE (a DDT metabolite).
- **Industrial organics:** Alkylbenzenes such as toluene, xylenes, ethylbenzene; alkanes such as decane through pentadecane; naphthalene and methylnaphthalenes; alkyl phthalates (used as plasticizers); chlorinated methanes and ethanes, especially $CHCl_3$; chlorinated benzenes and chlorinated phenols; benzaldehyde; dicyclopentadiene. Of these compounds, the hydrocarbons are typical of oil refinery operations, and the chlorinated benzenes and phenols are associated with herbicide and insecticide production.

Studies carried out on water from other major waterways such as the lower Great Lakes (Lakes Erie and Ontario) and the Rhine and Danube Rivers have shown similar findings. In Lake Ontario, the presence of Mirex has been of particular public concern, while in more recent studies of Rhine water, public attention has focussed more on PCB's and dioxins. More detail on these substances will be provided in Chapter 9.

These trace analyses are carried out by first separating the volatile and non-volatile organics. Volatile substances are recovered first, using equipment similar to that shown in Figure 7.4. The water sample is transferred from the collection flask A to the sparging bottle B by displacement by means of ultra-pure helium. Ultra-pure helium is then bubbled through the water sample, volatilizing the volatile organic fraction. The gas stream exits the flask and passes through a short column of solid having a high affinity for organics. When the volatilization is complete, this column is removed and is then attached to the front end of a gas chromatography

[14]L.H.Keith *et al*, "Identification of organic compounds in drinking water from thirteen United States cities" in *Identification and analysis of organic pollutants in water*, Ed. L.H. Keith, Ann Arbor Science, 1976, Chapter 22.

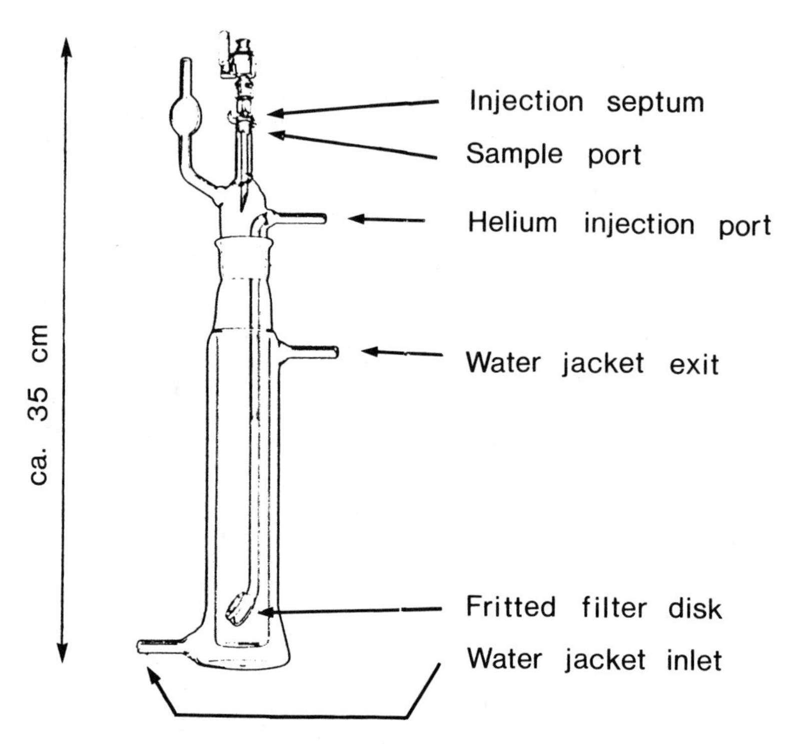

Figure 7.4: Apparatus for stripping volatile organic compounds from water. Reproduced from Reference 11, Chapter 6.

column. The volatile organics desorb from the "pre-column" upon heating and pass into the gas chromatographic (gc) column, where they separate and are identified by mass spectrometry (ms).

Non-volatile organics are extracted with an organic extraction solvent such as hexane. The hexane extract is then concentrated to a small volume in order for the solutes to be present at a detectable level, and the concentrated sample is analyzed directly by gc/ms.

A very important consideration in either of the analyses is that the results not be confounded through inadvertent contamination. The helium used for sparging and the hexane used for extraction must both be ultra-pure; since relatively large volumes of each are used, even a tiny amount of impurity would be concentrated to a level which would interfere with the analysis of the sample. Ordinary plastic tubing cannot be used because the alkyl phthalates present as plasticizers would be extracted from the tubing and contaminate the sample. Likewise, great pains must be taken to make sure that the collection bottles and the laboratory glassware do not contain contaminants that will interfere with the analysis. This is checked by running frequent "blank" analyses, in which the whole procedure is carried out except that no water sample is collected. A whole service industry has grown up dedicated to the provision of these ultra-pure materials, contaminant-free detergents for washing glassware, and reference standards having guaranteed concentrations of stated pollutants for calibration of the instruments.

The presence of anthropogenic—and potentially toxic—organic chemicals in lakes and rivers used as sources of drinking water raises the issue of possible adverse health effects. In the Great Lakes, which supply drinking water to many millions of people, concern has arisen because of the discovery of cancers in some of the fish taken from these waters. The concern follows from three specific observations: fish in the wild are only rarely cancerous; Great Lakes fish seem to have abnormally high incidence of cancer; toxic chemicals may be detected in Great Lakes water and in the fish taken from them.

It is an unproven—but worrisome—point that drinking water taken from the Great Lakes might expose neighbouring human populations to hazardous amounts of these chemicals. However, as with many environmental issues, cause and effect are not clear-cut. First, the exposure of a person who drinks the average two liters of treated Great Lakes water each day is very different from the exposure of the fish which actually lives in the water, and which presumably becomes equilibrated with the substances in it. A second point concerns the hazard of ingesting trace amounts of chemicals which are known to be toxic, but only at much higher levels. This point may be illustrated by means of a dose-response curve (Figure 7.5), which is constructed to show the effects on the test organisms of defined doses of toxicants. For aquatic toxicants, a commonly used protocol is the 96-hour test on immature rainbow trout.

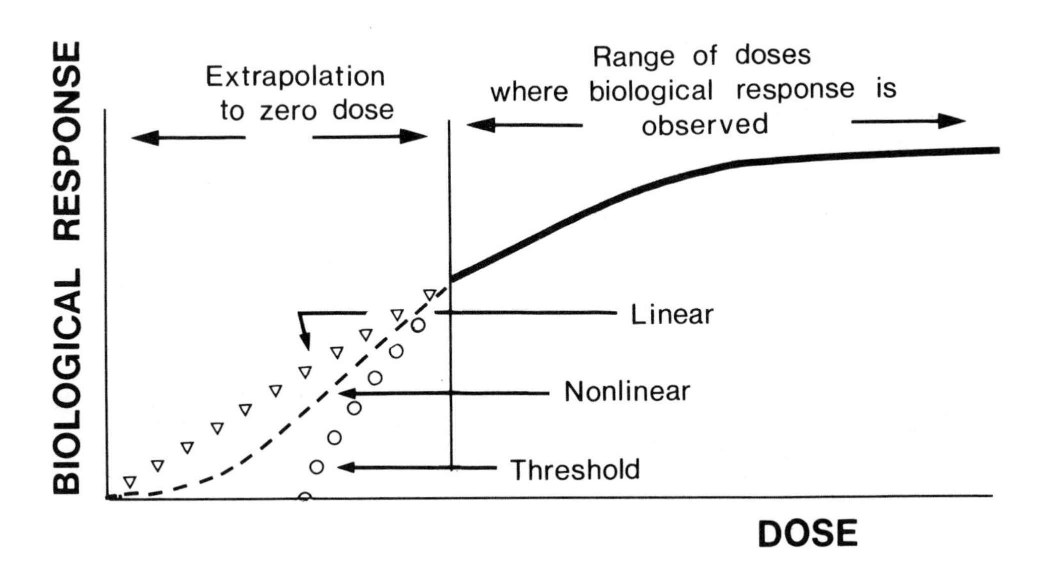

Figure 7.5: Sketch of a typical dose-response curve.

Unfortunately for scientific studies, the concentrations that are of interest in environmental toxicology are almost always those that come so close to the origin of the dose-response curve that no effect is observed in the laboratory (the dotted region of Figure 7.5). This presents a difficulty. Does the dose-response curve extrapolate linearly to the origin, or is there a "no effect" dose below which the toxicant has no deleterious effect? This question is controversial and not yet resolved. A linear relationship suggests that low levels of toxicants in our drinking water would cause harm to at least a few members of the human population, while a threshold would imply less cause for alarm, at least for the human population. Even so, we should still take very seriously the matter of whether human activities are polluting the environment to the point where other species—the fish in this case—are being endangered.

The foregoing discussion raises for the first time in this book the matter of toxic, or more specifically carcinogenic, chemicals in the environment. In the United States, the topic is usually discussed under the title of the "Delaney Amendment," which was introduced in 1960 to regulate the addition of colouring agents to foodstuffs. Foodstuffs may not be offered for sale if they contain any quantity of a colorant which has been shown to cause cancer in animals or humans. A recent court challenge to the amendment made the argument that there may be harmless levels of such substances—i.e., levels at which no harmful effects can be detected. This issue has arisen because analytical capability has now advanced to the point of the routine detection of parts per billion (1 in 10^9), parts per trillion (1 in 10^{12}), and even parts per quadrillion (1 in 10^{15}) in some analyses. Such concentrations were undetectable in 1960. The well known Canadian analytical chemist Dr Walter Harris has noted that we can probably now detect and quantitate **any** analyte of interest in any sample we choose. Thus legislators in all countries face this dilemma: we can probably detect any selected carcinogen in any substance destined for human consumption if we look hard enough; we do not know how toxicological dose-response curves behave at very low doses; so how should legislation as to the permissible levels of such substances be framed? Concerning the court challenge just mentioned, a 1987 ruling of the United States Supreme Court upheld the status quo by reaffirming the Delaney clause, namely that **no** amount of such substances is permissible[15]. Although this sidesteps the problem, the alternative of setting permissible levels for each and every carcinogen is simply not possible with present knowledge.

Before leaving the subject of problem organics in drinking water, we should note that not all problem chemicals are anthropogenic. The natural substance geosmin is an example of a particularly troublesome natural contaminant.

Geosmin, a monoterpene, is elaborated by *Actinomyces* species which occur in soil. The odour of geosmin is the smell of newly dug earth or

[15] J. Long, *Chem Eng News,* November 9, **1987,** 23.

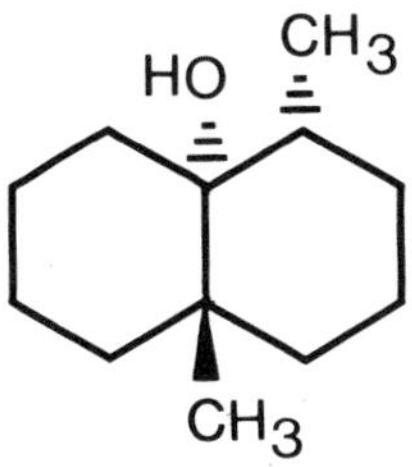

Geosmin

that of beets cooking; it is not a flavour which is desired by the consumers of drinking water. Because geosmin is a tertiary alcohol, it is resistant to oxidation, and survives unchanged during conventional water treatment, including the action of substances such as chlorine or ozone. The city of Regina, Saskatchewan has had a longstanding problem with geosmin, and other natural foul-smelling substances. Regina takes its water from a rather shallow lake (Buffalo Pound) some 100 km away, and the off taste of the water has been an annual summer event associated with algal blooms in the lake. The city installed an activated carbon treatment facility in 1985, in an attempt to resolve the difficulty[16].

Activated carbon can be used for the removal of both natural and anthropogenic organic compounds responsible for causing taste and odour problems. Because it is expensive, very few communities use it, although Angehrn[12] recommends activated carbon pretreatment of water that is to be disinfected by UV irradiation. The technique involves passing the partly finished water over a bed of carbon which has been prepared (or activated) by partly burning wood chips in a limited supply of air, affording material with a very large surface area. Absorption of organics by charcoal may be familiar from the organic chemistry laboratory, where charcoal is often used to remove coloured impurities from organic compounds. Of course, the activated carbon has only a finite capacity to adsorb impurities from the drinking water; when it is "spent" it may be reactivated, although with some loss, by further partial burning. This drives off and burns away the adsorbed organics.

Organic chemicals may also be removed from water by packed-tower aeration, in which air is used to strip organics from water, taking advantage of the equilibration between the dissolved and the gaseous phase. This technique only works, of course, if the substances in question are volatile. The method depends upon the Henry's law equilibration of the solute between the aqueous and gas phases. In this context, it is now recognized that public exposure to volatile organic compounds (VOC's) can occur not only by

[16] L. Gammie and G. Giesbrecht, "Operation of full scale granular activated carbon contactors for removal of organics," in *Treatment of drinking water for organic contaminants,* Ed. P.M. Huck and P. Toft, Pergamon, Elmsford, N.Y., **1987,** 67.

consuming drinking water but also through volatilization in confined areas such as shower stalls (compare comment on radon, Section 4.2.1). Other technologies under consideration for removing VOC's from domestic water are ozone oxidation, ultraviolet radiation (both of which can be combined with disinfection), ultrafiltration and reverse osmosis[17].

7.5 Other water treatments

Many parts of the world lack sufficient sources of potable water. The Middle East is one such region, where a growing population lives in an area which is largely desert. Such wells and waterholes as exist are often brackish, that is, rather high in salt and hence unpalatable. Therefore there is much interest in methods for the desalination of both these brackish waters and of seawater.

An obvious method of purifying salty water is by distillation. Because of the large energy requirement ($\Delta H°$ of vaporization for water = 44 kJ mol^{-1} at 25°C), the cost is high, about $4 per 1000 U.S. gallons (3800 L). The technology is practical in Middle Eastern countries where energy costs are low. Another promising technology is **reverse osmosis** which can be best appreciated by thinking first of conventional osmosis, which is familiar from biology. If a dilute solution and a concentrated solution are separated by a semipermeable membrane (which allows water molecules to pass through it, but not other solutes), water will flow from the dilute solution into the concentrated one. Under these conditions, a substantial difference in height between the two water columns may be established. The difference in pressure exerted on the semipermeable membrane by the two columns is called the osmotic pressure.

In conventional osmosis, the natural tendency of two solutions, initially of differing concentrations, to equalize their concentrations leads to a difference in pressure across the semipermeable membrane. In reverse osmosis, pressure is applied externally to one side of a semipermeable membrane which separates two solutions, initially of equal concentrations. Water flows through the membrane from the high pressure side to the low pressure side, establishing a difference in concentration of the solutions on either side of the membrane. Hence the technique of reverse osmosis consists of forcing a high-salt solution such as seawater through a semipermeable membrane; purified water passes through the membrane, leaving a more concentrated solution behind[18]. This technology is in use in the Bahamas, to provide potable water from a brackish lake, and in parts of the Middle East where the wells are brackish. The direct production of potable water from seawater by a single pass through a semipermeable membrane is now possible.

[17] For discussion, see R.M. Clark, C.A. Fronk, and B.W. Lykins, *Environ Sci Technol* **1988**, 22, 1126.

[18] "Prognosis on reverse osmosis," *Environ Sci Technol*, **1977**, 11, 1052.

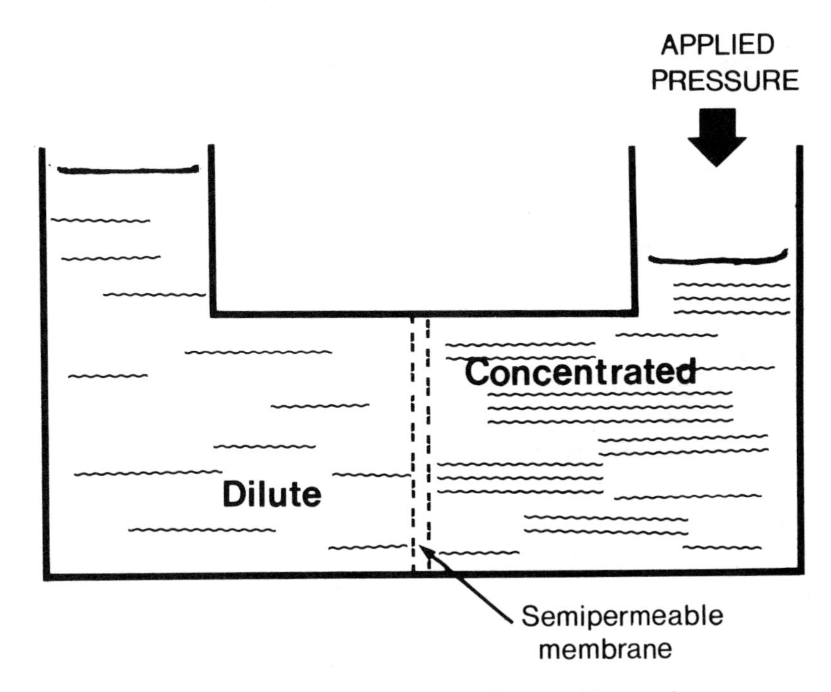

Figure 7.6: Osmotic pressure across a semipermeable membrane

7.6 Metals in drinking water

Iron has already been cited as a potential nuisance in water. It stains bathroom fixtures and gives an unpleasant "metallic" taste to the water, but it is rarely present in amounts that might be toxic since, upon oxidation, it precipitates from solution. Other metals which are of concern for their toxicity are **lead, cadmium, mercury,** and **aluminum.** Sources of the heavy metals lead, cadmium, and mercury in raw water include dissolution of these metals from the underlying rock, where they are usually present as their sulfides, from the use of water that has been contaminated by sewage or industrial effluents, or by leaching from unsecured dump sites. High **sodium** concentrations are suspected of being responsible for promoting cardiovascular disease, and can render the water unpalatable.

7.6.1 Lead

Lead in drinking water is to be avoided because lead is toxic, and has a long residence time in the body (see Chapter 10). Its long residence time causes it to be considered a cumulative poison. A feature of lead toxicity is that whereas adults typically retain less than 10% of the lead they ingest, the unborn fetus absorbs 50%, the proportion dropping as a child grows older. Lead can cross the blood-brain barrier, again in the order fetus > child >

adult, leading to mental retardation in severe cases, and IQ deficiency or behavioural problems at lower levels.

An important origin of lead in drinking water is the use of lead plumbing[19]. Nevertheless, some exposure to lead occurs even with copper piping because of the use of lead solder to join the pipes. Some years ago, a recall of electric kettles was prompted by the discovery of large amounts of lead in the solder inside them. Lead salts were formerly used to glaze pottery drinking vessels. The lead may be leached by acidic liquids such as fruit juices. A 1990 news report mentioned that wines and spirits can leach ppm levels of lead from crystal decanters upon long storage.

The word plumbing comes from the Latin word for lead; the Romans first introduced lead piping as a means of distributing water. Roman houses had indoor plumbing and hot water systems as long as two thousand years ago—some of these systems still survive. Lead is much easier to work and to repair than the copper piping that is used nowadays. Anyone living in an older home should check to determine whether the water system, or more likely part of it, is made of lead.

In hard water areas, lead plumbing causes few problems, since a deposit of scale ($CaCO_3$) quickly covers the surface of the lead piping and the water does not actually come in contact with the metal. Soft waters, which are low in minerals, do not provide this protection. Soft waters also tend to be more acidic, promoting the dissolution of the somewhat electropositive lead ($E^\circ = 0.12$ V). In addition, water from the "hot water" side of the system is likely to have a higher lead content than the "cold side" because: (i) more rapid dissolution will occur at higher temperature and (ii) scale tends to deposit in the hot water heater, leaving the hot water pipe without the protection of the scale. Householders with lead piping should not use the hot water tap for drinking or cooking, and should make a particular point of not using, even from the cold tap, the "first draw" water (the water that has been standing in the pipes overnight). Instead they should run the water for one or two minutes in the morning in order to flush this more contaminated water from the system.

A survey in the United States Northeast showed that 9% of the population was using drinking water having a lead concentration higher than the recommended upper limit of 50 μg of lead per litre. Those at risk were people not using municipal water, especially those with cisterns to collect rainwater, or shallow dug wells. In extreme cases (first draw water, lead piping, very acidic water) lead concentrations as high as 3 mg/L were recorded, 60 times the recommended limit. Clinical signs of retardation were evident in children born to women who used "first draw" water having $[Pb] > 0.8$ mg/L during pregnancy. Even for homes supplied with municipal water, it is quite common for 30 μg/L of lead to be found in older homes, with three times this in the first draw water. On account of this concern about lead, a

[19] M.E. McDonald, "Acid deposition and drinking water," *Environ Sci Technol* **1985**, 19, 772.

reduction in the allowable concentration of lead in United States drinking water from the present 50 μg/L has been proposed[20].

7.6.2 Cadmium and Mercury

Like lead, cadmium and mercury[19] are cumulative poisons. These metals generally contaminate the raw water, rather than being introduced from the piping system, as was discussed for lead. The aforementioned survey in the U.S. Northeast indicated that 2% of the population was exposed to cadmium concentrations greater than the recommendation of 10 μg/L, and that 22% were drinking water contaminated with more than the standard of 2 μg/L of mercury. The latter statistic originates both from the lower recommended limit for mercury, and from the widespread environmental contamination by mercury used in industry (see Chapter 10).

7.6.3 Aluminum

Until recently, aluminum has generally been considered to be innocuous in drinking water and in the diet. However, the current situation regarding the toxicology of aluminum is extremely confused[21]. Although aluminum is very abundant in the Earth's crust, it appears to have no natural biochemical function. Aluminum is kept out of the body by three barriers: the gastrointestinal tract, serum binding to transferrin (this can interfere with iron transport, however), and the blood-brain barrier[22].

Allegations have been made that patients suffering from renal failure are susceptible to aluminum toxicity. Anemia (interference with iron transport), softening of the bones (replacement of calcium phosphate by aluminum phosphate), and a form of senile dementia can result. As well, a link to Alzheimer's Disease (also a form of senile dementia) has been proposed, with the brains of dementia patients showing characteristic "tangles" of neurons and "hot-spots" of unusually high aluminum concentration; whether the aluminum is a cause of the disease or simply a side effect is not known. However the Aluminum Association, an industry organization, claims that this connection could not be substantiated[23].

Several studies have been reported in recent years in an attempt to determine whether there is a predisposition towards Alzheimer's disease among populations whose diets are high in aluminum, but so far the results are not clear cut. One British study showed an increase in the incidence

[20] *Chem Eng News,* August 8, **1988,** 23.

[21] S. Krishnan, "The case against aluminum," *Canadian Research,* March **1988,** 32–35; A. Prescott, "What's the harm in aluminum?" *New Scientist,* January 21, **1989,** 58–62.

[22] K. Thompson, "The aluminium controversy," *Chem in Britain,* May **1989,** 448.

[23] Press kit obtained from The Aluminum Association, 900 19th St., N.W., Washington, D.C., 20006, USA.

of Alzheimer's disease in parts of the country where the water naturally contained elevated levels of aluminum, but the effect was small[24].

The typical dietary intake of aluminum has been estimated at 22 mg/day in North America[19]. The highest concentration of aluminum measured in a survey of 200 drinking water systems was found to be 2.7 ppm. At an estimated 2 L of water drunk per day, this amounts to 5.4 mg of aluminum, or less than 25% of the average daily intake. As shown in Chapter 6, the solubility of $Al(OH)_3$ is much less than 2.7 ppm over the pH range 6–8 that is normal for drinking water (Problem 14); the explanation for these higher concentrations of aluminum is most likely the use of unnecessarily large amounts of filter alum as a clarifying and coagulating agent. Putting the matter in perspective however, buffered aspirin tablets contain as much as 35–200 mg of aluminum per tablet; no evidence has yet been presented that long term use of this medication is a health hazard. Most likely, healthy individuals eliminate dietary aluminum rather than absorb it.

Aluminum cookware is another potential source of aluminum in the diet[25]. Aluminum metal is very electropositive, but despite this, it does not usually dissolve easily because of a tightly held protective film of aluminum oxide. High concentrations (up to 100 ppm) of aluminum can be detected in fruits and fruit juices cooked in aluminum cookware. The combination of high $[H^+]$ and high [citrate] is responsible, since aluminum forms strong complexes with citrate ion.

7.6.4 Sodium

Sodium ion may naturally be present in drinking water if the supply is brackish. Above about 300–400 ppm of sodium chloride, the water is unpalatable on account of its chloride content, and special purification measures such as reverse osmosis may be required. Naturally soft water frequently contains sodium rather than the "hardness" cations calcium and magnesium, while water softened artificially in domestic water softeners has the hardness cations replaced by sodium (Chapter 5). Even though drinking water is only one source of dietary sodium, it is recommended that those who use domestic water softeners retain one outlet of unsoftened water for drinking. The use of sodium salts by industry and the practice of using sodium chloride as road salt de-icer contribute substantially to the pollution of surface and ground waters by sodium[26].

---

[24] "Aluminum in water puts kidney patients at risk," *New Scientist,* January 21, **1989,** 28.

[25] R.B. Martin, "The chemistry of aluminum as related to biology and medicine," *Clin Chem,* **1986,** 32, 1797.

[26] *Sodium, Chlorides, and Conductivity in Drinking Water,* World Health Organization, Copenhagen, Denmark, 1979; *Health Effects of the Removal of Substances occurring naturally in Drinking Water, with special reference to demineralized and desalinated water,* World Health Organization, Copenhagen, Denmark, 1979.

Excessive intake of sodium has been statistically associated with hypertension (high blood pressure) and cardiovascular disease. For this reason the United States National Academy of Sciences has recommended that the sodium content of drinking water be kept below 100 ppm of sodium. Controversy has surrounded the relationship between high sodium levels in drinking water and blood pressure. Research to date suggests that young children and adults are probably affected in this way, but teenagers are probably not[27]. Glasgow, Scotland is an area where deaths from cardiovascular disease are statistically high, and the water is naturally high in sodium.

7.7 Nitrates in drinking water

Concern has arisen about the presence of nitrate ion in drinking water. Agriculture is usually the major source of this contaminant, through manure seepage from feedlots, seepage from the holding tanks used to contain liquid manure from intensive hog production, and excessive use of fertilizer. The higher crop yields obtained today compared with a generation ago are largely due to increased use of chemical fertilizers. However, low crop prices combined with high land and machinery costs encourage farmers to cultivate fields right up to their margins, thus promoting run-off from fields to waterways[28]. The prevalence of nitrates in ground water has become an issue in countries such as Britain and the Netherlands, where high population densities are combined with intensive agriculture.

The suggestion has been made in Britain that it is unnecessary to protect the quality of each and every water supply. The logic of the argument is that people drink about 2 L of water daily, but use many times this quantity of water to wash clothes, dishes, and cars, flush toilets, run showers, and water lawns. Yet the proposal is a mischievous one, because if society accepts contamination of natural water by nitrate, other contaminants will surely follow.

Nitrate in drinking water is of concern because of its toxicity, especially towards young children. Actually, it is not the nitrate ion itself that is toxic, but rather the nitrite ion NO_2^-, which is formed from it by the reducing action of intestinal bacteria, notably *Escheriscia coli*. In adults, NO_3^- is absorbed high in the digestive tract before reduction can take place. In infants, whose stomach are less acidic, *E. coli* can colonize higher up the digestive tract and reduce the nitrate before it is absorbed.

Nitrite ion is toxic because it can combine with hemoglobin. The resulting complex between hemoglobin and nitrite ion is called **methemoglobin**;

[27] See several chapters in *Inorganics in drinking water and cardiovascular disease* Volume 9 of *Advances in modern environmental toxicology,* Princeton Scientific, Princeton, NJ, 1985. Proceedings of a conference held in 1984 at the University of Massachusetts.

[28] T. Addiscott, "Farmers, fertilizer and the nitrate flood," *New Scientist,* October 9, **1988,** 50.

the association constant for methemoglobin formation is larger than that for oxyhemoglobin formation, and so the nitrite ion ties up the hemoglobin, depriving the tissues of oxygen. Severe cases of **methemoglobinemia** can result in mental retardation of the infant.

H.H. Comly (1945) was the first person to draw a link between nitrates in drinking water and the incidence of methemoglobinemia. Research shortly afterwards showed that no cases of methemoglobinemia had been reported in any area of the United States where the water supply contained less than 45 ppm of nitrate ion. This value has become accepted as the upper limit for the nitrate concentration in drinking water. However, Winneberger[29] has questioned this standard, pointing out that in other studies no relationship between nitrate concentration in the water and methemoglobinemia has been found. He suggests that the two issues may be unconnected.

Although this chapter has drinking water as its focus, it may be noted here that for most consumers their nitrate/nitrite (daily intake ca. 95 mg) comes principally from foodstuffs. Vegetables tend to concentrate nitrate ion, especially if they are grown with the assistance of fertilizers high in nitrate. Lettuce, spinach, and celery contain up to 700 ppm wet weight of NO_3^-. Cured meats, notably bacon, contain nitrate as a curing agent and nitrite as a preservative to arrest the growth of *Clostridium botulinum* (which elaborates the highly toxic botulism toxin). The maximum amounts of these additives are regulated in most countries (e.g., 500 ppm of $NaNO_3$ in the U.K.)[30]. Because of the methemoglobinia issue, the addition of nitrates and nitrites to baby foods is now generally prohibited.

At stomach pH, nitrite ion is converted to $H_2NO_2^+$, which is capable of nitrosating secondary amines and secondary amides. The significance of this is that the resulting N–nitrosamines may be carcinogenic: the prototype compound N–nitrosodimethylamine (or dimethylnitrosamine) is carcinogenic in many animal species, although neither it nor any other nitrosamine has been proven to be a human carcinogen. Returning to the topic of drinking water, dimethylnitrosamine was in the news in February 1990, when the water supply in Elmira, Ontario was found to be contaminated with parts-per-trillion levels. A tire factory in the town was restrained from discharging water containing the nitroso compound. However, it is unclear whether the plant was the only, or major, source of contamination of the water supply, in that microbial action in sewage treatment plants produces the same substance.

---

[29] J.H.T. Winneberger, *Nitrogen, Public Health and the Environment,* Ann Arbor Science, 1982, Chapter 1.

[30] For discussion, see *The health effects of nitrate, nitrite, and N-nitroso compounds,* National Academy of Sciences, Washington, D.C., 1981; also C. Glidewell, "The nitrate/nitrite controversy," *Chem in Britain,* February **1990,** 137–140.

7.8 Fluoridation of drinking water[31]

No other public health issue approaches fluoridation in spurring passionate debate over the question, "to fluoridate, or not to fluoridate?" It was discovered in the 1930's that certain areas of the United States had an unusually low incidence of tooth decay (dental caries) and that these same regions had an unusually high concentration of natural fluoride ion in the water. The chemistry behind the argument in favour of fluoridation is relatively simple. Tooth enamel is composed of a mineral called **hydroxylapatite**, $Ca_5(PO_4)_3OH$, and because hydroxylapatite contains the basic anions OH^- and PO_4^{3-}, its solubility increases in acidic solutions. (Few things rot your teeth faster than excessive consumption of cola drinks, which have low pH to begin with and whose sugar is converted by bacterial action into organic acids.) Tooth decay occurs when the tooth enamel is damaged, allowing the entry of bacteria which destroy the material underneath.

Fluoride ion can replace the hydroxide ion in hydroxylapatite, yielding a new mineral, fluorapatite, which is intrinsically less soluble[32].

$$F^-(aq) + Ca_5(PO_4)_3OH(s) \longrightarrow Ca_5(PO_4)_3F(s) + OH^-(aq)$$

The suggestion was therefore made that it would be advantageous to add fluoride ion artificially to the water supply in areas where natural fluoride was low or absent, in order to reduce the incidence of dental caries, especially among children. The first community to pursue this course was Grand Rapids, Michigan in 1945. Today about half of all North Americans drink artificially fluoridated water, and in addition, pastes and solutions containing fluoride ion are routinely applied to patients' teeth during visits to the dentist's office. A level of 1 ppm in drinking water is usually recommended (Problems 13 and 15); higher concentrations lead to mottling of the teeth (dental fluorosis). The latter effect is seen at as little as 3–5 ppm; i.e., there is little "safety factor" from the doses used in drinking water. In Ontario, for example, drinking water may be supplemented with up to 1.2 ppm of fluoride, but must be rejected if the F^- concentration exceeds 2.4 ppm[33].

Opponents of fluoridation are likely to make the following arguments.

- Fluoride salts are poisonous. Sodium fluoride can be purchased as a rat poison. We should not be adding known poisons to our water. (Editorial comment: as Paracelsus noted four centuries ago, it is the dose that makes the poison. Just because sodium fluoride is toxic at high doses does not make it dangerous to ingest in small doses.)

[31] B. Hileman, "Fluoridation of water," *Chem Eng News,* August 1, **1988,** 26.

[32] The solubility properties of fluorapatite are interesting. No Ksp for this material has been reported, because dissolution from the crystal surface leaves the surface with a different composition from the bulk: D.R. Simpson, *Am Minerol,* **1969,** 54, 1711.

[33] *Drinking Water Objectives,* Ontario Ministry of the Environment, Toronto, Ont., 1976, 11.

- People should not be medicated against their will. Maybe fluoridation is a good idea, but those who do not want to be medicated cannot remove it from their water. On the other hand, fluoride preparations may easily be purchased by those who wish to medicate themselves or their children.

Until recently, the statistical (but not necessarily the ethical) case in favour of fluoridation seemed very convincing. The incidence of dental caries, especially among children, has declined dramatically during the forty years since fluoridation was first introduced, suggesting that a cause and effect relationship existed between fluoridation and healthier teeth. However as in many cases where a statistical correlation exists between two events (here, fluoridation and tooth decay), the one is not always a consequence of the other. While it is true that the incidence of dental caries has declined in North America, comparable declines have been found both in communities where the water is fluoridated and in those where it is not[34]. This suggests that we cannot definitively assess the value of fluoridation as a public health measure.

Finally, the question arose in 1990 as to whether fluoride ion might be carcinogenic in rodents, when four male rats drinking water with high levels of fluoride developed osteosarcoma, a rare bone cancer. An epidemiological study by the U.S. National Cancer Institute indicated that osteosarcoma in human males has increased over the past two decades, but could not link this trend to fluoridation[35].

7.9 Drinking water standards

Each jurisdiction has its own standards. Those quoted in Table 7.2 are the recommendations of the Canadian federal government, but they are similar to those approved elsewhere[36].

---

[34] "New studies cast doubt on fluoridation benefits," *Chem Eng News,* May 8, **1989,** 5–6.

[35] E. Marshall, "The fluoride debate: one more time," *Science,* **1990,** 247, 276–277. B. Hileman, "Use of fluoride given clean bill of health," *Chem Eng News,* February 25, **1991,** 6.

[36] *Guidelines for Canadian Drinking Water Quality,* Ministry of Supply and Services, Ottawa, Canada, 1979, 74.

Table 7.2: Drinking water standards.

Substance	M.A.C., ppm[a]	Substance	M.A.C.,ppm
Inorganics			
aluminum	0.2[b]	arsenic	0.05
cadmium	0.005	chloride	250.
copper	1.0	cyanide	0.2
fluoride	1.5	iron	0.3
lead	0.05	manganese	0.05
mercury	0.001	nitrate	45.
selenium	0.01	silver	0.05
sodium	lowest practical	sulfate	500.
uranium	0.02	zinc	5.0
Organics			
Aldrin/Dieldrin	0.0007	DDT	0.03
Lindane	0.004	Parathion	0.035
phenols	0.002	trihalomethanes	0.35

[a] M.A.C. = maximum acceptable concentration
[b] EEC standard

Further reading

1. G.E. White, *Handbook of chlorination,* Van Nostrand Reinhold,New York, 1972.
2. J. Katz, *Ozone and chlorine dioxide technology,* Noyes Data Corp., Park Ridge, New Jersey, 1980.
3. V.L. Snoeyink and D. Jenkins, *Water chemistry,* Wiley, New York, 1980.

7.10 Problems

1. Filter alum is used to coagulate the solids in a water sample having $[HCO_3^-] = 2.6 \times 10^{-3}$ mol L^{-1} (take $[CO_3^{2-}]$ as insignificant). The filter alum is used at a rate of 1.0 kg per 1.0×10^5 L of water. Calculate

   (a) the mass of $Al(OH)_3$ formed

   (b) the $Al^{3+}(aq)$ concentration of the water thus treated.

2. (a) Using the equilibrium constants in the text, calculate the concentrations of Cl_2, HOCl, H^+, and OCl^- when pure water is treated with 100 ppm of Cl_2.

   (b) Repeat this calculation for a hard water sample, taking the original water to contain 185 ppm of $Ca(HCO_3)_2$ as its only solute. Assume $[H_2CO_3] = 1.0 \times 10^{-5}$ mol L^{-1}.

3. (a) A water sample has pH 6.58 and total alkalinity 8.5×10^{-4} mol L^{-1}. Calculate its total alkalinity after 8.3 ppm of Cl_2 has been added.

   (b) Explain why the disinfecting power of the 8.3 ppm of Cl_2 would be different if the pH of the water were adjusted to pH 8.58 with NaOH prior to chlorination.

4. Calculate the dose of chlorine that would be required at pH 8.5 to achieve the same level of disinfection that would result from the use of 1.0 ppm of chlorine at pH 7.0.

5. Three 1.00 L water samples contain 1.0 ppm each of HOCl, O_3, and ClO_2 respectively. Each is acidified and treated with excess KI, and the I_2 liberated is titrated against 1.27×10^{-3} mol L^{-1} $Na_2S_2O_3$. What titer is expected for each sample?

6. (a) Calculate the rate of oxidation of 1.5 ppm of $Fe^{2+}(aq)$ by atmospheric oxygen over the pH range 5–8.

   (b) How does the half life of Fe^{2+} change with pH?

7. The figure below relates titratable chlorine to disinfecting power of different chlorinating agents.

   (a) Deduce the relative germicidal efficiencies of HOCl, OCl^-, and NH_2Cl.

   (b) Three water supplies have chlorine residuals of 3.1 ppm. They contain respectively HOCl (pH 6), OCl^- (pH 8.5) and NH_2Cl (pH 8.5). Are they all equally efficiently disinfected? Explain.

   (c) Do the data support the statement that for each of these agents (separately) disinfection is a process that is first order in titratable chlorine

   i.e. rate $= k$[titratable chlorine][microorganisms].

8. A chlorination facility is built to the specification that the residence time of the water in the chlorination tank should be 25 minutes, and that 2.0×10^6 liters per hour of finished water can be produced.

   (a) At what rate should chlorine be injected into the tank if the finished water is to have a chlorine residual of 1.2 ppm? Assume

   (i) zero chlorine demand

   (ii) a chlorine demand of 0.44 ppm.

   (b) How large a chlorination tank will be required?

9. The rate of decomposition of ozone in water is given by

$$\text{rate} = 2.2 \times 10^5 [OH^-]^{0.55} [O_3]^2 \text{ mol L}^{-1} \text{ s}^{-1}$$

The equilibrium solubility of ozone is given as

$$S = \frac{\text{mg } O_3 \text{ per L (aq)}}{\text{mg } O_3 \text{ per L (g)}} = 0.41 \text{ at } 20°C$$

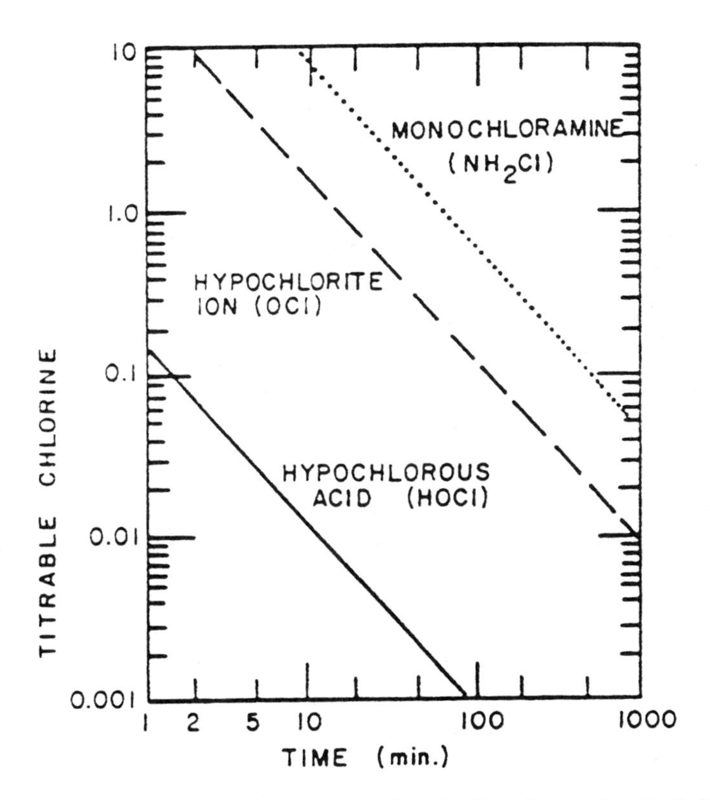

Figure 7.7: Figure reproduced from *Chemical and nonchemical disinfection,* N.P. Cheremisinoff, P.N. Cheremisinoff, and R.B. Trattner, Ann Arbor Science, 1981, p. 26.

(a) A water sample at 20°C and pH 7.55 is equilibrated with $O_3(g)$ at 2.6×10^{-3} atm partial pressure. Calculate the initial rate of decomposition of ozone as the water leaves the ozonization chamber.

(b) How long will it take for the concentration of $O_3(aq)$ to fall to 1.0 ppm?

10. Dry air at 1.0 atm is passed over an electric discharge, converting 0.85% of the oxygen to ozone. The ozonized air is then equilibrated with water at 25°C in a vessel which contains 12.0 L of ozonized air and 1.00 L of water. Calculate the concentration of ozone in the water at equilibrium.

$$K_H = 1.3 \times 10^{-2} \text{ mol L}^{-1} \text{ atm}^{-1}.$$

11. Chlorine dioxide has the following thermodynamic properties:

	ΔH_f° kJ mol^{-1}	S° J mol^{-1} K^{-1}
$ClO_2(g)$	102.5	256.7
$ClO_2(aq)$	74.9	164.9

(a) Calculate the Henry's Law constant for ClO_2 in water at 10°C

(b) What is the concentration of $ClO_2(aq)$ (in ppm) in equilibrium with 10 ppm of $ClO_2(g)$?

12. A 1.5 kW germicidal lamp converts electricity to 254 nm photons with 37% efficiency. Calculate the photon output of the lamp.

13. Calculate the volume of a 0.1000 mol L^{-1} sodium fluoride concentrate needed to fluoridate 5.0×10^5 liters of water to a fluoride concentration of 0.90 ppm. (The raw water analyzes for 0.15 ppm of fluoride.) Also express your answer in terms of the factor by which the concentrate should be diluted with the raw water.

14. Review Problem 10, Chapter 5 on the solubility of $Al(OH)_3(s)$, and calculate over the pH range 6.5–8.5 the concentration of dissolved aluminum in equilibrium with $Al(OH)_3(s)$ in pure water.

15. Calculate the concentration of dissolved aluminum in equilibrium with $Al(OH)_3(s)$ in the presence of 1.0 ppm of fluoride ion (a) at pH 7.00 (b) in a fruit juice at pH 4.00. The successive association constants for the complexation of the first two fluoride ions to $Al^{3+}(aq)$ are 2.5×10^6 and 1.6×10^5 L mol^{-1} respectively.

Chapter 8

Sewage and Waste Disposal

Introduction

Two topics make up this chapter: the return of "used" water to the environment, and the disposal of non-aqueous wastes (other than the special problems associated with the treatment of chlorinated wastes such as PCBs and dioxins: these are treated separately in Chapter 9).

With good management, sewage and other waste waters can be discharged without adverse effect on the quality of the receiving rivers and lakes. The treated water must be acceptably free of organic and inorganic toxic substances and of pathogenic microorganisms, and have a low enough BOD (Chapter 5) that aquatic life in the receiving water body is not threatened. Unfortunately, these standards may not be attained; industrial discharges imperil both wildlife and human sources of drinking water, while the contamination of drinking water supplies by sewage is a leading cause of diseases such as cholera and typhoid fever, especially in less developed countries.

Other wastes can be categorized as aqueous and non-aqueous industrial wastes—of which some are better described as hazardous wastes—and solid wastes, which include municipal garbage. We will discuss landfilling, and the methods of solidification of liquid wastes which render them safe for disposal by landfilling.

8.1 Sewage treatment

Even today, a large fraction of the world's sewage is discharged into rivers and oceans without any treatment whatever. As cities grow, the volume of sewage to be disposed of increases correspondingly. The example of Cairo, Egypt shows the magnitude of the problem[1]. Treatment facilities date from the time when the city's population was one-third of its present 12 million; much of the sewage flows directly into the Nile through open ditches. About $1.5 billion is being spent to develop a system which will serve the current

[1] J. Bedding, "Money down the drains," *New Scientist,* April 15, **1989,** 34–38.

population, but the population growth rate is such that the new system will be at capacity the day it is finished.

Lest we think that this is only a "Third World" phenomenon, we should remember that the city of Montreal, population ca. 2 million, has only recently constructed a municipal sewage treatment facility. Hitherto, human waste has been discharged directly into the St. Lawrence River. The need for sewage treatment facilities cannot be overemphasized, recognizing the location of many of the world's cities on major rivers such as the Rhine, Danube, Mississippi, Yangtze, and other rivers, and the multiple usage of this water for drinking, industry, transportation, and recreation.

Complicating the design of a sewage treatment plant for a major city is the issue of whether separate sanitary and storm sewer systems should be provided. Storm sewers take the run-off from city streets; sanitary sewers accept the waste water from homes and from industry. Additionally, certain industries with a high demand for cooling water may discharge used cooling water directly to lakes and rivers (see thermal pollution, Chapter 5). When the storm and sanitary sewer system is combined, much greater capacity must be provided at the sewage treatment plant to accommodate the extra flow following heavy storms or during snow melt. Should the load exceed the capacity of the plant on such occasions the surplus, comprising mixed storm water and sanitary sewage, may have to be discharged without treatment.

Sewage treatment may involve the following three phases.

1. Primary settling
2. Secondary treatment, and the related problem of disposal of sewage sludge
3. Tertiary, or advanced, treatment. We shall cover phosphate removal in most detail, and digress to include the related topic of soaps and detergents.

Before release to a river, the treated sewage is usually clarified ("polished") by passing it through a sand filter.

8.1.1 Primary settling

The sewage enters a lagoon, or clarifier, whose capacity is large enough to allow a residence time of several hours in the lagoon (Problem 1). A coarse screen at the entrance to the clarifier removes large objects such as pieces of wood, tree branches, and similar debris. A grit tank ahead of the main clarifier allows the deposition of grit, sand, and like material. Typically, the sewage enters and leaves at opposite ends of the lagoon, and moves through slowly enough that any solid particles settle out. Some greasy material may float to the surface, where it is removed by a skimming device. The effluent from the primary settler is almost clear, but has a high BOD (of the order of 500—1000 ppm). Solids removed at the primary stage are sent for landfilling.

Advanced primary treatment[2]

"Advanced" primary treatment has been put forward as an alternative to the combination of primary and secondary sewage treatment. The tiny particles which fail to separate in primary settling are the major source of BOD in the liquid entering the secondary stage. Experiments in the late 1970's showed that the use of coagulants such as filter alum or ferric chloride on this liquid could reduce its BOD almost as much as conventional secondary treatment. Since that time further advances to the technology have centred on the use of mixtures of charged ionic synthetic polymers and $FeCl_3$ to accomplish this goal.

The city of San Diego, California currently uses advanced primary treatment. Proponents of the technology cite reduced capital construction costs, compared with secondary treatment plants, and less sludge production (the inorganic/polymer/particulate sludge is actually less than the biomass accumulated in the conventional process). Opponents point to the higher coliform count of the effluent, and the undesirability of adding further chemical substances to the sewage stream.

Complicating the issue is a 1972 U.S. law—the Clean Water Act, passed before the design of advanced primary treatment—which requires U.S. cities which discharge sewage into waterways to install secondary treatment systems. San Diego is clearly in violation of the letter—even if not the spirit—of this legislation, and is being sued by the U.S. EPA to force the introduction of secondary treatment facilities.

8.1.2 Secondary treatment

The objective in secondary treatment is to reduce the BOD, perhaps from the range of 800 ppm to around 80 ppm, a 90% reduction (Problem 2). There are two common approaches. The **trickling filter** is a large round bed of sand and gravel, with coarse gravel on top and successively finer layers beneath. A rotating boom sprinkles the bed with water from the primary settler. The gravel bed quickly becomes colonized with microorganisms which use the carbon compounds in the water as an energy source. A well maintained trickling bed may remain operational for several decades. Its chief threat is the presence of toxic substances, e.g., from industry, which would kill the microorganisms.

Disadvantages of trickling filters are that they require a lot of space—an important consideraton where land is expensive—and that biological activity is much reduced at low temperatures—a problem during winter months in many countries. An alternative method of secondary treatment is the **activated sludge reactor,** which requires less land and which, being enclosed, can be maintained at the optimum temperature for biological activity. The reactor is a large tank in which the waste water is agitated and

[2] "Mud-slinging over sewage technology," News item in *Science*, **1989**, 246, 440–442; see also subsequent correspondence page 1374 of the same volume.

aerated to provide the oxygen required by the microorganisms (as opposed to the trickling filter where oxygen diffuses in naturally from the air). In order to keep the concentration of microorganisms high and in the maximal growth phase, some of the sludge of microorganisms is continuously removed. Following secondary treatment in an activated sludge reactor, a second period of settling allows solid material to precipitate. Alum (Chapter 7) may be used to assist settling.

Sewage sludge

Both primary and secondary sewage treatment involve settling of particulate matter, and thus produce sludge. The term sludge refers to a material having a high—more than 95%—water content, even when it has been air dried. (It looks solid enough, but you can't get it on a shovel!) Dewatering is aided by heating (digesting) the sludge, which causes the small particles coagulate; the analogous process is used in the laboratory and in industry to coagulate precipitates which are too fine to filter easily. Digestion is anaerobic, and is accompanied by the release of reduced gases such as methane, often in quantities sufficient to be worth collecting for use as a fuel to heat more of the sludge. The dewatered sludge, which now contains coarser particles, can be air-dried to a material that is recognizably solid. It may be disposed of by land-filling, by incineration, by ocean dumping, or by spreading it on the land for use as a fertilizer. The extent to which these methods are practised varies from country to country.

Sewage sludge is rich in organic matter, and includes compounds of nitrogen and phosphorus. These attributes make it an attractive material to use as a fertilizer and soil-conditioner, since its high organic matter content can compensate for the loss of natural organic material which occurs when land is used for intensive crop production. One innovative approach has been to use large amounts of sewage sludge (hundreds of tonnes per hectare) to remediate land which has been severely disturbed, such as areas where mining spoil has been dumped, abandoned strip mines, and the soil caps on municipal landfill sites, when they are full. In these cases, generous doses of sludge can transform barren terrain into an aesthetically pleasing landscape of grass or trees. In such situations, the organic matter in the sludge is particularly important for allowing the soil to retain moisture.

One of the great attractions of using sewage sludge as a fertilizer is to provide a method of disposal, and it has traditionally been free for the price of hauling it away, either as the raw sludge—to be applied as a liquid—or as the dewatered material which is about 50–60% water by weight[3]. The sludge is relatively low in fertilizer value (about $20 per tonne on a dry

---

[3] *Guidelines for sewage sludge utilization on agricultural lands,* Ontario Ministry of Agriculture and Food, revised 1986; see also M.D. Webber, *Phosphate fertilizer and sewage sludge use on agricultural land—the potential for cadmium uptake by crops,* Environment Canada Report EPS 4–WP–79–2, 1979.

weight basis[4], compared with $240 per tonne for bulk 15:15:15 fertilizer), so that it is not economic to haul it great distances. This constrains its use to the immediate vicinity of the point of production.

A potential drawback to the use of sewage sludge as a fertilizer is the presence of both organic and inorganic toxic substances. The former are the more oxidation-resistant organic substances such as organochlorine compounds which may be present in the raw sewage. These survive treatment, and become bound—non-covalently—to the organic matrix of the sludge. The inorganic toxicants are metals such as arsenic, cadmium, lead, mercury, and zinc, a proportion at least of which emanate from industrial operations which discharge their wastes into the sewerage system. These metals may be taken up by crops and thus be introduced into the food chain, or they may leach from the soil and contaminate ground water. In order to avoid excessive contamination of the soil, strict limits are placed in most countries on the amounts of sludge which may be applied to fields used for crop production. In addition, waiting periods between sludge application and harvest are required to avoid the transfer of pathogens.

Over a five year period, sludge from a Danish town showed the following average composition[5] on a dry matter basis:

Fertilizer elements, %:	N, 4.2	P, 1.8	K, 0.26	
Trace metals, ppm:	Cr, 70	Ni, 40	Cu, 510	Zn, 2800
	Ag, 4	Cd, 5	Hg, 3	Pb, 400

Of these elements, chromium is not considered to pose a threat to human or animal health, since it is not concentrated by plants, and nickel toxicity has been observed only on highly acidic soils. Copper is rarely a problem—although sheep are very susceptible to copper poisoning, and so fertilizing pastures with sewage sludge demands a longer waiting period for sheep than for cattle. Potential problems relate to zinc, which is toxic to many crop plants but only at higher levels. Cadmium, silver, mercury, and lead are all very toxic: cadmium is one of the elements of most concern. It has no known biological function, and the normal North American diet contains cadmium at 70–90% of the World Health Organization's recommendation of 70 μg per person per day. Recent evidence suggests that heavy metals may significantly reduce the yields of leguminous crops because of their toxicity towards nitrogen-fixing bacteria[6].

The rate of movement of metals through soil depends strongly on how tightly the metal binds to the soil, which in turn hinges upon the chemical nature of the soil. Sandy and gravelly soils are highly porous to both water

[4] This value is courtesy of Dr. Tom Bates of the Ontario Agriculture College, University of Guelph.

[5] Data in this section are taken from "Utilization of sewage sludge on land: rates of application and long term effects of metals," proceedings of a conference held in Uppsala, Sweden, 1983, D. Reidel Publishing Co., Dordrecht, Holland, 1984.

[6] K. Giller and S. McGrath, "Muck, metals and microbes," *New Scientist,* November 4, **1989,** 31–32.

and dissolved metals. They do not retain metals, but their porosity allows metals applied to them to contaminate ground water. "Muck" soils, which are very high in organic matter, and which are ideal soils for growing vegetables, bind metals but the association is relatively loose; uptake into the crop can occur if the concentrations of metals become high. Clay soils bind metal ions most tightly; movement of metal ions through clays is very slow (of the order of only cm/yr). The low mobility can be understood by reference to the structures of clays, which are complex aluminosilicates, with the aluminum/silicon/oxygen backbone arranged in sheets. These sheets are sometimes cross-linked to form a three dimensional structure. As with the zeolites (Chapter 5), clays can act as ion exchangers, with the backbone carrying a net negative charge which is balanced by the presence of interstitial cations. Trace metals introduced with the sewage sludge are scavenged, replacing Na^+, K^+, Ca^{2+}, Mg^{2+} etc. from the interior of the clay[7].

The availability of metals to a crop depends also on the pH of the soil. Heavy metals have insoluble hydroxides and carbonates, and become more soluble as the acidity increases. Liming the soil increases the pH and hence reduces the availability of these metals. A further advantage of liming clay soils is that high $[H^+]$ inside the clay limits the exchange of heavy metals into the clay.

The availability of metals from sludge depends on the character of the sludge as well as that of the soil to which it is applied.[8] The sludge itself binds cations sufficiently tightly that they are not in an "available" form. As a result, the guidelines which have been promulgated in most countries to regulate the amounts of sludge which can legally be applied to agricultural land may offer a greater margin of safety than had been previously supposed.

In summary, the following considerations govern the safe use of sewage sludge as a fertilizer.

- The amount of sludge which may be safely applied varies with the soil type.

- Agricultural soil that is fertilized with sewage sludge should be analyzed regularly for build-up of toxic metals. Such an analysis should consider the availability of the metals, rather than their total concentrations.

- Sludge should never be applied to a growing crop, otherwise the crop will absorb any toxic materials before they have been immobilized by the soil.

[7]For more on the interaction of metals with soils, see L.J. Evans, "Chemistry of metal retention by soils," *Environ Sci Technol,* **1989,** 23, 1046–1056.

[8]A.L. Page, T.G. Logan, and J.A. Ryan, *Land application of sludge,* Lewis Publishers Inc., Chelsea, Michigan, 1987.

8.1.3 Tertiary treatment

Most cities in the developed world employ both primary and secondary sewage treatment. There are several different kinds of tertiary treatment, each designed to reduce the concentration of a specific substance or group of substances in the sewage. Consequently, not all the following processes are practised at any one plant. Although tertiary treatment is far from universal, phosphate reduction is now in widespread use, and will be the main focus of this section.

Phosphorus

The addition of phosphorus, in the form of phosphate anion, to natural waters is associated with **eutrophication,** or accelerated aging of lakes. The two chief sources of anthropogenic phosphate in the environment are sewage—as a result of the use of phosphates in detergents—and agriculture. Agricultural run-off carries phosphate fertilizers from fields into drainage ditches and drainage tile; run-off also occurs as a result of spreading manure—especially in winter when the ground is frozen—and of seepage of animal wastes from feedlots.

Figure 8.1: Algal bloom in Lake 226 in the Experimental Lakes Area, Ontario. The lake was dammed with a temporary barrier. The far basin, fertilized with P, N, and C, was covered with an algal bloom within 2 months. No increases in algae were seen in the near basin, which received similar quantities of N and C but no P.

The natural life span of any lake is finite. Silt washes in from streams and rivers; biomass is produced in the lake, settling out to form sediment. The lake gradually becomes shallower, until it forms first a marsh and then dry land. Depending upon the size and depth of the original lake, these natural processes occur with a time-span of thousands, or tens of thousands, of years.

Under conditions of eutrophication, excessive levels of nutrients cause drastically increased populations of algae during the summer. The algae form floating mats, also called "blooms." When the algae die, their decomposition places a greatly increased BOD on the lake, and the attendant reduction in the dissolved oxygen concentration may compromise other aquatic life in the lake. At the same time, the increased deposition of sediment accelerates the natural aging of the lake.

In natural waters, phosphorus is usually the "limiting nutrient" i.e., algal growth is limited by the supply of this nutrient but not, for example, by the supply of carbon or nitrogen. The following graphs, which refer to lakes in Oregon[9], show clearly a linear correlation between algal growth and the concentration of phosphate in the water, but no such link with the concentrations of carbon or nitrogen (Problem 3)

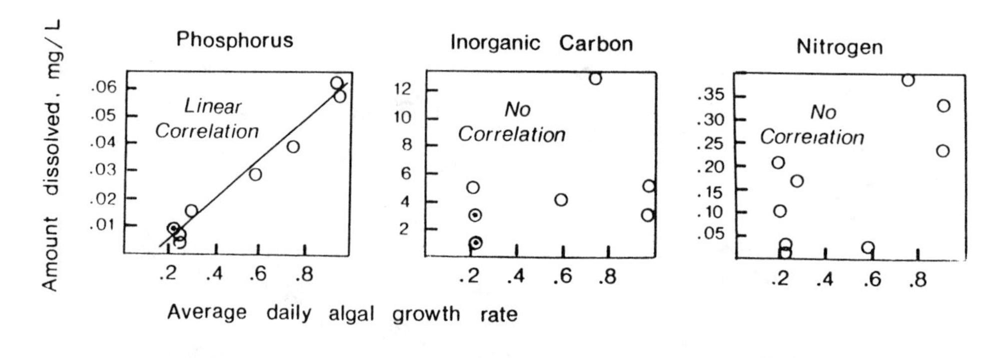

Actually, the aqueous concentrations may be misleading: lakes which apparently have carbon as the limiting nutrient are often limited in practice by phosphorus. This apparent inconsistency arises because carbonate can be resupplied by atmospheric CO_2. Likewise, it has been found that in lakes apparently limited by nitrogen, nitrogen-fixing blue-green algae can increase the available nitrogen concentration. The uptake of nutrients into biomass occurs in the approximate ratio C:N:P = 100:15:1, but as the graphs show, phosphorus concentrations in natural waters are usually so much lower than those of carbon and nitrogen that phosphorus can be the limiting nutrient,

[9]G.E. Likens, Ed. "Nutrients and eutrophication, the limiting nutrient controversy," Proceedings of the symposium on nutrients and eutrophication, Michigan State University, 1971, 139. For more on eutrophication, see J. Emsley, "The phosphorus cycle" in *The handbook of environmental chemistry,* Ed. O. Hutzinger, Volume 1A, Springer-Verlag, Berlin, West Germany, 1980.

even though only 0.01 times as much phosphorus as carbon is needed for growth[10].

In the 1960's very large amounts (50% and more by weight) of inorganic phosphates were added to domestic detergents as "builders" (see Section 8.1.4 on soaps and detergents). A high proportion of these phosphates was carried through the sewage treatment process and discharged with the finished sewage into waterways. Measures taken in the past two decades to alleviate the formation of algal blooms include tertiary treatment of sewage to remove phosphate, and limiting the amount of phosphate which may be included in detergent formulations. For example, the Canadian government limited the phosphate content (expressed as percent P_2O_5 by weight) to 20% in 1970 and to 5% in 1973. The average phosphorus content of raw Ontario sewage (i.e., what arrives at the treatment plant) dropped from nearly 10 mg phosphorus per liter in 1969 to 5.2 mg/L in 1974[11]. In the Great Lakes basin today, phosphate contamination from agriculture is probably the greater problem, especially in Lakes Huron and Erie[12] (Problem 4). Tertiary treatment of sewage can therefore solve only part of the phosphate loading problem; fortunately however, the unit cost of removing phosphorus from this source is many times less than preventing seepage from feedlots and run-off from fertilized land.

One of the chief phosphate materials in detergents is sodium tripolyphosphate[13] (STP) $Na_3H_2P_3O_{10}$. The term polyphosphate refers to linear phosphate anhydrides while metaphosphates are their cyclic analogues. Both linear and cyclic phosphates hydrolyze to monophosphate (also called orthophosphate) in the course of sewage treatment. This reaction is analogous to the biochemical hydrolysis of ATP to ADP and inorganic phosphate. For example, the first step in the hydrolysis of STP can be represented by equation [1].

$$\text{(1)} \quad \text{HO–PO}_2\text{–O–PO}_2\text{–O–PO}_2\text{–OH}^{3-} + \text{H}_2\text{O} \longrightarrow \text{HO–PO}_2\text{–O–PO}_2\text{–OH}^{2-} + \text{HO–PO}_2\text{–OH}^{-}$$

The species $\text{HO–PO}_2\text{–OH}^-$ is recognized as another way of writing $H_2PO_4^-$.

Raw sewage typically contains 5–15 ppm of phosphorus, both monophosphate and polyphosphate. Phosphorus is removed[14] in tertiary treatment

[10] D.W. Schindler, "Evolution of phosphorus limitation in lakes," *Science,* **1977,** 195, 260–262.

[11] S.A. Black, "Experience with phosphorus removal at existing Ontario municipal wastewater treatment plants," Chapter 13 of *Phosphorus management strategies for lakes,* Eds. R.C. Loehr, C.S. Martin, and W. Rast, Ann Arbor Science, Ann Arbor, Michigan, 1980.

[12] N.A. Berg, "Control of phosphorus from agricultural land in the Great Lakes basin," Chapter 19 of previous reference.

[13] STP is sometimes called sodium triphosphate instead of sodium tripolyphosphate. Be sure to distinguish it from trisodium phosphate Na_3PO_4.

[14] V.L. Snoeyink and D.L. Jenkins, *Water Chemistry,* Wiley, New York, **1980,** Section 6.8.

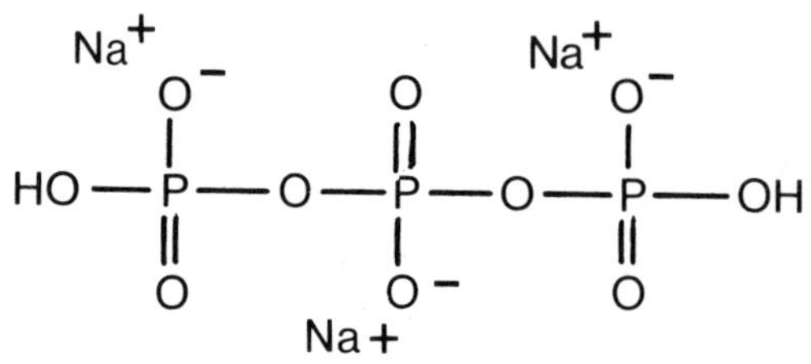

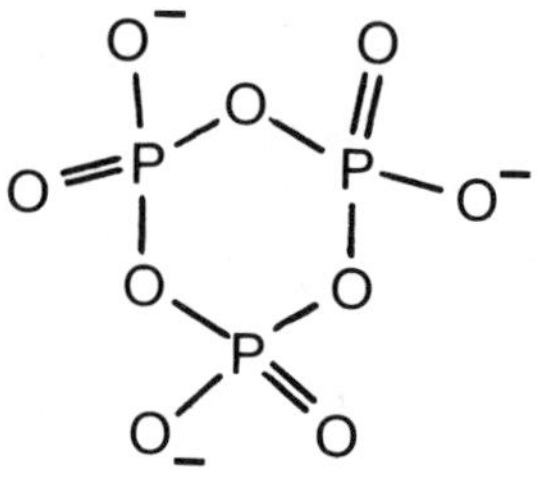

sodium tripolyphosphate (STP) trimetaphosphate anion

by precipitation with lime, $Ca(OH)_2$, which also raises the pH, usually to about 9. Low pH is to be avoided because phosphate $PO_4^{3-}$is a basic anion, and so phosphate salts become more soluble the lower the pH, equation [2].

$$(2) \qquad PO_4^{3-} \xrightarrow{H^+,\ pK_a 12.3} HPO_4^{2-} \xrightarrow{H^+,\ pK_a 7.2} H_2PO_4^-$$

The chemistry of calcium/phosphate systems is complex. Besides $Ca_3(PO_4)_2$, which has $K_{sp} = 1 \times 10^{-24}$ (mol $L^{-1})^5$, there are the much more soluble[15] $CaHPO_4$ and $Ca(H_2PO_4)_2$, and the less soluble hydroxylapatite $Ca_5(PO_4)_3OH$, which has $K_{sp} = 1 \times 10^{-56}$ (mol $L^{-1})^9$. In thermodynamic terms, solutions containing Ca^{2+} and PO_4^{3-} ions should precipitate hydroxylapatite, the least soluble solid phase. In practice, an amorphous phase corresponding more closely to $Ca_3(PO_4)_2$ is formed first, and this only slowly transforms to the less soluble hydroxylapatite. Precipitation of hydroxylapatite can be stimulated if seed crystals of preformed hydroxylapatite are present, but even then, the process requires many hours. For this reason, phosphate removal is optimized if the waste stream is partly recycled through the reaction tank, so that preformed hydroxylapatite is present at all times (Problem 5).

Even natural waters are not necessarily at equilibrium with respect to hydroxylapatite. For example, consider the Oregon lakes referred to earlier. They had phosphate concentrations in the range 0.01–0.06 ppm. Typical concentrations of Ca^{2+} in natural waters are 10–100 ppm. Thus, these natural waters were not at equilibrium with respect to hydroxylapatite, otherwise the phosphate concentration would have been much lower, and eutrophication of lakes due to phosphate would not normally be a problem (Problem 6).

[15]The higher solubility of $CaHPO_4$ and $Ca(H_2PO_4)_2$ is why rock phosphate, which consists of $Ca_3(PO_4)_2$ and hydroxylapatite, is treated with sulfuric acid in the manufacture of fertilizer. Untreated rock phosphate is too insoluble to release its phosphate fast enough for plant uptake and growth.

Phosphate can also be removed from sewage or other waste waters by precipitation with either Al^{3+} or Fe^{3+}; a lower pH is needed compared with precipitation using lime.

$$(3) \qquad M^{3+}(aq) + PO_4^{3-}(aq) \longrightarrow MPO_4(s)$$

This equation somewhat oversimplifies the situation because in both cases the metal phosphate is actually a hydroxyphosphate $M_x(OH)_y(PO_4)_z$. Recall from Chapter 7 that both iron and aluminum can precipitate from aqueous solutions as their hydroxides. Careful pH control is essential if it is the phosphate rather than the hydroxide which is to precipitate. In the case of aluminum phosphate, the waste water is treated with filter alum $Al_2(SO_4)_3$ as a source of $Al^{3+}(aq)$. The minimum solubility of aluminum occurs near pH 6.5, while—as discussed above for the calcium phosphates—the solubilities of all phosphates increase with increasing acidity, since the phosphate anion is basic. In practice, there is a pH "window" near pH 5 where the least soluble phase in the $Al^{3+}/PO_4^{3-}/H_2O$ system is $AlPO_4$, whose K_{sp} has the value 1×10^{-21} (mol $L^{-1})^2$. Above about pH 5.4, $Al(OH)_3$ is the least soluble component of the system, and hence phosphate removal is impracticable (Problems 7 and 8). For iron as precipitating agent, the optimum pH for phosphate removal is lower.

When iron, in the form of $FeCl_3$ or $Fe_2(SO_4)_3$, is used to precipitate phosphate an additional chemical reaction arises at the stage of sludge digestion which, as described above, is an anaerobic process. Unlike aluminum, iron has two common oxidation states, Fe^{2+} and Fe^{3+}. During sludge digestion, the initially formed iron (III) phosphate undergoes reduction, and the material which is finally formed has a composition close to $Fe_3(PO_4)_2$, whose K_{sp} is 8×10^{-34} (mol $L^{-1})^5$.

The foregoing approaches to phosphate removal are chemical. As such, they require expenditures for the chemicals needed for precipitation and disposal, for example by land-filling, of the chemical phosphate sludge which is produced. An alternative strategy is **biological phosphate removal**, in which the phosphorus is incorporated into the biomass of microorganisms. Biological phosphate removal makes use of essentially the same biological processes as those which cause eutrophication in the environment at large, but under controlled conditions within the sewage treatment plant. In activated sludge plants, part of the phosphorus load becomes fixed in the biomass of the microorganisms, and hence becomes part of the sludge (this accounts for the fertilizer value of the sludge). It is found that the phosphorus-to-carbon ratio corresponds to the removal of 1 ppm P for every 100 ppm reduction in the COD of the sewage ($P/C = 0.01$). If, as is common, P/C for incoming sewage is ca. 0.03, only one-third of the phosphorus can be removed even if the COD of the effluent is reduced to zero, which is unattainable in practice. Food industry wastes have P/C close to 0.01, and from these, biological phosphate removal can be very efficient. Some removal of phosphate into sludge will occur even if biological phosphate removal is not consciously practised; optimization of the conditions, notably the times during which the sludge process is aerobic and anaerobic, can improve the efficiency of the process[16]. Mixed biological/chemical treatments are also in operation.

Further BOD reduction

Water leaving the secondary sewage treatment has BOD in the range of 50 ppm of O_2, mostly associated with very fine particles. Two methods in use to reduce the BOD below this level are **microstraining** and **coagulation.**

Microstraining involves forcing the water through very fine (μm) screens made of stainless steel. The particles are trapped on the screens, and have to be removed by back-flushing whenever the pressure needed to filter the water becomes excessive.

Coagulation (compare advanced primary treatment) involves the use of filter alum, and hence the same principles that are used in the clarification of drinking water (Chapter 7). A gelatinous precipitate of $Al(OH)_3(s)$ brings down with it any particles suspended in the water. This process is most effective in the pH range 6–7, where the solubility of aluminum is near its minimum. Since alum is also useful in the pH range 4.5–5.5 for the removal of phosphate, coagulation and phosphate removal can be combined, with suitable adjustment of the pH.

Approaches to discharging sewage after secondary treatment have been described by Hocking[17]. The receiving water may be vigorously aerated in order to counter the BOD of the added sewage. A passive method for achieving this objective has been used at Red Deer, Alberta, where the receiving water, the Red Deer River, is subject to very variable flow. The system is illustrated in Figure 8.2.

When the river flow is heavy (e.g., spring run-off), the water simply flows over the weir, and the treated sewage presents no significant BOD burden. When the river is low (summer and winter: in winter it is also ice-covered), water flows through the U-tube. This pulls in air through the aspirator, as in a laboratory aspirator. The water downstream is thereby aerated to a level which will not injure aquatic life.

Disinfection

In the 1960's and early '70's, a number of municipalities installed facilities to treat finished sewage with a heavy dose of chlorine ("superchlorination") to kill any remaining pathogens before discharging the water. This procedure served also to oxidize ammonia formed by microbial reduction during

[16] M. Florentz and J. Sibony (Eds.), "Enhanced biological phosphorus removal from wastewater," *Water Sci Technol,* **1985,** 17: special issue on the proceedings of a conference in Paris, France, 1984.

[17] M.B. Hocking, *Modern chemical technology and emission control,* Springer-Verlag, Berlin, West Germany, 1985, 86.

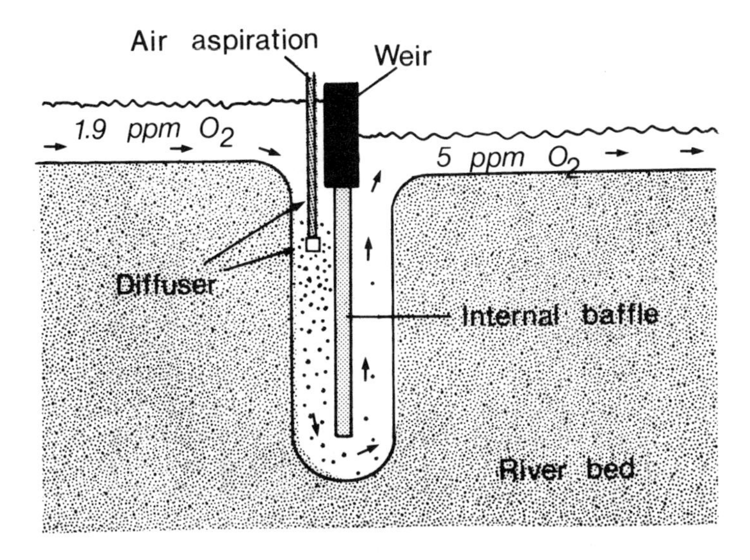

Figure 8.2: U-tube river aerator, of the design used at Red Deer, Alberta, Canada. Reproduced from Reference 17.

the secondary treatment; the reaction between Cl_2 and NH_3 is discussed under waste-water treatment. Because side products such as $CHCl_3$ result when water is chlorinated (Chapter 7), the advisability of practising superchlorination is now questioned, and low-dosage chlorination is preferred.

8.1.4 Soaps and detergents

It is convenient to discuss this topic in the present chapter because of the problems which soaps and detergents can cause in sewage treatment. The mechanism of action of soaps and detergents in cleaning is to sequester greasy materials in a form which makes them, for practical purposes, water soluble.

Surfactants

Chemically, both natural soaps and synthetic detergents contain **surfactants** which are substances having the structural characteristic of a long hydrocarbon chain to which is attached a polar "head group" which is frequently ionic. In the case of natural soaps, the head group is a carboxylate anion; laundry detergents are most often based on surfactants having sulfonate anions as head groups. As purchased in the supermarket, a detergent is a complex formulation of a surfactant and **builder,** which is a basic substance whose role is to provide a source of hydroxide ion. Here we discuss only the surfactant (surface active compound) and the builder. Other ingrediants include: bleaches, optical brighteners, foam regulators, and sometimes proteases and anti-static compounds. Here we discuss only the surfactant (surface active compound) and the builder.

$CH_3(CH_2)_{10}CH_2OSO_3^- \; Na^+$

Dreft, an alkyl sulfate

$CH_3CH(CH_3){-}(CH_2CH(CH_3))_n{-}C_6H_4{-}SO_3^- \; Na^+$

1

alkylbenzenesulfonate

Soaps are the oldest surfactants. The first complete detergent, Persil (1907), comprised soap, builder (sodium carbonate + sodium silicate), and bleach (sodium perborate)[18]. Long chain alkyl sulfates were introduced in the 1930's (e.g., Dreft, 1933)[19], and alkylbenzenesulfonates in the 1940's (Tide, 1946).

The special feature of a surfactant molecule is the combination of a very polar, hydrophilic head group with a non-polar, hydrophobic hydrocarbon tail. At very low concentrations the surfactant dissolves normally in water, to give a solution containing individual hydrated molecules. Above a certain solute concentration called the "critical micelle concentration" (CMC), the surfactant aggregates into structures called **micelles,** each containing of the order of 10^2 solute molecules. The micelle is approximately spherical, and is arranged so that all the polar head groups can be solvated by water, while all the hydrocarbon tails are entangled in the centre of the micelle, where they are not exposed to water.

The driving force for micelle formation is mainly entropic. Recall from Chapter 5 that non-polar solutes have low solubility in water because they order the water molecules around them, and this greatly reduces solvent entropy. The loss of entropy in the water/detergent system is greater when each surfactant molecule must be solvated individually by water molecules

[18]G. Jakobi and A. Lohr, "Detergents and textile washing: principles and practice," VCH Verlagsgesellschaft mbH, Weinheim, West Germany, 1987.

[19]Dreft, $C_{12}H_{25}OSO_3^-$ Na^+—chemical names sodium lauryl sulfate or sodium dodecyl sulfate (SDS)—is also encountered in the laboratory as a denaturing agent for proteins, for example in SDS-polyacrylamide gel electrophoresis.

than when the surfactant molecules aggregate to form the micelle. Over a rather large range of the total dissolved solute concentration, the size of the micelles remains constant, and what changes is the number of micelles per unit volume. Mutual repulsion inhibits the micelles from coalescing.

Hydrophilic surface of ionic or polar moieties

Hydrophobic core of hydrocarbon chains

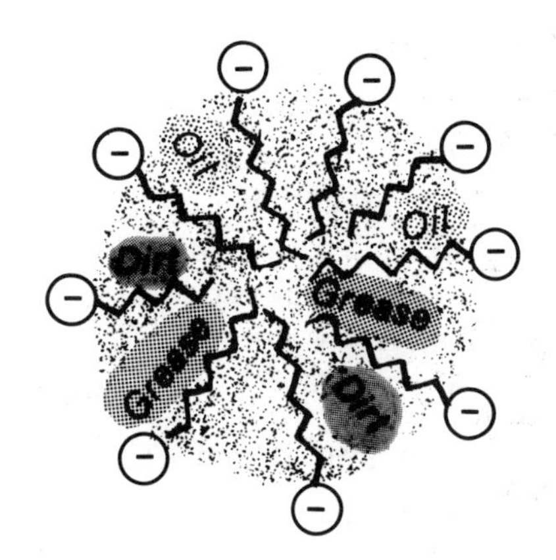

The cleaning action of the detergent solution involves the uptake of "grease" i.e., non-polar substances, into the interior of the micelles, where they dissolve in the hydrocarbon-like milieu. This has the effect of solubilizing these substances. Thus when you wash greasy hands, the grease migrates from your hands into the aqueous phase, where it disappears down the drain. In reality, the grease is dissolved in the hydrocarbon-like interior of the micelles, rather than in the water itself. Detergent action is assisted because the surfactant lowers the surface tension of water; this allows the soap or detergent solution to "wet" surfaces more effectively, providing better contact between the aqueous phase and the object being cleaned.

Soaps

Soaps are made by the base hydrolysis of long chain triglycerides of animal or vegetable origin: indeed, the basic hydrolysis of esters is still known as "saponification," which means soap-making. Take tristearin—the triglyceride of stearic acid, $C_{17}H_{35}CO_2H$—as an example of saponification.

$$\begin{array}{l} C_{17}H_{35}CO_2\text{-}CH_2 \\ \qquad\qquad\quad | \\ C_{17}H_{35}CO_2\text{-}CH \\ \qquad\qquad\quad | \\ C_{17}H_{35}CO_2\text{-}CH_2 \end{array} + 3NaOH \longrightarrow 3C_{17}H_{35}CO_2^-Na^+ + \begin{array}{l} CH_2OH \\ | \\ CHOH \\ | \\ CH_2OH \end{array} \qquad (4)$$

In pioneer days, soap was made by heating animal fat such as lard or tallow with ashes from the fire. These ashes contain the base potassium carbonate (potash), which in the presence of water provides the OH^- required for hydrolysis[20].

When lye (NaOH) is used to hydrolyze the triglycerides, the soap thus formed is the mixture of sodium salts of whatever carboxylic acids were incorporated into the triglycerides. Ordinary hand soaps are of this type. Salt is added to assist in precipitating the raw soap (Problem 9), which is then filtered, and appropriate colorants and scents are added. Potassium soaps have lower solubility and are used in applications such as shaving soaps. Transparent soaps are made by adding a little ethanol to the raw soap curds. Special soaps for use in seawater are made from shorter chain carboxylic acids (about C_{12} rather than about C_{16}–C_{18}) because ordinary soaps are too insoluble to use in seawater, where the high concentration of Na^+ reduces their solubility by the common ion effect.

The major drawback to the use of natural soaps is that in hard water areas they give a precipitate (scum) of insoluble calcium or magnesium salts with Ca^{2+} and Mg^{2+}. For example:

$$(5) \quad 2RCO_2^-Na^+(aq) + Ca^{2+}(aq) \longrightarrow (RCO_2)_2Ca(s) + 2Na^+(aq)$$

This reaction removes the soap from solution with consequent loss of cleaning power, and produces an unsightly scum, which gives the familiar bathtub ring and which turns white clothes grey (Problem 10).

Synthetic detergents

Synthetic detergents based on sulfonate (or, less commonly, sulfate half-ester) salts provide efficient cleansing action even in hard water areas. Their calcium salts are more soluble than those of soaps, so precipitation of scum is less likely.

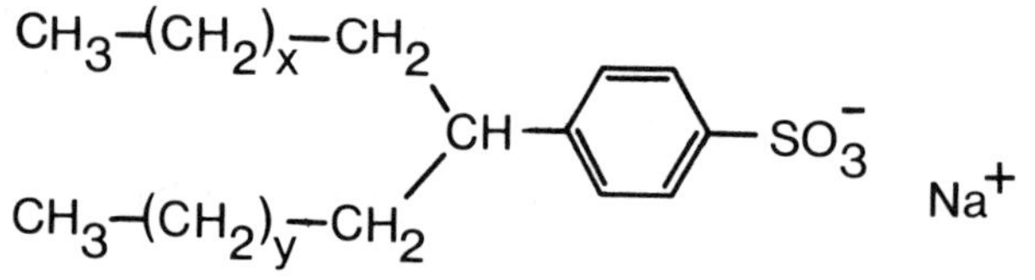

$CH_3-(CH_2)_x-CH_2$
$CH-C_6H_4-SO_3^-$ Na^+
$CH_3-(CH_2)_y-CH_2$

2

Sulfonate detergents are manufactured by polymerizing a simple alkene precursor (ethylene or propylene) to low molecular weight (oligomeric) products, followed by an acid-catalyzed Friedel-Crafts alkylation of benzene. Following sulfonation of the long chain alkylbenzene, the sulfonic acid is neutralized to give the sodium salt.

[20]The pioneers used an iron kettle for making soap. Should you decide to make your own "pioneer" soap, do not use an aluminum pot. Why?

$$CH_2{=}CH_2 \xrightarrow{H^+} CH_3(CH_2CH_2)_nCH_2CH{=}CH_2$$

$$\xrightarrow[\text{benzene}]{H^+} CH_3(CH_2)_xCH(C_6H_5)(CH_2)_yCH_3 \xrightarrow[\text{2. NaOH}]{\text{1. } H_2SO_4} CH_3(CH_2)_xCH(C_6H_4SO_3^-\ Na^+)(CH_2)_yCH_3$$

The early sulfonate detergents used in the 1940's and '50's were branched chain structures (e.g., **1.**) Such molecules are biodegraded very slowly, since each branch point acts as a block to microbial oxidation. Branched chain surfactants can survive secondary sewage treatment without being broken down; the result was spectacular mountains of foam at weirs and waterfalls on rivers downstream from sewage treatment plants. The detergent properties of the undegraded surfactants were also damaging to aquatic life in the receiving waters (Problem 11).

Linear alkylbenzene sulfonates (**2**) were introduced in the 1960's. They are much more easily biodegraded, and hence do not survive secondary sewage treatment.

Cationic detergents are usually based on long chain quaternary ammonium salts, such as alkyltrimethylammonium chlorides, $RN(CH_3)_3^+\ Cl^-$, where R is typically C_{12}–C_{16}. This feature mimics the structures of phospholipids found in the membranes of living cells. Cationic surfactants tend to disrupt cell membranes, and are used as biocides, or antifouling agents (for example, to retard the growth of algae in the water used in industrial heat exchangers). Long chain cationic detergents of structure $R_2N(CH_3)_2^+Cl^-$ are used as fabric softeners. Adsorption of the alkylammonium salt by the fabric gives the anti-static property because the fibres acquire positive charges and repel each other.

Non-ionic surfactants are used both in liquid laundry detergents and in products such as hair shampoos. Generally they contain the usual long hydrocarbon chain, which is oligomerized with ethylene oxide to produce —$(CH_2O)_n$–CH_2OH as the polar head group. (Other liquid laundry detergents are based on the anionic "alcohol ether sulfates," R–$(CH_2O)_n$–$CH_2OSO_3^-Na^+$.) The hydrophobic part of the molecule may be derived from a long chain alcohol or alkylphenol, giving R–$(CH_2O)_n$–CH_2OH, or the amide of a long chain carboxylic acid, giving R–CO–N($(CH_2O)_n$–$CH_2OH)_2$. Non-ionic surfactants are useful in detergents to be used with synthetic fabrics and at low wash temperatures.

Amphoteric surfactants are zwitterionic i.e., they contain both a cationic and an anionic group. $R(CH_3)_2N^+-(CH_2)_nCO_2^-$ and $R(CH_3)_2N^+-(CH_2)_nSO_3^-$ are typical structures (Problem 12). They are used only in specialty detergents, as they are expensive to produce.

Builders

Hydroxide ion plays an important role in detergent action. In the case of a natural soap, hydroxide ion is automatically present, because carboxylate salts are noticeably basic ($K_b \approx 10^{-9}$ mol L^{-1}, corresponding to $K_a \approx 10^{-5}$ mol L^{-1} for the corresponding carboxylic acids).

$$(6) \qquad RCO_2^-(aq) + H_2O(l) \rightleftharpoons RCO_2H(aq) + OH^-(aq)$$

Sulfonic acids are strong acids ($pK_a < 0$), and so the conjugate sulfonate anions are very feeble bases which contribute almost no OH^- to their aqueous solutions (Problem 13). For effective detergent action, synthetic laundry and dishwashing detergents, for example, must also include a source of OH^-, known as a **builder.** In the case of industrial degreasers and restaurant strength dishwasher detergents, the builder may be NaOH, and the pH of the detergent solution in use may be as high as pH 12–13. This is unacceptable for household use, where young children may eat the detergent or get it in their eyes. In consumer products, the builder is a weak base, which provides a safe source of hydroxide ion.

Until about 1970, complex phosphates were the preferred choices as detergent builders, since they are basic, and hence provide a source of OH^-. For example, for STP:

$$(7) \qquad P_3O_{10}^{5-}(aq) + H_2O(l) \longrightarrow P_3O_9OH^{4-}(aq) + OH^-(aq)$$

A second desirable attribute of phosphate builders is that they form soluble complexes with Ca^{2+}. This seems odd, remembering that Ca^{2+} is used to precipitate phosphate ion. Precipitation indeed occurs with calcium ion and monophosphate anion, but the complex phosphate anions form soluble association complexes. The linear polyphosphates form "wrap-around" complexes with Ca^{2+}, while the cyclic metaphosphates sequester the calcium cation in a central cavity where it is surrounded by $-O^-$ ligands. Sodium hexametaphosphate has long been marketed as a water softening aid for use with laundry detergents under the tradename Calgon (**cal**cium **gone**).

With the restriction on the amount of phosphate builder which may be used in a detergent formulation, manufacturers have looked to other builders. Sodium carbonate (soda ash, $pK_b = 3.68$) is an inexpensive source of OH^-, but it does not complex calcium ions; instead it precipitates them. Soda ash can therefore replace part, but not all, of the phosphate in the detergent. Aluminosilicates such as the zeolites (Chapter 5) are useful builders, usually in combination with STP. They are, of course, an insoluble component of the detergent. Their ion-exchanging properties allow them to bind calcium and magnesium ions. Aluminosilicates are especially valuable for high temperature applications because, unlike STP, they bind Ca^{2+} more strongly at high than at low temperatures.

The sodium salt of nitrilotriacetic acid (NTA, $N(CH_2CO_2Na)_3$, pK_b's 3.72, 11.05, and 12.34) has the desired properties in terms of providing OH^- with water and complexing Ca^{2+} by chelation (Problem 14). It can thus replace all the phosphate in a detergent formulation, but it is excluded from the large U.S. market because of toxicological concerns. These were raised in the early 1970's following experiments which involved repeated administration of very large amounts of NTA to experimental animals. However, NTA is used in Canada and in Europe, and nearly two decades of use have failed to show any undesirable effects. The adverse effects seen in the U.S. studies are now thought to be the result of interference with the metabolism of Ca^{2+} and Mg^{2+} due to their complexation with these large doses of NTA. Another suggested drawback to the use of NTA is that it might solubilize heavy metals from rocks and sediments by reactions such as:

$$(8) \quad NTA^{3-}(aq) + Pb^{2+}(aq) \rightleftharpoons (NTA{\bullet}Pb)^-(aq) \quad K_{ass} = 3 \times 10^{13}\ L\ mol^{-1}$$

This seems an exaggerated fear (Problem 15), in addition to which, NTA is biodegraded fairly rapidly, i.e. it does not persist in the environment.

8.2 Other aqueous wastes

The types of waste water are almost as numerous as the types of industry, so it is impossible to give a comprehensive treatment. Most aqueous industrial wastes are very dilute, with solute concentrations usually in the parts-per-million range. This is understandable in that the disposal of any waste product represents two costs to industry: first the actual cost of disposal; second, the cost of the raw materials and other resources that went into generating the waste. Unreacted starting materials and valuable products and catalysts are recovered from process streams to the maximum extent possible. Wherever possible, byproducts are used or sold for profit rather than having to pay for their disposal.

8.2.1 Specific industries and specific wastes

The **food processing industry** and the **pulp and paper industry** both generate large quantities of high-strength wastes. Examples from the food industry include the manufacture of butter and cheese, meat packing and processing, and the preparation of fruits and vegetables for canning, freezing, jam making, juices, sauces, ketchups etc. Although the substances present in these wastes are entirely "natural," their effect on the environment is very damaging if they are simply discharged into rivers and lakes, on account of their high BOD. Problems can also arise if food wastes are discharged directly into a municipal sewage system, because the BOD is so much higher than that of the rest of the sewage stream. Many municipalities levy a charge on industrial users of the sewage system based on the BOD of the material discharged into the municipal sewers. Where aqueous streams are to be discharged to the environment, strict legal limits are placed on the levels of solutes they may contain.

Thus we see that industrial operators are involved to a major extent with waste processing—especially BOD reduction—whether their waste is discharged directly to the environment or to a municipal sewerage system. Modern facilities for the treatment of aqueous wastes represent multi-million dollar investments for the industries concerned. Although the design of these facilities is more an engineering problem than a chemical one, we may note that their are two basic strategies for BOD reduction: aerobic and anaerobic treatment. In aerobic treatment, oxidizing microorganisms convert the organic waste to CO_2, H_2O, and biomass; anaerobic treatment converts waste—principally carbohydrate, $(CH_2O)_n$—to CH_4, CO_2, H_2O, and biomass. Either system has the effect of reducing the BOD of the treated aqueous stream.

Waste may be treated either in an open lagoon or in a closed bioreactor. When the waste is discharged into a lagoon, the liquid is constantly aerated vigorously in order to provide the oxygen needed by aerobic organisms, and to prevent the lagoon from becoming completely anaerobic. Anaerobic degradation is a reductive process which produces malodorous sulfides and amines as byproducts, as well as complaints from the neighbours. The recognition that volatilization from open lagoons can contribute substantially to air pollution is the reason for the trend towards activated sludge systems—whether aerobic or anaerobic—rather than open lagoons for waste treatment. The sludge byproduct is digested, dewatered, and land-filled. When bioreactors are used to treat industrial waste, an important consideration is any possible toxicity of the waste stream towards the microorganisms in the reactor. Since the treatment process is a biological one, the health of the organisms must always be maintained.

As has been noted already, food industry wastes have a higher ratio C:(N,P) than sewage, and so conversion to biomass can be very effective at reducing the dissolved nutrients. If necessary, ammonium phosphate may be added to optimize this process if the proportion of carbon is too high.

Oily wastes are generated in several industries, notably petroleum refining. These wastes have such a high water content that it is uneconomic to recover the oil. They pose a particular environmental threat because oil can cover such a large area of the surface of a natural water body, thereby interfering with oxygen transport across the air-water interface. The organic compounds present in oily waste are hydrocarbons, mostly alkanes, which are much less easily oxidized than food industry wastes. In addition, since oil-water mixtures may be used in applications where lubrication is needed, they frequently contain emulsifiers, and separation of the oil from the water is difficult[21]. Various mechanical separation methods are in use, as well as evaporation, to remove the water. Chemical methods for the destruction of the oil component are sometimes used, including combustion and ozonization.

An approach to the disposal of oily wastes is **land farming,** where an area of land is set aside to spread the waste. Biological action will eventually oxidize even alkanes, as has been shown by the natural clean-up, over only a few years, of ocean beaches which suffered disastrous oil spills. In order to accelerate the natural process of biological oxidation, the land is tilled regularly (hence the term farming) with ordinary agricultural equipment, and nitrogen and phosphate fertilizers are added in the proper proportions to allow the soil microorganisms to make optimum use of the carbon source provided in the oily waste. Analysis of the site determines the maximum rate at which the waste can be applied. In northern latitudes, land farming, like conventional agriculture, is restricted by the length of the "growing season." In addition, the vagaries of the weather determine the rate of waste degradation in the land farm, as in conventional agriculture. Whether land farming will ultimately survive as a treatment method for oily wastes is problematical, because volatilization of organic compounds from land farms is bringing the practice under increasing scrutiny from government regulators. Eventually, land farms may be replaced by advanced closed bioreactors.

Problem organics include substances that are toxic and/or resistant to oxidation. The aqueous wastes from process streams in pesticide manufacture, for example, will inevitably be contaminated with a few ppm of the product being manufactured. Very odorous compounds such as the chlorophenols are a special problem if the discharge water is to be used downstream for drinking (Chapter 7). If the product is highly toxic, land farming will be inappropriate because the organic compound will be toxic towards the soil microorganisms. The development of new technologies to decontaminate waste streams is a very active area of industrial research, especially as government regulations become more and more restrictive as to what may pass beyond the plant boundary. The use of adsorbing agents such as activated charcoal suffers from problems such as the high cost of the adsorbent, its finite capacity, and what to do with the spent adsorbent.

Many industries generate **acidic wastes** e.g., spent pickling liquors containing hydrochloric and sulfuric acids, which are used to remove oxide films from metals before they are painted or electroplated; spent acid from explosives manufacture. These acids are neutralized with limestone. World-wide it is estimated that 2.5 million tonnes of waste sulfuric acid are released to the environment every year. Alternatively, it may be economical to re-

[21] V.V. Pushkarev, A.G. Yuzhaninov, and S.K. Men, *Treatment of oil-containing wastewater,* English translation, Allerton Press Inc., New York, 1983.

cover the waste acid through concentration (removal of excess water) for re-use. Waste pickling liquors are sometimes used to precipitate phosphate from waste water, since they are able simultaneously to lower the pH, and provide dissolved iron with which to form iron phosphate.

Ammonia presents a problem in both sewage and waste water treatment because of its toxicity towards aquatic life. The two methods commonly used to remove ammonia from waste water are breakpoint chlorination and air stripping. In some cases biological oxidation of ammonia is employed. This process, called nitrification, yields nitrate ion as the final product rather than elemental nitrogen.

Breakpoint chlorination (Problems 16 and 17): The overall reaction between chlorine (or equivalently HOCl) and ammonia is:

$$(9) \qquad 2NH_3(aq) + 3HOCl(aq) \longrightarrow N_2(g) + 3HCl(aq) + 3H_2O(l)$$

In this reaction chlorine is reduced from the +1 to the −1 oxidation state, while ammonia is oxidized to elemental nitrogen.

This reaction occurs in several stages.

$$(10) \qquad NH_3(aq) + HOCl(aq) \longrightarrow NH_2Cl(aq) + H_2O(l)$$

Chloramine NH_2Cl is a mild disinfecting agent which has been used to disinfect drinking water in small scale operations. It is not much used now; it is a rather slow acting, weak disinfectant, and there are uncertainties about its toxicology. Dissolved ammonia is therefore a problem in the context of drinking water treatment, because it adds to the chlorine demand, and the chloramines thus formed are much weaker disinfectants than HOCl or ClO^-.

Chloramine next reacts with further HOCl:

$$(11) \qquad NH_2Cl(aq) + HOCl(aq) \longrightarrow NHCl_2(aq) + H_2O(l)$$

The chloramine and the dichloramine react together, giving the overall stoichiometry of Equation 9.

$$(12) \qquad NH_2Cl(aq) + NHCl_2(aq) \longrightarrow N_2(g) + 3HCl(aq)$$

Both NH_2Cl and $NHCl_2$ are "active chlorine" compounds, in the sense defined in Chapter 7. Consequently, as chlorine (or HOCl) is gradually added to a solution containing ammonia the solution initially shows a chlorine residual due to NH_2Cl. As more chlorine is added and $NHCl_2$ forms, Reaction 12 intervenes, and the chlorine residual drops again. The "breakpoint" is where the chlorine residual drops to zero again (Figure 8.3), at which point all the ammonia originally present has been oxidized to N_2.

Air stripping:

For volatile compounds, air stripping is an effective method of removing the organic contaminants from water, but leads to air pollution. An alternative strategy is volatilization from the aqueous waste stream followed by

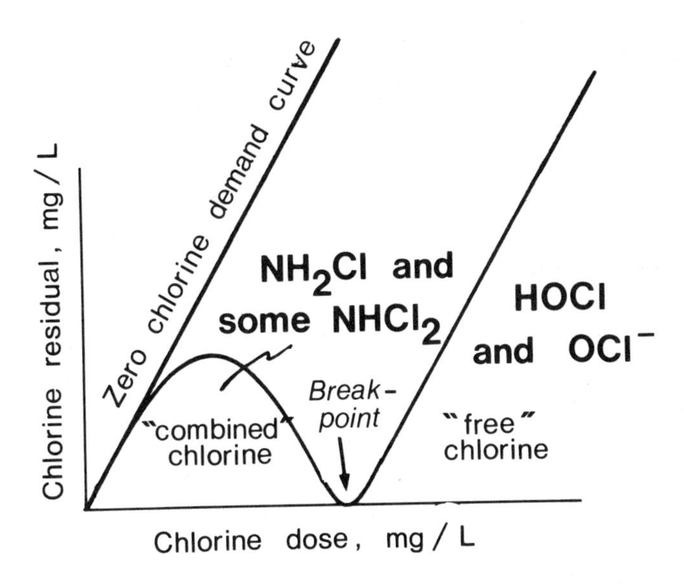

Figure 8.3: Schematic of the progress of a breakpoint chlorination reaction.

catalytic oxidation to carbon dioxide, using air as the oxidant. Gaseous and low boiling substances such as ammonia and Volatile Organic Compounds (VOC's, Section 7.4) can be removed from aqueous solution by air stripping. Ammonia may be used as an example, since it is very soluble in water but can nevertheless equilibrate between the aqueous and vapour phases: $K_H = 5.3$ mol L^{-1} atm^{-1}. If an aqueous solution containing ammonia is sprayed down a tower against an updraft of air, ammonia will transfer from the aqueous to the gas phases until equilibrium is reached, thereby removing the contaminant from the waste water stream. VOC's can be treated similarly. The advantage of air stripping is that it is very inexpensive; its disadvantage is that it exchanges pollution of the water for pollution of the air.

The acceptability of air stripping is increasingly being questioned. One possible technology that is under development is to pass the exhaust air over an oxidation catalyst so as to convert the VOC's to CO_2 before the air is returned to the environment. Another possibility is oxidation in the aqueous phase by means of UV irradiation in combination with the use of powerful oxidants such as ozone and hydrogen peroxide[22]

Cyanide is a component of several waste streams. Cyanide is used to complex heavy metals in electroplating and in other metal finishing pro-

[22]N. Lewis, K. Topudurti, G. Welshans, and R. Foster, "A field demonstration of the UV/oxidation technology to treat ground water contaminated with VOCs," *J Air Waste Management Assoc*, **1990**, 40, 540–547.

cesses. Simultaneous contamination by cyanide and by heavy metals is therefore common. Cyanide is also used in the mining industry, in certain oil flotation processes. Oil flotation is used to separate materials on the basis of density (see also Chapter 6), for example in separating the valuable ore from the unwanted rock (gangue), or in separating ores of different density (e.g., copper, lead, and zinc sulfides). The addition of sodium cyanide to the water/oil/surfactant mixture has been found empirically to improve the efficiency of the separation, but it leaves concentrations of NaCN up to 50 ppm in the waste stream.

The gold extraction industry also consumes large amounts of cyanide, since the ability of dilute solutions containing CN^- to dissolve gold makes for economical mining of very low grade gold ores (see also the section on Hydrometallurgy, below). Usually the ore is preconcentrated by flotation prior to cyanidation. The dissolution of gold by cyanide is remarkably efficient (Problem 18), the dissolution depending in practice not on the equilibrium constant for Reaction 13, but on factors such as the rate of diffusion of oxygen and the physical accessibility towards the reagents of the gold atoms in the crushed rock.

$$(13) \quad Au(s) + 2CN^-(aq) + H^+(aq) + \tfrac{1}{4}O_2(g) \longrightarrow Au(CN)_2^-(aq) + \tfrac{1}{2}H_2O(l)$$

The release of cyanides into the environment is clearly undesirable on account of the great toxicity of cyanides.

The usual method for removing cyanide from aqueous waste streams is to oxidize it using chlorine (or, on a smaller scale, calcium hypochlorite).

$$\begin{array}{lrcl} (14) & CN^-(aq) + HOCl(aq) & \longrightarrow & OCN^-(aq) + HCl(aq) \\ \text{or:} & CN^-(aq) + Cl_2(aq) + H_2O(l) & \longrightarrow & OCN^-(aq) + 2HCl(aq) \end{array}$$

The cyanate ion thus formed hydrolyzes quite readily:

$$(15) \quad OCN^-(aq) + H^+(aq) + H_2O(l) \longrightarrow NH_3(aq) + CO_2(aq)$$

Heavy metals are present in numerous wastes. The term heavy metals is usually used by chemists to indicate metals beyond about iron in the periodic table, but since toxic metals in the environment are usually heavy metals, the term heavy metals is often used inclusively for toxic first transition series elements such as chromium and vanadium.

Many metals with important commercial uses are toxic, and hence undesirable for indiscriminate release into the environment. Copper, chromium, silver, nickel, and cadmium are all used in electroplating; vanadium, nickel, mercury, and the precious metals are all used as the active ingredients in catalysts; mercury is used in batteries and still to some extent in the electrolytic manufacture of chlorine and sodium hydroxide (Chapter 10); lead is used for storage batteries and to a decreasing extent in gasoline additives (Chapter 10).

Removal of these metals from aqueous waste streams can be accomplished by precipitation of the metal cation with a suitable anion. Most metal sulfides are very insoluble, and so the metals may be precipitated by passing hydrogen sulfide into the solution containing the metal cation(s). For a dipositive metal:

$$H_2S(g) + M^{2+}(aq) \longrightarrow MS(s) + 2H^+(aq) \quad (16)$$

The chemical sludge of precipitated sulfides must be collected and disposed of. Sulfide is a basic anion, so precipitation is more efficient at high pH, and the pH may therefore be raised by adding lime prior to treatment with H_2S.

8.2.2 Acid mine drainage

Acid mine drainage is an unsightly environmental problem associated with the mining industry. It involves the acidification of creeks downstream from a mine, and the deposition of iron as an unsightly slimy orange precipitate of the hydroxide on the rocks of the stream bed. Although acid mine drainage is technically an example of contamination of natural waterways rather than one of waste water disposal, it is convenient to discuss it here.

Acid mine drainage is actually a biological problem. As noted elsewhere in this book, many important metals are mined as sulfides, while coal also contains metal sulfides such as FeS_2. Some of these materials will be present in the waste material ("spoil") that is discarded at the mine site. Bacteria of the *Thiobacillus* family are able to use these sulfur compounds for the reduction of molecular oxygen; some species of *Thiobacillus* oxidise the sulfur right through to sulfuric acid. In the immediate vicinity of the mine pH values as low as 1–2 may be reached, and under these acidic conditions, iron compounds—both Fe(II) and Fe(III)—in the mine spoil may also dissolve. The dissolution and oxidation of FeS_2 is important in this context, because oxidation of the Fe(II) in FeS_2 to Fe(III) occurs simultaneously with oxidation of S_2^{2-}.

The oxidation of Fe(II) occurs both chemically and microbially. The chemical reaction follows the rate law[23] shown in Equation 17.

$$\frac{-d[\mathrm{Fe(II)}]}{dt} = k\,[\mathrm{Fe(II)}]\,[\mathrm{OH^-}]^2\,p(\mathrm{O_2}) \quad (17)$$

This rate equation suggests that the rate increases with $[OH^-]^2$, or equivalently that the $\log_{10}$(rate) increases proportionately with (2 × pH): see also Chapter 7. Below about pH 3 chemical oxidation of Fe(II) should be extremely slow. The experimental points (Figure 8.4) fit the rate law for chemical oxidation of Fe(II) above about pH 5 (the straight line portion of the curve, and its dotted extrapolation). Below pH 4, the reaction is

[23] W. Stumm and G.F. Lee, "Oxygenation of ferrous iron," *Ind Eng Chem,* **1961**, 53, 143.

noticeably faster than the chemical rate law predicts, showing the incursion of biological oxidation at low pH.

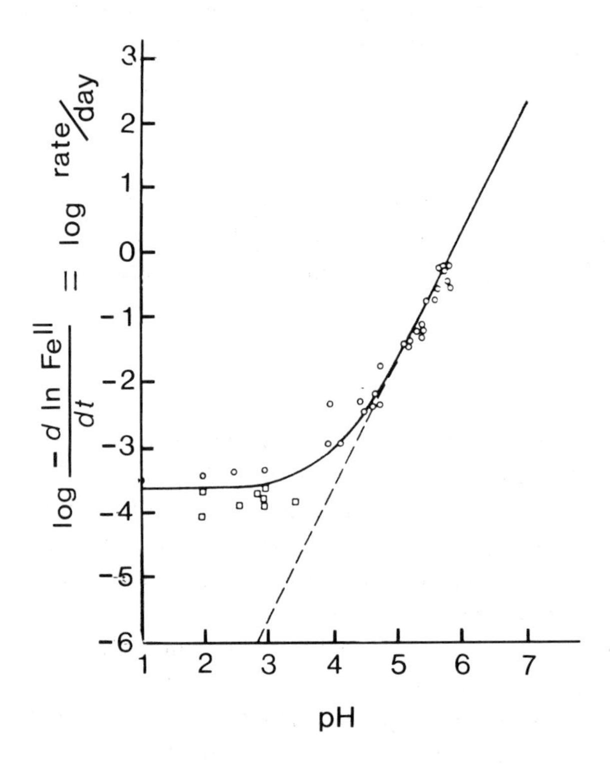

Figure 8.4: Oxidation of Fe(II) as a function of pH.

Near the mine site, bacterial oxidation of sulfide causes the pH of the water to be very low. As the stream travels away from the mine site, it becomes diluted with uncontaminated water. The pH rises, and the dissolved iron precipitates when the pH exceeds about 3.5 (compare discussion in Section 6.5.3). Thus the manifestations of acid mine drainage are (i) the loss of aquatic life close to the mine site due to acidity, (ii) the deposition of hydrated Fe_2O_3 some distance downstream (Problem 19).

In the case of ores containing metals other than iron, acidity in the vicinity of the mine causes these other metals (e.g., Cu, Pb, Zn) become leached in soluble form. Since their concentrations are often > 100 ppm, it may be economically feasible to recover them. In the case of copper, this may be done by passing the leachate over a large amount of scrap iron, a process called **cementation** (not to be confused with the use of cement to immobilize solid hazardous wastes). The more electropositive iron displaces copper from solution (Problem 20).

$$Cu^{2+}(aq) + Fe(s) \longrightarrow Fe^{2+}(aq) + Cu(s) \tag{18}$$

8.2.3 Hydrometallurgy

The example of cementation just discussed provides a link between conventional mining operations—in which ores are taken from the ground, concentrated, and processed chemically—and hydrometallurgy, in which the valuable metal is removed from its orebody by an extraction process. Hydrometallurgy makes possible the utilization of low grade ores, those containing so little of the valuable metal that the handling and separation of the ore by conventional methods would be uneconomic. For example, very low grade gold ores may be leached in place by treating a heap of the crushed ore with aqueous sodium cyanide solution. Experiments have been conducted in which the ore is leached *in situ,* but difficulties arise in the collection of the leachate, and the inaccessibility of the reagent to the gold when the ore is not crushed.

Low grade copper ores can also be worked using hydrometallurgy. Unlike gold, which occurs as the element and must be oxidized to get it into solution, copper is normally found in the +2 oxidation state, as the insoluble CuS (sometimes Cu_2S) or a basic carbonate such as malachite, $CuCO_3.Cu(OH)_2$. One technique which may be used is extraction with a large organic ligand having a high selectivity for copper. Designating the ligand as L:

$$\text{(19)} \qquad CuS(s) + 2L(aq) \longrightarrow CuL_2^{2+}(aq) + S^{2-}(aq)$$

The Cu(II) cation, now surrounded by two large organic ligands, can be extracted from the aqueous phase into an organic solvent, the complex broken down, and the ligand recycled. One such ligand and its Cu^{2+} complex are shown below.

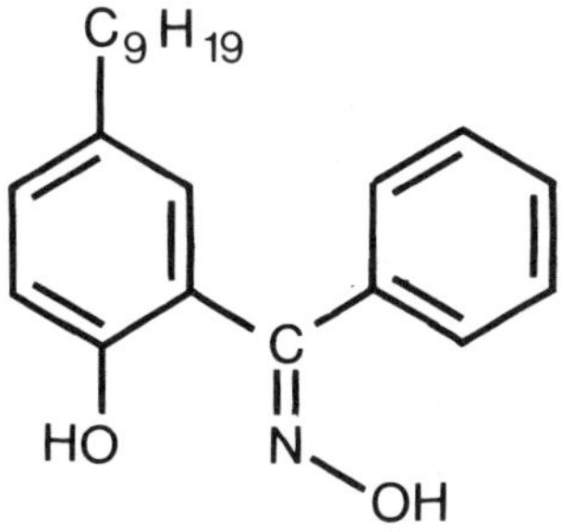

Ligand

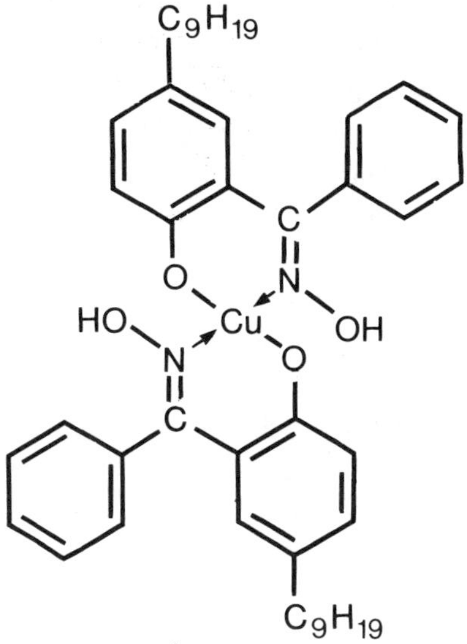

Copper complex

8.2.4 Hazardous and toxic wastes

These terms have become a source of confusion, because they now have a regulatory meaning in addition to their everyday use. In the U.S., for example, under the provisions of the Resource Conservation and Recovery Act (RCRA, pronounced "Rickra" in the trade), the EPA may designate materials as "toxic." In this regulatory sense chlorobenzene, for example, is a toxic waste, and may not be discharged at levels exceeding 100 ppm in either solid waste or waste water. "Hazardous waste" or "toxic waste" therefore does not imply that the particular waste material under discussion is specifically a hazard, or poses a specific threat of toxicity, since the amount of material is not defined. Unfortunately, the regulatory use of these words is sure to increase confusion, and the sense of "chemophobia" among the general public. Current EPA regulations under RCRA cover eight metals such as arsenic and mercury, as well as approximately 30 pesticides and other organic chemicals, including various chlorinated phenols and chlorinated aliphatic solvents[24].

8.3 Non-aqueous wastes

So far, we have considered wastes that are mainly composed of water. Among non-aqueous wastes, municipal garbage comprises one of the largest volumes of waste (140 million tonnes per year in the U.S. alone[25]). Liquid wastes include used solvents, and special problems such as polychlorinated biphenyl (PCB) wastes (Chapter 9). Sludges include the "still bottoms" that are left over from the distillation of all kinds of liquid products such as monomers to make plastics, herbicides and pesticides. Solids include the precipitates formed when toxic metals are precipitated from aqueous waste streams (a special case of which is radioactive waste), and sewage sludge, already discussed. In the past, many of these wastes have been disposed of improperly and have ultimately come to public attention as environmental disasters such as Love Canal (improper land-filling), Times Beach (improper spraying of waste oil), and "Valley of the Drums" (land-filling of incompatible chemicals in unprotected steel drums). Emerging issues are the export of hazardous wastes from Europe to impoverished West African nations, and the lack of controls on waste disposal in industrializing Third World nations such as India and China. With more openness in the news media, similar revelations are likely from Eastern Europe, where preliminary information suggests that waste disposal has hitherto had a low priority.

---

[24] News item in *Chem Eng News,* March 12, **1990**, 4.

[25] P.R. O'Leary, P.W. Walsh, and R.K. Ham, "Managing solid waste," *Sci Am,* **1988**, 259, 36–42.

8.3.1 Recycling

The disposal of waste is a monumental problem for society, because of the enormous amounts involved and their variety. Although industry receives much bad press over the waste disposal issue, a great deal of the waste generated by industry is actually recycled: defective product, manufacturing overruns, machined scrap, etc. There is no particular altruism in this; it is generally cheaper to recycle scrap than to make new product. Today, major corporations are very conscious of the adverse corporate image of poor waste disposal practices.

In this section we explore some possible options for handling wastes, both hazardous and (nominally) non-hazardous. We shall also examine a few well known cases where improper waste disposal has led to problems. At the outset, we must recognise that few options exist for handling large volumes of waste. As a slogan "Recycling" sounds excellent, and indeed should be practised wherever possible; but it can by no means eliminate waste. The Ontario government has promoted an industrial "4-R's" program: waste Reduction, Reuse, Recycling, and resources Recovery[26]. Advantages to industry of the 4-R's program are indicated to be: savings in raw materials costs; savings in time and energy; lower treatment and disposal costs; less risk of legal liability; improved corporate image; and less employee exposure to hazardous materials. Central to the program is a Waste Audit, to establish the identities and quantities of the wastes being generated. In order to assist Canadian industry manage its waste, Waste Exchanges have been set up both within Ontario and interprovincially; listings of wastes currently available are maintained so that Company A, having waste acid to neutralize for example, may be able to use waste alkali from Company B, rather than using virgin base.

Estimates are that full implementation of a 4-R's program could reduce industrial wastes by up to 75%. Even so, this would still leave 25% which would require treatment, so that waste management will continue to be a difficult problem. On the municipal waste front, at the time of the report (1983), the authors noted that the actual participation of individuals in recycling programs fell far short of their stated willingness to participate when measured in surveys. This situation has changed—quite suddenly—and very successful recycling programs have now been introduced in many Ontario municipalities. Ironically, the success of municipal recycling programs is being undermined at the present time by insufficient capacity to treat and sell the material collected, especially newsprint. In the Netherlands, a small densely populated country where landfill sites are hard to find, two thirds of all newsprint and half of all glass are now recycled[27].

[26] "Waste reduction, reuse, recycling and recovery," Appendix 2 of *Blueprint for waste management,* Ontario Ministry of the Environment, 1983. See also the brochure "Industrial waste reduction: 4Rs," Ontario Waste Management Corporation.

[27] L.J. Brasser, "Solid waste disposal in the Netherlands," *J Air Waste Management Assoc,* **1990,** 40, 1364–1366.

Figure 8.5: An example of improper disposal of chemical wastes.

8.3.2 Landfilling

Landfilling refers to dumping waste on the ground. Landfill sites may be secured or unsecured. By far the majority are unsecured; that is, there is minimal management of the site once the waste has been dumped. In the case of garbage dumps, the site may be termed a "sanitary landfill site" if the garbage is regularly covered with soil to minimize access by vermin and the consequent spread of disease. Management of the site is normally limited to planting green cover once the site is full. Underneath the green cover, anaerobic decomposition of the garbage proceeds for many years, releasing methane and other gases, and causing the ground to settle slowly. This restricts severely the uses to which the land may subsequently be put, and this restriction is permanent, from the perspective of normal city planning.

In the past, municipal landfill sites—and even hazardous waste dumps—have been located with little regard to the underlying soil structure. The "Valley of the Drums" in the U.S. is a particularly poor example of chemical waste disposal. Until the 1970's thousands of tonnes of chemical waste were deposited in this location using old metal drums to contain the waste, but with no regard to segregating different kinds of waste or to preventing spillage from broken or rusty drums. Contamination of ground water was the inevitable result. This example and others like it led to the creation of the "Superfund" in the U.S.A.; the Superfund is financed partly by the chemical industry, and is used to clean up abandoned chemical waste dumps, often at a cost of many millions of dollars.

As noted earlier, soils vary greatly in their ability to retain waterborne material; clays are impervious, while sand and gravel are highly porous. This means that rain and melted snow may seep into the landfill and leach (dissolve) soluble material that has been buried. The leachate may find its way into underground water courses and hence contaminate sources of drinking water, possibly at a considerable distance from the landfill site, thereby causing contamination by toxic metals and organic compounds such as pesticides. The problem is compounded by the fact that many municipal garbage dumps also accept industrial wastes, with the regulation of wastes often being a matter of local ordinance.

Contamination of aquifers from unsecured landfills is such a problem that on-land dumping is now considered a least favoured method of disposal. A 1984 U.S. Act of Congress stated, "Certain classes of land disposal facilities are not capable of assuring long-term containment of certain hazardous wastes, and to avoid substantial risk to human health and the environment, reliance on land disposal should be minimized or eliminated, and land disposal, particularly landfill and surface impoundment, should be the least favored method for managing hazardous wastes." Once contamination of an aquifer has occurred, clean up is virtually impossible[28]. This is illustrated for a site in New Jersey which was contaminated with ca. 2 tonnes of VOC's (mostly 1,1,1–trichloroethane and tetrachloroethylene). Remediation of the site (*cf.* Section 8.4.3) consisted of pumping out the water at an average rate of 1200 L min^{-1}. The following data show the lack of success of remediation: numerical values are concentrations of tetrachloroethylene in the aquifer.

- before remediation: 10 ppm
- after pumping for 6 years < 0.1 ppm
- 4 years after pumping ceased: 13 ppm

Besides acting as sources of water pollution, unsecured municipal and industrial landfill sites may also cause air pollution through volatilization of substances dumped there. Odour problems are an obvious manifestation of this. As an example of chemical emissions, most of the airborne load of PCB's over the United States is believed to originate from old transformers which have been dumped (mostly in hazardous waste dumps rather than municipal garbage dumps) and leaked their contents. At any time some 900 tonnes of PCB's, mostly from dumps, circulates in the troposphere above the U.S., with a residence time of one week[29]. However, risk analysis

[28] C.C. Travis and C.B. Doty, "Can contaminated aquifers at Superfund sites be remediated?," *Environ Sci Technol,* **1990,** 24, 1464–1466

[29] R.G. Lewis, B.E. Martin, S.L. Sgontz, and J.E. Howes, "Measurement of fugative atmospheric emissions of polychlorinated biphenyls from hazardous waste landfills," *Environ Sci Technol,* **1985,** 19, 986.

indicates minimal hazard to the general population as a result of these and similar emissions[30].

Secured landfill sites are much more carefully planned. A prerequisite is that the underlying soil structure should be an impervious clay. A similar impervious layer is placed as a "cap" on the filled site to prevent the entry of rain and melted snow. Any leachate is pumped from the bottom of the landfill and treated before release to the wider environment. In addition, the ground water is monitored around the site by passing boreholes down into the aquifers, and testing the quality of the water regularly. Secured landfill sites are expensive to construct, in addition to which there is the continued cost of maintenance and aquifer monitoring. Although this approach does provide a safe burial for hazardous wastes, we must question from an ethical point of view the desirability of passing on to future generations the responsibility of maintaining our generation's waste.

8.3.3 Degradable plastics

Plastic wastes now contribute 7–8% by weight of North American municipal garbage, up from less than 3% in 1970. About half this amount represents discarded packaging material (the total of packagings forms nearly one-third of all municipal garbage). The littering of roadsides, the countryside, and ocean beaches by plastic rubbish calls attention to the near-indestructibility of synthetic plastics such as polyethylene, polystyrene, polyvinyl chloride etc. As with most other pollutants, many of the virtues of plastics (light weight, inertness to chemical and microbial attack, low cost) become their weaknesses when they are discarded[31]. At present, the proportion of plastics recycled is very small, mostly polyethylene terephthalate from soft drink bottles and polyethylene from plastic milk jugs[32].

Two approaches to degradable plastics are now available: photodegradable plastics and biodegradable plastics. However, it must be emphasized first, that these products form a very small proportion of all plastics and second, that they do not degrade away literally to nothing. Furthermore, the true effectiveness of these measures is probably limited[33]. Photodegradation does not work on buried plastic, as in a landfill; even biodegradation is exceedingly slow under these conditions: witness the success of archeologists in recovering organic materials from ancient historic sites.

---

[30] T.F. Wolfinger, "Screening-level assessment of airborne carcinogen risks from uncontrolled waste sites," *J Air Pollut Control Assoc,* **1989,** 39, 461–468.

[31] In this context, the disposable coffee cup has become a very visible symbol of plastic litter. As a result, there is pressure to replace it with more "environmentally friendly" paper cups. However, Hocking (*Science,* **1991,** 251, 404–405) has argued that in terms of the resources needed to manufacture each type of cup, the paper cup may actually be the less friendly environmentally.

[32] C.H. Kline, "Plastics recycling takes off in the USA," *Chem & Ind,* July 17, **1989,** 440–442.

[33] A.M. Thayer, "Degradable plastics generate controversy in solid waste issues," *Chem Eng News,* June 25, **1990,** 7–14

Photodegradable plastics are copolymers of conventional polymers such as polyethylene with a small amount of carbon monoxide. The presence of carbon monoxide in the feed affords a polymer containing a small proportion of randomly located carbonyl groups, the key feature of which is their ability to undergo chain scission upon absorption of light in the 300–330 nm range. This is a very well known reaction in organic photochemistry[34], and proceeds by the reaction mechanism below:

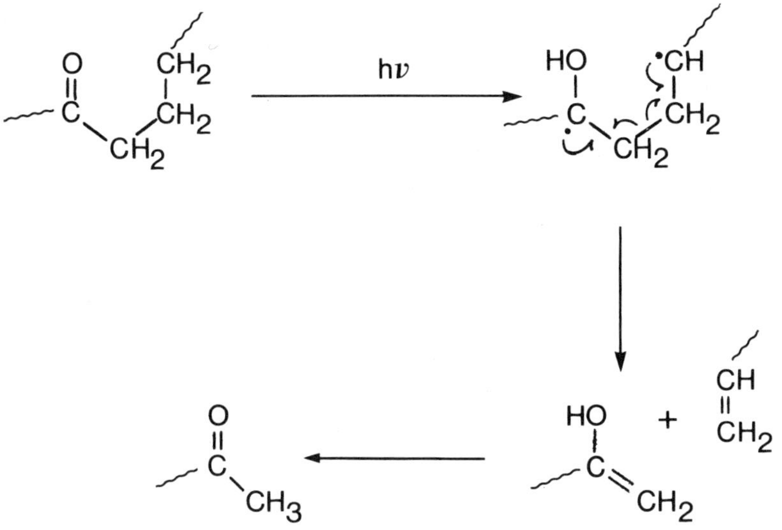

The photoreaction creates breaks in the polymer chain, which leads to loss of structural integrity as follows. The structure of the plastic is maintained by van der Waals attractions of neighbouring chains; although van der Waals interactions are usually thought of as weak, they can be substantial when large numbers of atoms are mutually in contact, as in an organic polymer. Breaks in the polymer chains therefore lead to the plastic disintegrating physically, although in the chemical sense little of the material has actually decomposed. The chief advantage of the photodegradable plastic is that unsightly litter "disappears" to powder upon sunlight exposure. Legislation has been introduced in Italy to require the use of photodegradable plastic in applications such as supermarket grocery bags.

As already mentioned, the carbonyl chromophores present in photodegradable plastics absorb near 300 nm, the extreme short wavelength of tropospheric solar radiation, also called UV-B. Artificial lighting—both incandescent tungsten bulbs and fluorescent lamps—produces very little UV-B, and window glass filters out this radiation from sunlight. This gives the advantage that plastic items made from this material will not disintegrate indoors, before they are used. Of course, they will only degrade outdoors if they are in direct sunlight, and not, for example, if they are buried in a landfill site.

[34]N.J. Turro, *Modern Molecular Photochemistry,* Benjamin-Cummings, Menlo Park, California, 1978, Chapter 10.

A different strategy for producing a photodegradable plastic is to incorporate a photosensitizer into the formulation. The photosensitizer is a light-absorbing substance which initiates chemical attack on the polymer molecules—for example, hydrogen atom abstraction, followed by attack of atmospheric oxygen. The redox properties of a number of transition metals make their compounds suitable photosensitizers[35].

Biodegradable plastics are conventional plastic materials which are formulated with 5–10% of starch as a binder. They have been introduced to the North American market as "environmentally friendly" grocery bags. Starch, an isomer of polyglucose, is readily biodegradable. When the binder degrades, the bag loses its structural integrity. However, only the starch component of the plastic is biodegradable; the remaining synthetic polymer remains unconsumed, but as a less noticeable powder. Plastics with starch binders are incompatible with excessive humidity, otherwise microbial degradation may be initiated before the plastic is used.

In the future is the possibility of a genuinely biodegradable series of plastics. The bacterium *Alcaligenes eutrophus* produces the polyester polyhydroxybutyric acid which has physical properties suitable for making soft drink bottles. When the bacteria are grown on a mixture of glucose and organic acids, a more flexible copolymer of 3-hydroxybutanoic acid and 3-hydroxypentanoic acid, acceptable for plastic bags, film, and certain medical applications. Such polymers would, of course, be completely biodegradable[36].

In conclusion, current photodegradable and biodegradable plastics are not **completely** degradable. Each contains a chemical feature which allows discarded items to lose their structural integrity, although the bulk of the synthetic polymer remains chemically intact.

An exciting development in the treatment of plastic wastes is their recycling into items such as fence posts, park benches and other structural items, which have the advantage over wood and steel that they do not rot. Even incompatible plastics (that is, different polymers which are immiscible) can be blended together to form articles having the strength needed for these applications.

8.3.4 Industrial wastes: the Love Canal

The Love Canal is one of the most notorious examples of environmental mismanagement. To understand the situation we must present a brief history[37].

In the late 1800's William Love proposed to build an 11 km canal to bypass Niagara Falls, and simultaneously to provide inexpensive hydroelec-

[35] G. Scott, "Polymers with enhanced photodegradability," *J Photochem Photobiol (A)*, **1990,** 51, 73–79.

[36] R. Pool, "In search of the plastic potato," *Science,* **1989,** 245, 1187–1189.

[37] S.E. Manahan, *Environmental Chemistry 4th ed,* Willard Grant Press, Boston, Mass., 534–537.

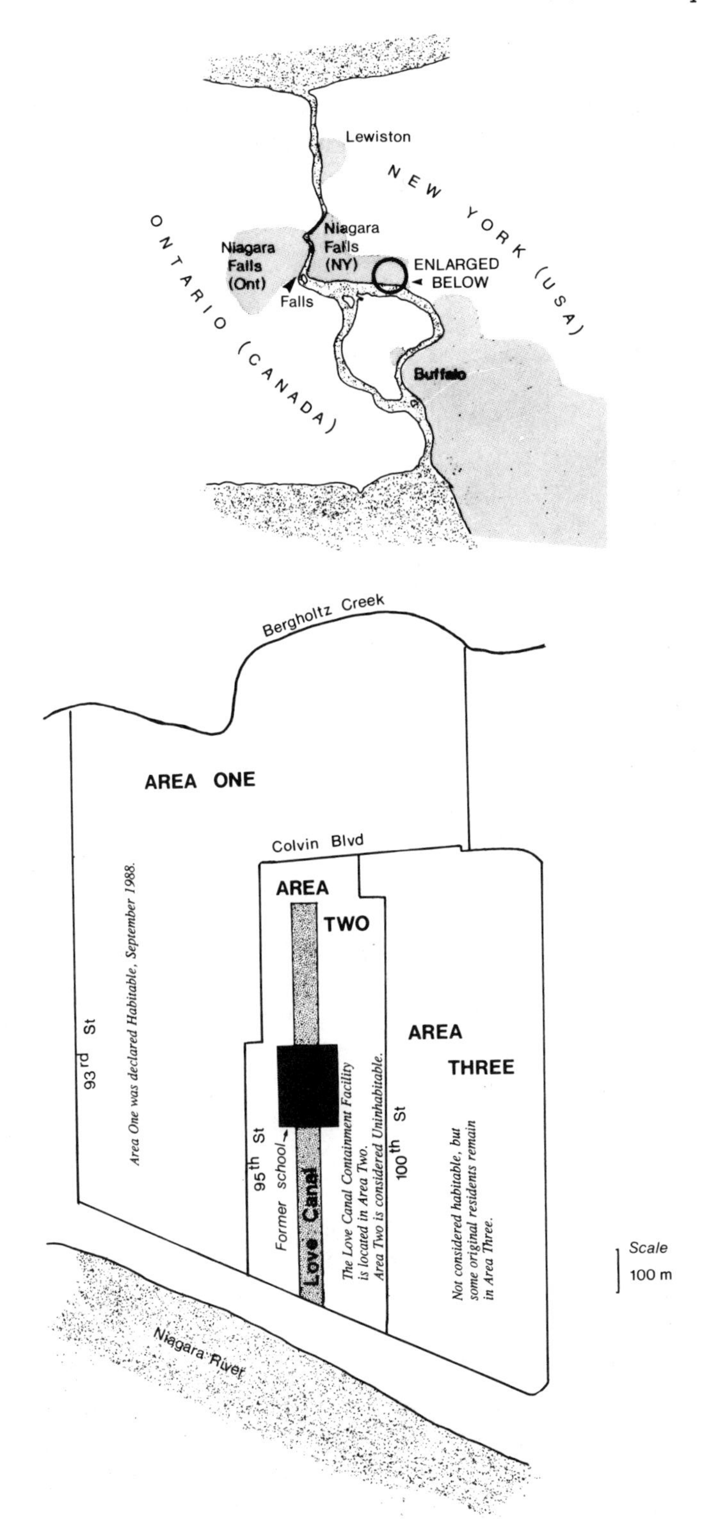

Figure 8.6: Area around Love Canal, near Niagara Falls, N.Y.

tric power. The project failed financially, leaving an open ditch about 15 m wide, 1.6 km long, and varying in depth between 3 and 13 m. Hydroelectric power came to the area a little later, attracting industry, including the Hooker Chemical Company, a manufacturer of chlorine and organochlorine derivatives[38]. Beginning in the 1940's the uncompleted canal was used for the disposal of chemical wastes, and by the time the site was filled, some 20,000 tonnes of these wastes had accumulated. The filled dump was sealed with a clay cap.

Subsequent events transformed the site into a disaster area. Homes were built around the abandoned canal; roads crossed it, breaching the clay cap; and a school was built on the property, the schoolyard of which was actually sited on the old dump. During the 1970's, rain and melted snow infiltrated the breached cap, allowing leaching of some of the contents of the dump. Toxic chemicals were found in the schoolyard and in the basements of the nearby homes. Eventually, many residents had to be evacuated and relocated, their homes were expropriated, and the site sealed off. Well over $140 million has been spent in an attempt to rehabilitate the area, which has been designated a "Superfund" site. The question of the financial liability by the present owner of Hooker Chemical, namely Occidental Chemical, represents an important challenge to environmental and corporate lawyers[39].

The Love Canal is not the only abandoned dumpsite in the vicinity of Niagara Falls, New York[40]. The area was one of the early locations for the manufacture of toxic organochlorine compounds, including DDT, Aldrin, Dieldrin, and Mirex: see also Chapter 9. As recently as the early 1960's environmental pollution did not appear to be a serious concern, and organochlorine wastes were dumped without any special precautions. The marriage of gas chromatography to mass spectrometry (gc/ms analysis) showed the world how pervasive was environmental contamination, particularly by organochlorine compounds. These substances are unreactive chemically; like other unreactive, long-lived pollutants, they have become widely distributed in the environment. Leachate from the many abandoned dumps around Niagara Falls, New York has been, and still is, a major source of the trace quantities of organochlorine compounds in the Niagara River and Lake Ontario. This contamination continues to be of public concern, since many millions of Americans and Canadians take their drinking water from Lake Ontario.

The critical event in the Love Canal story seems to be the development of the area around the former dump for residential purposes. In this respect "Love Canal" is by no means an isolated incident. Residential development on former garbage dumps has frequently occurred, leaving the unfortunate

---

[38]The manufacture of chlorine is discussed in Chapter 10.

[39]See, for example a news item in *Chem Eng News,* June 19, 1989, 20.

[40]V.A. Elder, B.L. Proctor, and R.A. Hites, "Organic compounds found near dump sites in Niagara Falls, New York," *Environ Sci Technol,* **1981,** 15, 1237–1243.

homeowners with cracked basements due to settling, and dangerous concentrations of methane in their basements. A partial solution to the problem is to bore holes into the ground, and vent (and if necessary flare) the methane into the atmosphere. Alternatively, the methane can be collected and piped to a point of use[26].

8.3.5 *Other dumping procedures*

Ocean dumping involves taking chemical or other wastes out to sea into international waters, and consigning the barrels to the deep. The 1972 "International convention on the prevention of marine pollution" (the "London dumping convention") prohibited this practice, as well as the custom of oil tankers flushing out their bilges at sea, thereby contributing to oil pollution. Ocean dumping of sewage sludge is common, the U.K. and Ireland in particular using this method to dispose of a substantial proportion of their sewage sludge. The U.S. EPA will no longer permit U.S. cities to practise ocean dumping of sewage sludge after 1992.

Coastal waters become polluted by receiving untreated sewage and discharges from coastal industrial plants. In addition, ocean-going ships still heave their garbage overboard, and with the increase in shipping activity, the littering of coastline with plastic has become a significant issue.

Waste is also buried deep underground. Liquid waste, especially aqueous, may be injected deep into the ground. Although the idea is that injection should occur deeper than any aquifer, any misjudgement of the underlying geology poses a substantial threat of ground water contamination. The Hanford nuclear reprocessing site, Washington State is the source of one such current controversy. Deep well injection of liquid wastes, including those from war gas manufacture, is thought to have triggered a number of minor earthquakes in the vicinity of Denver, Colorado. This would most certainly indicate that deep well injection is unsafe in seismically unstable areas.

The use of former mines to store waste is another possibility. It is more expensive than deep well injection, and so is usually reserved for treated waste, as discussed below. For wastes that are dangerous, or very slow to degrade, the issues to be addressed are groundwater contamination and the geological stability of the site.

Waste output per capita in the developed world is very high, and the combination of two factors—greater environmental consciousness and shortage of suitable new sites to replace filled dumps—is putting great pressure on municipalities to find ways of disposing of waste. This has led to the phenomenon of exporting waste, chemical as well as municipal, to Third World countries, which are short of cash and where environmental protection may not yet have become a matter for public concern. West Africa has become the repository of much of this waste, especially from Europe. An important step towards rectifying this situation is the 1989 "Basel convention on the control of trans-boundary movements of hazardous wastes and their disposal"[41], signatories to which agree to restrictions on such trans-boundary movements, and also to guidelines on how such waste should be handled. Critics of the convention have argued that it does not go far enough, and point to the fact that several major waste-producers, including the U.S., U.K., U.S.S.R., and the "two Germanys," had not signed (as of late 1989).

8.4 Ultimate treatment methods for hazardous wastes

"Ultimate" methods are designed as permanent disposal methods for waste. A distinction must be made between organic and biological hazardous materials—which can be destroyed—and metallic wastes, where the hazard is associated with the metallic element itself. The latter can be immobilized, but not destroyed. In this situation, the ultimate solution is to immobilize the waste in a non-toxic, non-migrating form, so that it can be landfilled safely.

8.4.1 Incineration

The objective of incineration is to convert organic and/or biological materials to CO_2 and H_2O. Incineration may be applied to domestic waste (garbage), chemical (industrial) waste, and biological (e.g., hospital) waste. Incineration has several potential advantages. First, combustion greatly reduces the volume of the waste. Industrial chemical waste can usually be reduced to negligible residual volume, while the residue to be landfilled from municipal garbage comprises about 25% of original volume. Prior separation and recovery of items of value such as metal and glass can minimize the volume of the residue. For example, an incinerator in London, England produces 2000 tonnes of ash from the 7000 tonnes of refuse incinerated weekly[42]. Further, once the residue is landfilled, it does not settle later—unlike raw garbage—allowing the possibility of subsequent development on the site.

Incineration also provides sterilization. This is important for preventing the spread of infectious disease from hospital waste; likewise it rids garbage of pathogenic microorganisms, and of potential food value for vermin. Combustion also offers the possibility of recovering the waste heat and putting it to use, whether the waste is industrial or domestic. Additionally, it may in the future become economical to recover trace metals from the **fly**

[41] D.A. O'Sullivan, "International controls on transport, disposal of wastes agreed upon," *Chem Eng News,* April 3, 1989, 21–22. See also a previous article: *Chem Eng News,* September 26, **1988,** 24–27. For specific examples, see A.K. Vir, "Toxic trade with Africa," *Environ Sci Technol,* **1989,** 23, 23–25.

[42] D.C. Ayres, "Organochlorine waste disposal—cremation or burial?," *Chem in Britain,* January **1987,** 41–43.

ash produced from municipal garbage. Finally, there is the possibility of combining fly ash with cement to make structural concrete (compare Section 8.4.2 below). To date this has been inhibited because of the presence of trace metals such as lead and cadmium in the ash, and the concern that they might subsequently enter the wider environment.

The chief disadvantage of incineration is air pollution: this includes NO_X production, as in all combustion processes, and the emission of particulate matter. The fly ash from municipal incinerators is very finely divided; it is of the micrometer dimension associated with ingestion into the lungs. It is catalytically reactive and has been found to be mutagenic in the Ames assay. Curbing fly ash emissions is feasible using, for example, electrostatic precipitators. Other trace emissions are the polychlorinated dibenzo-*p*-dioxins and dibenzofurans (PCDD's and PCDF's); these form in trace quantities whenever chlorine-containing organic material is burned (Chapter 9).

PCDD's and PCDF's are emitted mostly as adsorbates on fly ash, rather than as free gaseous species. Although the most toxic PCDD and PCDF congeners are only minor components of the total, their presence raises considerable public opposition to incineration of municipal waste. The actual amounts of PCDD's and PCDF's are very small; surprisingly, hospital incinerators can be among the more significant sources of these pollutants, mainly because of the lack of pollution control equipment compared with municipal solid waste incinerators[43].

Various schemes have been devised to make use of the energy value of waste. For example, municipal waste can be used to raise steam for heating purposes, or to make cement. The ash becomes incorporated into the product, and so requires no disposal. In the case of waste plastics and used automobile tires, the energy value represents much of the energy content of the petroleum from which the product was originally made. This issue came to prominence in 1990, when a huge fire at a Canadian storage site for tires burned for two weeks, and caused some damage to groundwater[44]. As in so many issues surrounding waste handling and disposal, storage carries with it the risk of fire or other accident. Many jurisdictions now permit the use of scrap tires as part of the fuel in cement kilns.

Incineration at sea has been a very controversial issue. This process involves taking hazardous wastes out into international waters in specially designed incinerator ships. Combustion takes place well away from land, and hence also from people. Any acidic gases (e.g., HCl) released are absorbed and neutralized by the ocean (pH 8.1). Combustion at sea began in Europe in 1969. The incinerator ship Vulcanus has incinerated a variety of hazardous wastes since 1972, including excess Agent Orange herbicide from the Viet Nam War, and test burns of PCB's in the Gulf of Mexico[45].

---

[43] C.C. Lee, G.L. Huffman, and R.P. Nalesik, "Medical waste management," *Environ Sci Technol,* **1991,** 25, 360–363.

[44] News Focus item, *J Air Waste Management Assoc.,* **1990,** 40, 900.

[45] P.S. Zurer, "Incineration of hazardous wastes at sea," *Chem Eng News,* December 9, **1985,** 24–42.

Critics have contended that the design of the incinerator ships is inadequate to ensure combustion at the efficiency that would be required on land, and have also cited spills, leaks, and the lack of adequate policing and inspection of ocean-going incinerator ships. For example, incineration of 10,000 tonnes of Agent Orange by Vulcanus, at an efficiency stated to be 99.9%, left 10 tonnes of the herbicide unaccounted for[46]. By 1981, the European countries had already agreed to phase out ocean incineration by 1990.

8.4.2 Solidification of hazardous wastes[47]

Most wastes are in the form of liquids or sludges. From a disposal point of view, a solid waste is easier to deal with, because it is immobile, and is also less likely to become dispersed through the environment.

The immobilization of wastes is variously known as solidification, fixation, and stabilization. Where a distinction is made, solidification has the connotation of forming a solid mass having sufficient structural integrity to allow it to be transported in convenient sized pieces without the need for secondary containment; stabilization suggests immobilization through a chemical reaction, or entrapment in an impermeable and inert structure. In either case, a solidified waste is less likely to pose a threat of toxicity, as the toxic component is unable to reach a target, even though it has not actually been destroyed. An objective of solidification is the production of a material which is sufficiently innocuous chemically that it may be safely disposed of by some method such as landfilling. One point to be kept in mind is that these fixation processes all increase the volume of waste.

Fung[48] classifies three kinds of toxic waste suitable for fixation: flue gas cleaning sludges, waste streams containing toxic inorganics, and waste streams containing toxic organic substances. Radioactive wastes are a special case of inorganic wastes. Organic wastes presents different problem from inorganic materials, because of possible incompatibility of the waste with the solidification matrix.

There are several quite different approaches to waste fixation:

- Cement and lime based technologies
- Thermoplastic and polymer techniques
- Encapsulation techniques
- Vitrification

---

[46]P.E. des Rosiers, "Remedial measures and disposal practices for wastes containing dioxins and furans," *Chemosphere,* **1983,** 12, 727–744.

[47]R.B. Pojasek, "Using solidification as a waste detoxication process," Chapter 11 of *Detoxication of hazardous waste,* Ed. J.H. Exner, Ann Arbor Science, Ann Arbor, Michigan, 1982.

[48]R. Fung, *Protective barriers for the containment of toxic materials,* Noyes Data Corp., Park Ridge, New Jersey, 1980, 269–287.

Cement and lime based technologies

Ordinary concrete is made by mixing together cement, sand and water in an appropriate ratio. Over the course of hours, the mixture hardens to a solid. In cement-based fixation, the waste is used in place of sand (if the waste is solid, e.g., fly ash), or in place of sand and water if the waste is an aqueous slurry. The result is a concrete in which the waste is incorporated.

Cement is manufactured by heating limestone with a silicate mineral, together with a lower proportion of aluminum oxide, until the mixture just begins to fuse into a glass. The limestone component decomposes:

$$CaCO_3(s) \longrightarrow CaO(s) + CO_2(g) \tag{20}$$

The resultant **clinker**, as it is called, is ground finely to give cement. Incidentally, the grinding operation at a cement plant is a source of pollution by particulates unless electrostatic precipitators are used to control the dust. The concrete sets when water is added to the anhydrous mixture to give an amorphous hydrated calcium silicate. Subsequent hardening of the initial "set" involves the growth of interlaced silicate fibers, as well as the absorption of CO_2 from the air, regenerating calcium carbonate. Organic wastes are incompatible with cement-based solidification because they inhibit the concrete from setting.

Many different types of cement exist; the table below shows the typical range of analyses for different types of Portland cement, which are the most commonly used for structural work. Rapid setting cements, which are sometimes useful in waste disposal, are made by the addition of a few percent of CaF_2. Any halides, sulfides, and alkali metals in the feedstocks are vaporized at the high temperature required to "calcine" the raw materials into clinker.

Table 8.1: Composition of Portland Cement[49]

CaO	62–65%	SiO_2	21–25%	Al_2O_3	3–5%
Fe_2O_3	2–4%	MgO	2–3%	SO_3	2–3%

The composition is given as a mixture of oxides, but because CaO is a basic oxide and SiO_2 is acidic, in practice a major constituent of the cement is $CaSiO_3$. Likewise, the inclusion of SO_3 in the analysis implies the presence of metal sulfates rather than free sulfur trioxide.

Flue gas cleaning sludges provide an example of the use of cement-based solidification. Those produced at coal-burning power stations (Chapter 6) result from acid emission abatement strategies. They contain finely

[49] *Kirk-Othmer Encyclopedia of chemical technology,* 3rd Ed., Wiley-Interscience 1975, Volume 5, 163–191.

divided $CaSO_3/CaSO_4$ together with some fly ash. When acid gas scrubbing is done under aqueous conditions, the sludge is largely water, and the small size of the crystals makes "dewatering" very difficult. The amounts are vast: tens of millions of tonnes annually in the United States alone. Fly ash from incinerators and the fly ash/calcium salt mixtures obtained by dry scrubbing are also very finely divided, as noted above, but they are dry, and hence less voluminous. Stabilization of these wastes is best effected by cement-based technology.

Inorganic wastes, such as the sludges produced by sulfide precipitation of aqueous waste streams, are generally suited to cement-based solidification. Since cement—and thus concrete—are basic materials, the sludges of sulfides, hydroxides, phosphates etc of heavy metals can be maintained under alkaline conditions. Recall that these basic anions are solubilized as the pH drops, thereby releasing the cations also. Concrete therefore preserves the waste in a chemical state which minimizes leaching into the environment. Treatment of the concrete with a sealant may be beneficial if the concrete is likely to become exposed to acidic conditions.

Lime based solidification depends on forming a "pozzolanic concrete" by admixture of lime ($Ca(OH)_2$), water, and what are called pozzolanic silicate minerals. This kind of concrete has been known and used since Roman times. Wastes which can be solidified in pozzolanic concrete include fly ash, cement kiln dust, and ground blast furnace slag. The advantages of this technology are the low cost of the materials and the simplicity of the mixing equipment. Like regular concrete, lime fixing adds greatly to the bulk of the waste. However, uncoated pozzolanic concrete is more susceptible to acid leaching than ordinary concrete.

Not all metals are immobilized equally efficiently. For example, laboratory studies with model sludges showed that lime treatment followed by solidification with cement was substantially more effective at immobilizing cadmium than lead. In part, this was because the initial lime treatment converted Cd^{2+} completely to $Cd(OH)_2$, which formed nucleation sites for concrete setting, while the corresponding Pb system still contained soluble mixed salts[50].

Thermoplastic and polymer techniques

Inorganic materials can be mixed with a matrix such as bitumen, which provides protection against acid leaching. Other thermoplastic materials that can be used include paraffin wax and polyethylene. A rather high proportion of the thermoplastic (e.g., 1:1 with the waste) is needed. In the case of bitumen, the mixture must be blended at temperatures above 100°C to soften the bitumen, and to drive off any water that is present.

---

[50] F.K. Cartledge *et al*, "Immobilization mechanisms in solidification/stabilization of Cd and Pb salts using Portland cement fixing agents," *Environ Sci Technol,* **1990**, 24, 867–873.

The treated mix is normally contained in a steel drum before disposal by landfilling.

Thermoplastic-based fixation is practical only for dry inorganic wastes. Strong oxidants such as perchlorates and nitrates are incompatible, as are salts such as sodium sulfate, which have easily dehydrated hydrates. Heating converts the latter to the anhydrous form, but if the matrix later comes in contact with water, rehydration can cause the matrix to split apart, increasing the likelihood of water-soluble substances leaching from the ruptured medium. Aqueous sludges are incompatible with a thermoplastic matrix because of the energy requirement to drive off the water; organics are unsuitable because bitumen is soluble in organic solvents, and because volatile organics would present a fire hazard at the heating/mixing stage.

In thermoplastic fixation, the thermoplastic substance, which is composed of large organic molecules, forms a barrier between the solid waste and the environment. The same goal may be achieved if the waste is mixed with a polymer precursor, and the polymer is generated *in situ*. Urea-formaldehyde has been extensively used for the latter purpose. Both wet and dry wastes can be treated; however, the polymer does not bond chemically to the waste, but rather entraps it in the polymer matrix. This means that any liquid will remain after polymerization, and the polymer mass must be dried before disposal by landfilling. A disadvantage of the system is that urea-formaldehyde is cured with the use of acidic catalysts. These tend to mobilize toxic metals, especially if the waste contains any water (and this aqueous phase, containing high levels of metals, will remain after polymerization).

Encapsulation techniques

Encapsulation is the formation of an impermeable jacket around the waste by use of an organic polymer such as polyethylene. The compacted waste is heated under pressure with a small proportion of a binder such as a polybutadiene to improve adhesion between the jacket and the waste. The polyethylene jacket is then applied, and adheres to the polybutadiene. The process is applicable to water-soluble (but dry) wastes, since the treated waste is physically kept out of contact with water. However, immobilization depends on maintaining the integrity of the jacket. This latter limitation can be overcome—albeit at greater expense—by **microencapsulation**, in which the individual waste particles are encapsulated rather than the whole block. The encapsulated particles are fused into a block by means of heat and pressure, but the barrier between the waste and the environment remains intact even if the block should for any reason crack or break.

Vitrification

Vitrification means formation of a glass. Ordinary glasses are mixtures of metallic oxides which fuse together at high temperature. The composition of a glass is expressed in terms of the constituent oxides (CaO, Na_2O, SiO_2, B_2O_3, etc), although the raw materials which go into the glass are mainly carbonates, which break down to oxides at vitrification temperatures: e.g.,

$$CaCO_3(s) \longrightarrow CaO(s) + CO_2(g) \quad (20)$$

Glasses vary greatly in their susceptibility to leaching. Borosilicate glasses (e.g., the tradename Pyrex) contain substantial amounts of boric oxide, and are resistant both to leaching and to fracture through thermal shock. Aluminosilicate glasses are very strong, and retain their strength at high temperatures.

Table 8.2: Composition of some common glasses[51]

	SiO_2	B_2O_3	Al_2O_3	CaO	MgO	Na_2O
soda-lime	73	1	2	5	4	15
borosilicate	81	13	2	–	–	4
aluminosilicate	64	5	10	9	10	2

A waste disposal method based on vitrification is possible because other metal oxides may substitute in part for the Na_2O, CaO, MgO etc in the structure. Vitrification is very energy intensive, since melt temperatures above 1300°C are needed, and on this account is applicable only to particularly hazardous inorganic wastes, which must under no circumstances leach into the environment. For example, vitrification has been studied extensively for treating the radioactive wastes formed as byproducts in nuclear fission power reactors. These wastes are first treated using aqueous and non-aqueous solution chemistry to recover uranium, plutonium, and other valuable materials. The residual metallic elements could be converted to a mixture of oxides and then incorporated into a glass.

The concern about leakage of radioactive wastes is so great that the following measures have been proposed.

a. select a site such as a deep mine in a geologically inert formation—such as the American Southwest or the granite rocks of the Canadian Shield

b. vitrification

c. seal the glass into stainless steel drums

d. encase the drums in concrete

Experiments with such systems have shown that the vitrification step alone affords a glass which is so resistant to leaching that prolonged contact with water leaves the water fit to drink—in the sense of being within

[51] R.H. Doremus, *Glass science,* Wiley-Interscience, New York, 1973, 102.

guidelines for radioactive content[52]. The stainless steel drums have been estimated to last at least 500 years, and the concrete case would further immobilize any leachate both by physical containment and because of its basicity. Finally, since the containment needed would be for such a long time, the mine site would be sealed to prevent accidental access, perhaps by members of future generations. The great depth of rock between the storage site and the surface would ensure that no radioactive emissions reached the surface. Another advantage of such a scheme is that, should the radioactive wastes later prove to have economic value they could be "mined."

Schwoebel and Northrup[53] have pointed out that all glasses eventually devitrify, and that the crystalline phase is often more susceptible to leaching than the glass. They suggest that incorporation of radionuclides into titanate ceramics would give even greater safety from leaching than conventional vitrification.

8.4.3 Soil remediation

This term refers to cleaning up the soil at a former manufacturing facility or waste site[54]. Such soils may be heavily contaminated with inorganic and/or organic compounds, depending on the previous use of the site. Clean up is often very expensive on account of the large volumes of soil involved. Because of this, corporate legal departments look very closely at deals involving acquisition of another company's plant or site, and in some cases the former owner is required to retain the responsibility for maintaining part of the property.

Space limitations allow us only to list some of the technologies under consideration in this active area of waste management.

a. biological remediation, using mixed populations of microorganisms ("bugs in a bag"). Generally, the organisms capable of metabolizing a particular waste are rather specific to that waste, and the best route to go is to optimize the consortia of microorganisms already present at the site. "Imported" microorganisms tend not to survive. Enclosed bioreactors may be used, especially in northern climates, where the outdoor growing season for the microorganisms is short.

b. vitrification has been suggested for some especially hard to treat soils e.g., those contaminated with PCB's.

---

[52]P.A.H. Saunders and J.D. Wilkins, "Radioactive waste disposal: chemical control," *Chem in Britain,* May **1987,** 448–452.

[53]R.L. Schwoebel and C.J. Northrup, "Nuclide stabilization in ceramics," Chapter 13 in *Toxic and hazardous waste disposal,* Volume 1, Ed. R.B. Pojasek, Ann Arbor Science, Ann Arbor, Michigan, 1982.

[54]R.C. Sims, "Soil remediation techniques at uncontrolled hazardous waste sites," *J Air Waste Management Assoc,* **1990,** 40, 704–732.

c. solvent extraction is possible in principle, but leaves the soil contaminated with the solvent. This problem may be circumvented by the use of supercritical fluid extraction[55].

Site remediation is a gigantic task for the mining industry. The issue of acidic drainage has been considered in Section 8.2.2. This is referred to as secondary contamination by Moore and Luoma, who have published a case study of a site in Montana[56]. The acidic water may carry metal ions tens of kilometers down river, thereby contaminating wide areas. Primary contamination is the presence on site of waste rock from oil flotation, called tailings, and flue dust and slag from smelters. These latter may contain toxic metals at concentrations thousands of times greater than the original ore. To give an idea of the scale of these operations, the Montana site has ponds covering 35 km^2, which could contain millions of tonnes of unrecovered metals such as arsenic, cadmium, copper, lead, silver, and zinc, but all at ppm concentrations in the water.

8.5 Problems

1. A sewage treatment plant is designed to process 3.0×10^6 L of sewage daily. What capacity is required for the primary settling lagoon if the residence time is to be 6 hours? Suggest possible dimensions for this lagoon if the water is to be no more than 1.0 m in depth.

2. Suppose that the raw sewage at the plant in Problem 1 has BOD of 850 ppm.

(a) If a 90% reduction in BOD is achieved during secondary treatment, what volume of oxygen will be required (assume 15°C)?

(b) At what average rate must oxygen be transferred from the atmosphere to the sewage?

3. A lake water sample has the following partial analysis: total carbonate, 86 ppm; nitrate, 0.12 ppm, ammonia, 0.04 ppm, phosphate (as PO_4^{3-}), 0.08 ppm. Which is the limiting nutrient?

4. The residence time of the water in Lake Erie is 2.7 years. If the input of phosphorus to the lake is halved, how long will it take for the concentration of phosphorus in the lake water to fall by 10%?

5. (a) Calculate the equilibrium concentrations of soluble phosphate produced when (i) hydroxylapatite (ii) $Ca_3(PO_4)_2$ dissolve in pure water.

[55] X. Yu, X. Wang, R. Bartha, and J.D. Rosen, "Supercritical fluid extraction of coal tar contaminated soil," *Environ Sci Technol,* **1990,** 24, 1732–1738.

[56] J.N. Moore and S.N. Luoma, "Hazardous wastes from large scale metal extraction," *Environ Sci Technol,* **1990,** 24, 1278–1285.

(b) A sewage sample contains, after secondary treatment, 8.8 ppm phosphorus in the form of ortho-phosphate. It is brought to pH 9.0, and $[Ca^{2+}] = 4.7$ mmol L^{-1} by the addition of lime. What fraction of the phosphate is precipitated if the precipitate is (i) hydroxylapatite; (ii) $Ca_3(PO_4)_2$?

6. (a) A lake has pH 7.25 and contains 0.04 ppm phosphorus and 75 ppm of $Ca^{2+}(aq)$. Is it saturated with respect to hydroxylapatite?

(b) Under these conditions, calculate ΔG for the reaction:

$$5Ca^{2+}(aq) + 3PO_4^{3-}(aq) + OH^-(aq) \longrightarrow Ca_5(PO_4)_3OH(s)$$

7. Using the solubility data for $Al(OH)_3$ from Chapter 6, and K_{sp} for $AlPO_4 = 1.0 \times 10^{-21}$ (mol $L^{-1})^2$, plot the solubilities, separately, for $Al(OH)_3$ and $AlPO_4$ between pH 4.0 and 7.0. Over what pH range is filter alum most effective in precipitating phosphate?

8. Filter alum $Al_2(SO_4)_3$ is often used to remove phosphate ion from waste water. A wastewater of pH 5.62 containing 25 ppm total phosphate is treated with alum until the equilibrium concentration of Al^{3+} is 4.0×10^{-9} mol L^{-1}. What fraction of the phosphate is precipitated as $AlPO_4(s)$? Consider only the equilibria below:

$$AlPO_4(s) \rightleftharpoons Al^{3+}(aq) + PO_4^{3-}(aq) \qquad K_{sp} = 1.0 \times 10^{-21}\ (\text{mol L}^{-1})^2$$
$$H_2PO_4^-(aq) \rightleftharpoons HPO_4^-(aq) + H^+(aq) \qquad K_a = 6.2 \times 10^{-8}\ \text{mol L}^{-1}$$
$$HPO_4^{2-}(aq) \rightleftharpoons PO_4^{3-}(aq) + H^+(aq) \qquad K_a = 4.8 \times 10^{-13}\ \text{mol L}^{-1}$$

9. (a) Explain in your own words why salt is used to help precipitate a soap.

(b) Sodium stearate has an aqueous solubility of ≈ 0.5 mol L^{-1}. A student prepares sodium stearate from tristearin (10 g) and an almost stoichiometric amount of NaOH, then pours the mixture into water (100 mL). What is the recovery of sodium stearate (i) if no salt is added (ii) if 10 g of salt is added to the solution?

(c) The CMC for sodium stearate is ≈ 0.001 mol L^{-1}. Why would a soap made from sodium stearate be of little value in sea water? Take sea water to be a 0.5 mol L^{-1} solution of NaCl.

10. You have decided to wash your clothes in ordinary soap (e.g. "Ivory" soap). Your water supply contains 47 ppm of $Ca^{2+}(aq)$. You will be using 50 g of soap in a 60 L washing machine.

(a) Assuming the soap to be sodium stearate ($C_{17}H_{35}CO_2Na$), use K_{sp} for the calcium salt $= 1.0 \times 10^{-12}$ (mol $L^{-1})^3$ to calculate the mass of scum produced.

(b) How much soap would be required to get the same detergent action if the Ca^{2+} concentration were 2 ppm?

11. (a) Calculate the BOD resulting from the discharge of 1.0 kg of a soap $C_{17}H_{33}CO_2Na$ into a pond of capacity 1800 m^3, assuming that the soap is completely degraded to CO_2 and water within 5 days.

(b) Calculate the COD when a detergent $NaSO_3C_6H_4(CH_2)_{11}CH_3$ is oxidised to $NaSO_3C_6H_4CO_2H$. The initial concentration of the detergent is 0.14 g L^{-1}.

12. An amphoteric detergent has the structure $CH_3(CH_2)_{11}NH_2^+CH_2CH_2CO_2^-$ in aqueous solution. The $R_2NH_2^+$ group has $K_a = 1.2 \times 10^{-11}$ mol L^{-1} and the RCO_2H moiety has $K_a = 1.4 \times 10^{-5}$ mol L^{-1}. Calculate the pH of a 0.010 mol L^{-1} solution of this detergent. (Note: the derivation is quite long: consider the release into the water of H^+ from RCO_2^- and of OH^- from R_2NH^+.)

13. (a) Calculate the pH of a 0.010 mol L^{-1} solution of a long chain soap $RCO_2^-Na^+$ for which $K_b = 8.0 \times 10^{-10}$ mol L^{-1}.

(b) Repeat this calculation for the case of a long chain detergent $RSO_3^-Na^+$. The conjugate acid RSO_3H has $K_a = 1.0 \times 10^{+2}$ mol L^{-1}.

(c) Repeat calculation (b) for the situation where in addition to 0.010 mol^{-1} RSO_3Na, the detergent also yields 0.0020 mol L^{-1} Na_2CO_3 as a builder.

In each case assume that the solvent is pure water.

14. For the reaction:

$$Ca^{2+} + HT^{2-} \longrightarrow CaT^- + H^+$$

K_c has the value 7.8×10^{-3} (H_3T = nitrilotriacetic acid). Work out, at 25°C, the equilibrium concentration of dissolved calcium

(a) when pure water is in equilibrium with $CaCO_3(s)$.

(b) when water containing 1.0×10^{-3} mol L^{-1} NTA at pH 7.5 is in equilibrium with $CaCO_3(s)$.

15. The association of Pb^{2+} with NTA may be represented as:

$$Pb^{2+}(aq) + HT^{2-}(aq) \rightleftharpoons PbT^-(aq) + H^+(aq)$$

for which $K_c = 13$. For a (i) 1.0 ppm (ii) 0.02 ppm concentration of NTA in a water body, calculate the additional Pb^{2+} that would dissolve. Take the concentration of lead in the water as 3.0 ppm, in the absence of any contamination by NTA. Assume a pH of 7.22.

16. A water sample contains 15 ppm of ammonia. What chlorine dose should be added in order to produce a chlorine residual of 0.85 ppm?

17. The rate of the reaction between HOCl and NH_3 to form NH_2Cl has the form:

$$\text{rate} = k[HOCl][NH_3]$$

Note that it is HOCl (not ClO^-) and NH_3 (not NH_4^+) that react together. Given that HOCl has $K_a = 3.0 \times 10^{-8}$ mol L^{-1} and NH_4^+ has $K_a = 5.6 \times 10^{-10}$ mol L^{-1} at 25°C, find the pH at which the rate of NH_2Cl formation is at a maximum.

18. Gold ores are frequently leached with cyanide, dissolving the gold according to the equation

$$Au(s) + \tfrac{1}{4}O_2(g) + 2CN^-(aq) + \tfrac{1}{2}H_2O(l) \longrightarrow Au(CN)_2^-(aq) + OH^-(aq)$$

In order to prevent undue environmental contamination by CN^-, you wish to operate this process under conditions such that at least 98% of the CN^- is converted to $Au(CN)_2^-$. Your process operates at pH 9.0 and the O_2 pressure inside the ore body is constant at 0.032 atm. Calculate K for the reaction above, and use it to determine the CN^- concentration you should use. Comment on your result. Thermodynamic data:

substance	ΔG_f° (298K), kJ mol^{-1}
$CN^-(aq)$	172.3
$H_2O(l)$	−237.2
$OH^-(aq)$	−157.3
$Au(CN)_2^-(aq)$	285.8

19. A stream in the vicinity of a mine has flow rate 7.7 m^3 per minute. Its pH is 1.82; it contains 17 ppm of dissolved iron, and on the average, the stream flow increases by 15 m^3 min^{-1} for every km downstream from the mine. Assuming that the rate of oxidation to $Fe^{3+}(aq)$ is not limiting, calculate the concentration of dissolved iron in the water as a function of distance from the mine, and show the result graphically. Hence plot the mass of $Fe(OH)_3$ precipitated per day, also as a function of distance from the mine.

20. Calculate the maximum theoretical removal of copper from a leachate containing 145 ppm of Cu^{2+} by the use of cementation with scrap iron

(a) if Cu^{2+} is the only cation present

(b) if the solution also contains 350 ppm of Fe^{2+}.

Chapter 9

Chlorinated Organic Compounds

Introduction

Other than ionic chlorides, chlorinated compounds are rare in nature, at least on land. Chlorinated and brominated organic compounds are elaborated by some marine organisms; the study of the chemistry and the properties of these unusual compounds has only been undertaken quite recently.

Because terrestrial organisms do not naturally encounter organochlorine compounds, the introduction of chlorine into an organic structure can interfere with metabolism. This has been exploited in the use of organochlorine compounds—of very varied structures—as herbicides and pesticides (insecticides, algicides etc). The presence of chlorine in an organic molecule reduces reactivity towards oxidation. Many organochlorines are also resistant to hydrolysis, especially if the chlorine is substituted into an aromatic ring. On this account, organochlorine compounds tend to be persistent if they are released into the environment. As well, organochlorine compounds tend to be insoluble in water, but soluble in fats and oils, leading to the phenomenon of **biomagnification,** whereby tissue concentrations of these compounds tend to increase as one studies organisms higher in the food chain (Problem 1).

In this chapter, we shall discuss representative examples of organochlorine pollutants. DDT will be introduced because of its historical importance. PCB's and dioxins will form the bulk of the chapter, with some reference being made to the chlorophenols.

9.1 DDT

The initials DDT stand for DichloroDiphenylTrichloroethane, which is an incorrect name for 2,2-bis-(*p*-chlorophenyl)-1,1,1-trichloroethane, structure **1**. DDT, which is both an alkyl chloride and an aryl chloride, was first synthesized in 1874. Its uncritical acceptance into the marketplace in the

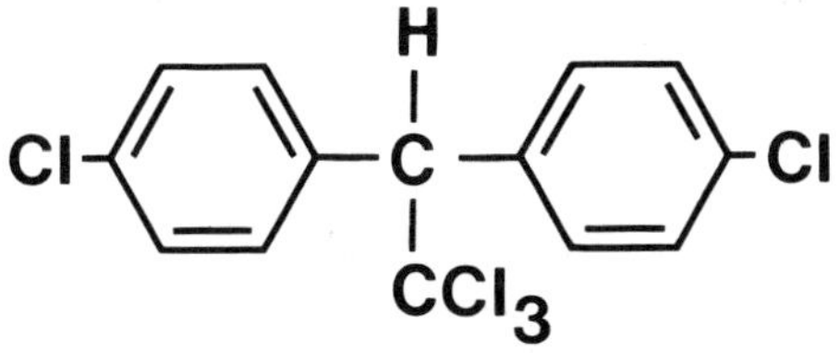

H

Cl–C–Cl

CCl_3

1

1940's was the result of circumstance[1]: during World War II, insect-borne disease was a serious threat to American and British forces. In Naples, typhus (which is carried by lice) reached epidemic proportions; in the malaria-ridden jungles of the Pacific islands and Far East, disease transmitted by mosquitos threatened to make any advance by troops impossible. The efficacy of DDT against insects was demonstrated experimentally in 1942, and full scale production was instituted immediately. The compound soon gained the reputation of a wonder chemical; it was more active against insects than any insecticide hitherto known, it was cheap to make (the price dropped to 60¢/kg by the mid 1950's), it had low acute toxicity to mammals, and was persistent. Persistence seemed an advantage at the time, because it reduced the frequency of application. Moreover, application rates were low (about 0.2–0.3 kg/ha), and since the material could be sprayed by aircraft, pests could be eradicated even in locations inaccessible from the ground.

Under these circumstances, when DDT became available for civilian use in 1945, its enthusiastic acceptance by farmers was hardly surprising. Early targets of DDT were insects which attacked cotton, but in the heady atmosphere of the times, DDT was soon tried against almost any pest, often with scant regard for the recommended application rates.

The early invstigators were aware of negative properties of DDT. Laboratory rats suffered fatty degeneration of the liver and kidneys after prolonged high exposure to DDT. Highly exposed soldiers in World War II had suffered aching joints, tremors, and depression—symptoms which indicate that DDT affects the nervous system. A possible explanation of the symptoms, which are slow to disappear, is that the lipid-soluble DDT accumulates in the insulating myelin sheaths around the nerves. The lipid solubility of DDT caused it to appear in milk, and the United States Food and Drug Administration recommended as early as 1947 that DDT not be used on feed or forage for dairy cattle, nor on vegetable crops for human consumption.

The low cost of DDT and its ease of application encouraged its use beyond the confines of the farm. Swamps, forests, and suburbs were sprayed to eradicate insect pests. By 1946 it was known that excessive use of DDT could cause the death of fish, birds and other wildlife. Residues of the

[1]Historical information in this section is based on T.R. Dunlap, "DDT: scientists, citizens, and public policy" Princeton University Press, Princeton, New Jersey, 1981.

lipid-soluble DDT were discovered in a wide range of wildlife, mainly in the fat, with carnivores having higher levels than herbivores. This was the first example of environmental biomagnification.

Carnivorous birds appeared to fare particularly badly; reduced hatchability of eggs and physical deformities in the chicks were linked statistically to high DDT levels in the parents[2]. Humans are also at the top of the food chain, and by the late 1960's citizens of several countries were horrified to learn that the levels of DDT in mothers' milk ranged up to 130 ppb, which would have been classified as unfit for human consumption (Problem 2).

Rachel Carson's book *Silent Spring* (1962)[3] was the first popular work to bring environmental contamination by pesticides to public attention. Carson focussed on the damage which uncontrolled use of persistent pesticides could cause to wildlife. Well publicized and well organized campaigns were mounted in several countries to prohibit the use of DDT and other persistent chlorinated insecticides such as Aldrin and heptachlor. Governments in many developed countries (Canada, Italy, Scandinavian countries, U.K., and U.S.) proclaimed bans on DDT or severely restricted its use around 1969–1970.

Despite the problems with its use, there is no doubt that DDT has saved countless lives in regions where malaria is endemic. Tropical countries which have discontinued its use have seen malaria incidence increase; some Third World countries still cling to the use of DDT because it is cheap and effective. Tissue levels in North American wildlife began to decline about 10 years after DDT was banned; however, it may routinely be detected even today because its long lifetime permits migration to North America from countries which still permit the use of DDT. As a result, DDT may still be found in all parts of the world, even in very remote regions.

The safety factor between the dose of DDT that will kill target insects and that which will harm non-target organisms is relatively small: DDT is described as an unselective insecticide. Research since DDT was introduced has focussed on the development of pesticides that are more selective and less persistent. For example, organophosphate insecticides are hydrolyzed within a few days into harmless products. Organophosphates are more acutely toxic however (their chemical structures are similar to those of nerve gases), and workers should be absent from fields and orchards during spraying operations.

As an example of a highly selective insecticide, pyrethroids are almost completely non-toxic towards mammals: they are familiar to us in household fly sprays. Pyrethroids are related to the natural insecticide pyrethrin, which is obtained from *Pyrethrum,* a daisy-like flower grown in large quan-

---

[2]C.A. Edwards, *Persistent pesticides in the environment,* CRC Press, Cleveland, Ohio, 1970. This book provides extensive tables, with data drawn from many countries, of DDT levels in wildlife, soil, food, and of bioconcentation factors between water and aquatic organisms.

[3]R. Carson, *Silent spring,* Houghton-Mifflin Co., Boston, Mass., 1962.

tities in Kenya[4]. Although another major thrust in pesticide research has been the development of pesticides which are less persistent in the environment, the natural pyrethrin suffers from the drawback of being too labile, especially in sunlight; cattle barns would need to be sprayed several times daily if pyrethrin were used. The challenge of the synthetic chemist therefore has been to prepare derivatives of pyrethrin which retain high selectivity towards insects and low mammalian toxicity, but which have a somewhat greater persistence (a few days) in use.

9.1.1 Chemistry of DDT

DDT is synthesized from trichloroacetaldehyde and chlorobenzene by a modified Friedel-Crafts reaction, Equation 1.

$$CCl_3-CHO \xrightarrow{H^+, PhCl} CCl_3-CH(OH)-C_6H_4-Cl \xrightarrow[-H_2O]{H^+, PhCl} Cl-C_6H_4-CH(CCl_3)-C_6H_4-Cl \quad (1)$$

Since chlorobenzene substitutes electrophilically at the *o*– and *p*– positions, the isomer shown (*p*,*p*′-DDT) is accompanied by small amounts of the *o*, *p*– and *o*, *o*′– isomers.

Until the advent of gas chromatography, DDT was analyzed by a colorimetric method, in which the sample was extensively nitrated with concentrated nitric acid and then treated with CH_3ONa/CH_3OH. Polynitro compounds give intensely coloured radical anions under these conditions, and these can be quantitated from the absorbance of the sample by means of calibration curves. Gas chromatographic analysis of DDT was developed in the early 1960's; analysis became simpler and detection limits lower, and as a result the wide dispersal of the compound through the environment and in wildlife became generally recognized.

DDT undergoes dehydrochlorination metabolically and also in the environment (probably photochemically) to 1,1-dichloro-2,2-bis-(*p*-chlorophenyl)-ethene, also known as DichloroDichlorophenylEthylene, DDE. DDE is separable from DDT by gas chromatography, but since DDE

[4]The natural insecticide pyrethrin is nòt new; its use was already established before DDT was introduced. The new developments in this area have been the synthetic pyrethroids. See J.P. Leahy, *The pyrethroid insecticides,* Taylor and Francis, London, England, 1985.

always accompanies DDT, the sum of the two is usually reported. In the early days of gas chromatography, there was confusion between DDT and some of the members of the PCB family (which had not yet been recognized as environmental pollutants), but this analytical problem was resolved by the introduction of gc/ms (gas chromatography coupled to mass spectrometry).

We finish this section with a look at some of the other early, persistent insecticides. Their use has now been discontinued, voluntarily or otherwise. All these compounds are highly chlorinated, persistent, and neurotoxic, but rather unselective.

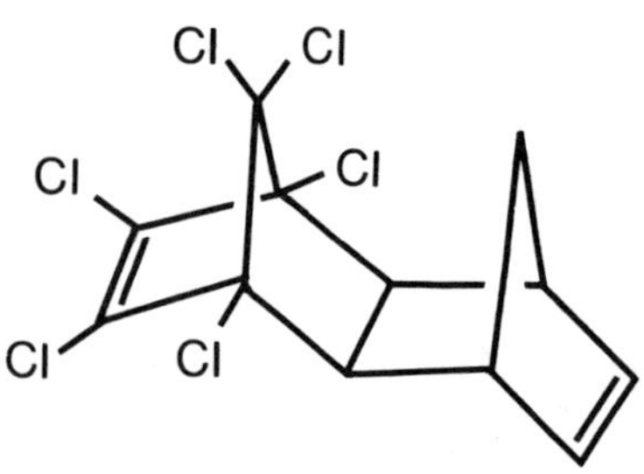

Aldrin

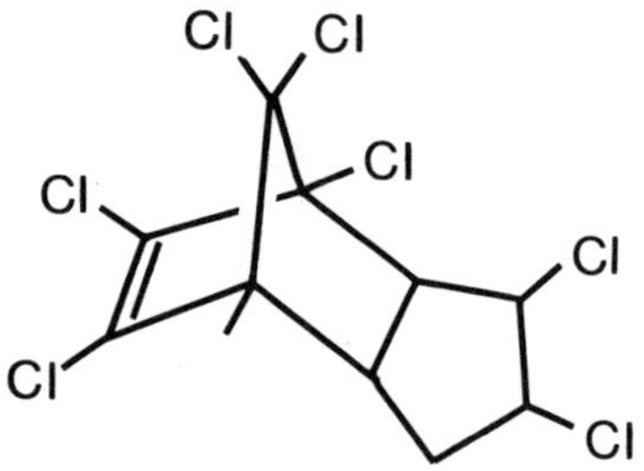

Chlordane

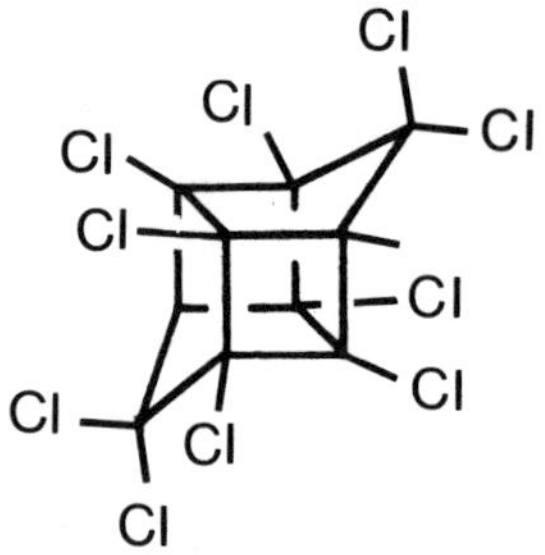

Mirex

9.2 PCB's

The initials PCB stand for PolyChlorinated Biphenyls. Unlike DDT, which contains one major isomer, commercial PCB preparations are mixtures. The mixtures contain both isomers and substances differing in the number of chlorine atoms. Excluding the parent compound biphenyl **2**, there are 209 chlorinated biphenyls.

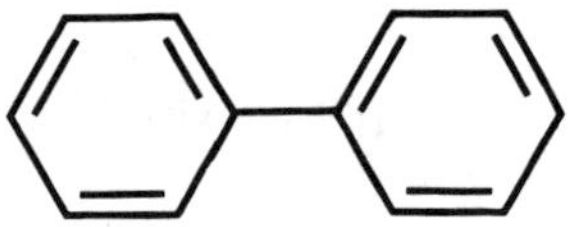

2

The term **congeners** is used to point up the relationship amoung members of a chemical family such as the PCB's. Thus all 209 chlorinated biphenyls can be described as congeners, irrespective of the number or location of the chlorine substituents.

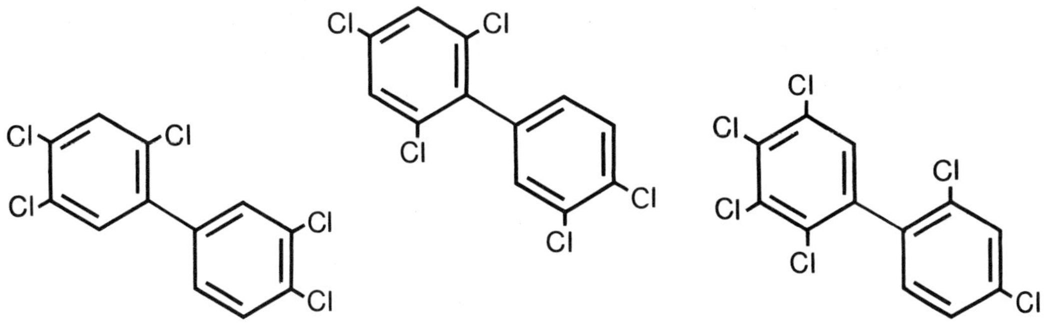

PCB's were first prepared in 1881; they were manufactured commercially, beginning in 1929, by chlorinating biphenyl under electrophilic conditions, Equation 2.

(2) Cl_2 / Lewis acid; Cl_x, Cl_y

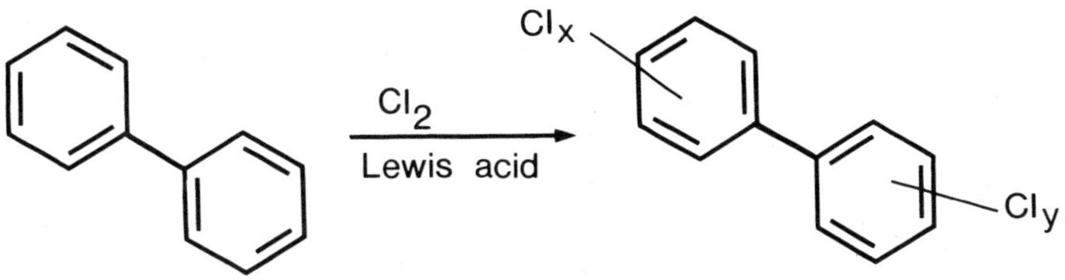

They were sold under various tradenames: Aroclor, Kaneclor, Santotherm, Clophen, and Phenoclor among others. PCB's were produced from the 1930's until the 1970's, with a peak production[5] of about 100,000 tonnes per year in 1970. In North America the leading manufacturer was Monsanto, whose tradename Aroclor was used for a series of formulations which were used as dielectric fluids in power transformers and capacitors (the device which levels out voltage fluctuations between one's domestic power supply and the utility supply). Each Aroclor product carried a four digit code, such as Aroclor 1242. The first two digits (12) represent the 12 carbon atoms of biphenyl and the last two (42 in this example) indicate the percent by weight of chlorine in the mixture (Problem 3).

Most PCB congeners are liquids or low-melting solids, and the Aroclor—or other PCB—formulations having 60% chlorine or less are liquids at room temperature. The great advantage of PCB's as dielectric fluids is their lack of flammability, coupled with excellent thermal and electrical insulation. Lack of flammability also implies resistance to oxidation, and this explains why the PCB's are so persistent when they are released into the environment. Transformer fluids were formerly formulated from PCB's alone, or

[5] R.F. Addison, *Can Chem News,* "PCBs in perspective" February **1986,** 15–17.

in admixture with either mineral oils or polychlorinated benzenes. Today, high-boiling mineral oil or silicone fluids are gradually replacing PCB's in these applications.

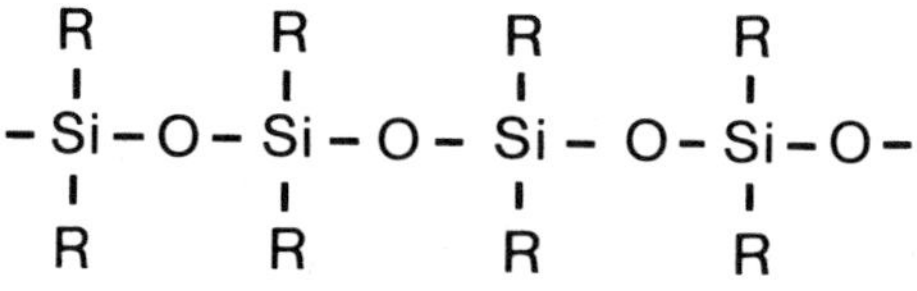

```
    R       R        R        R
    |       |        |        |
  -Si-O-  Si-O-   Si-O-   Si-O-
    |       |        |        |
    R       R        R        R
```

Silicone

9.2.1 Occurrence and analysis of PCB's

The detection of PCB's in wildlife was made by the Swedish scientist Jensen in 1966. The technique of gas chromatography combined with mass spectrometry (gc/ms) enabled analysts to detect and quantitate PCB's at the parts per billion level. Because PCB's are complex mixtures, the separation capability of gas chromatography is essential to their analysis (unlike DDT which is a single compound, apart from its ubiquitous transformation product DDE). One way in which quantitation has been done is to select one (or a few) of the more abundant congeners as a marker, quantitate it (them) by gas chromatography, and then use a conversion factor to obtain an estimate of the total PCB concentration. The chromatographic pattern of the relative amounts of the different congeners permits the chromatogram to be used as a "fingerprint" for a particular Aroclor or other commercial mixture. However, a problem with fingerprinting environmental samples is that the material being analyzed may have undergone change since its release to the environment. For example, the less chlorinated PCB congeners are more volatile and also more reactive metabolically under most circumstances. Consequently, an "aged" environmental sample of, say, Aroclor 1242 may resemble 1254 or 1260 by the time it is retrieved from the environment. Additionally, typical environmental samples contain other chloroaromatic compounds—such as DDT, dioxins, and dibenzofurans—besides PCB's, and painstaking chromatographic separations prior to gas chromatographic analysis are needed to isolate the PCB fraction[6] (Problem 4).

[6]Because of inevitable losses during clean-up procedures such as chromatographic separation, the sample is usually "spiked" either with a substance which is chemically similar to the mixture to be analyzed or with an isotopic (e.g., ^{13}C) variant of one of the congeners, to serve as an internal standard.

Like DDT, PCB's are lipophilic, and were quickly found to be both very widespread and capable of biomagnification (Problem 5). PCB's are more volatile than DDT, and are readily transported through the troposphere. Much of the atmospheric loading of PCB's comes from unsecured municipal landfills and hazardous waste dumps[7], although no emissions above background ensue from a well-designed PCB waste landfill[8] (Problem 6). PCB's have become globally distributed as a result of atmospheric transport, and this probably explains their provenance in remote locations such as Lake Superior[9] and the polar regions[10]. A U.S. survey of municipal incinerators showed that PCB emissions from this source did not contribute substantially to the atmospheric background[11]; similar conclusions have been reached in the United Kingdom[12].

9.2.2 Production of PCB's

Figure 9.2 shows the production of the Aroclors by Monsanto after their introduction in the 1930's. The total U.S. production between 1929 and 1977 has been assessed at about 600,000 tonnes (worldwide, about 1 million tonnes). In 1982, half of the U.S. total was estimated still to be in service, 21% was buried in landfills, 11% had been exported, and 11% had escaped into the environment[13]. In the U.K. 36,000 tonnes of a total production of 67,000 tonnes of PCB's from 1954–1977 had been landfilled by 1988[12].

Environmental contamination by PCB's became a problem both because of transformers and capacitators taken out of service and dumped, and because in later years PCB's began to find more varied uses as plasticizers, de-inking fluids for recycling newspaper, and in the production of non-carbon copy paper. The latter are called "open" uses: that is, escape

---

[7]T.J. Murphy, L.J. Formanski, B. Brownawell, and J.A. Meyer, "Polychlorinated biphenyl emissions to the atmosphere in the Great Lakes region. Municipal landfills and incinerators" *Environ Sci Technol,* **1985,** 19, 942–46; M.H. Hermanson and R.A. Hites, "Long-term measurements of atmospheric polychlorinated biphenyls in the vicinity of Superfund dumps" *Environ Sci Technol,* **1989,** 23, 1253–258.

[8]R.G. Lewis, B.E. Martin, D.L. Sgontz, and J.E. Howes, "Measurement of fugative emissions of polychlorinated biphenyls from hazardous waste landfills" *Environ Sci Technol,* **1985,** 19, 986–991.

[9]B. Hileman, *Chem Eng News,* February 8, **1988,** 22–39; also D.L. Swackhamer, B.D. McVeety, and R.A. Hites, "Deposition and evaporation of polychlorinated biphenyl congeners to and from Siskiwit Lake, Isle Royale, Lake Superior," *Environ Sci Technol,* **1988,** 22, 664–672.

[10]Antarctic: R.W. Risebrough, W. Walker, T.T. Schmidt, B.W. de Lappe, and C.W. Connors, "Transfer of chlorinated biphenyls to Antarctica," *Nature,* **1976,** 264, 738–739; Arctic: D.J. Gregor and W.D. Gummer, "Evidence of atmospheric transport and deposition of organochlorine pesticides and polychlorinated biphenyls in Canadian Arctic snow," *Environ Sci Technol,* **1989,** 23, 561–565.

[11]J.J. Richard and G.A. Junk, "Polychlorinated biphenyls in effluents from combustion of coal/refuse," *Environ Sci Technol,* **1981,** 15, 1095–1100.

[12]G.H. Eduljee, "PCBs in the environment," *Chem in Britain,* March **1988,** 241–244.

[13]S. Miller, "The persistent PCB problem" *Environ Sci Technol,* **1982,** 16, 98A–99A.

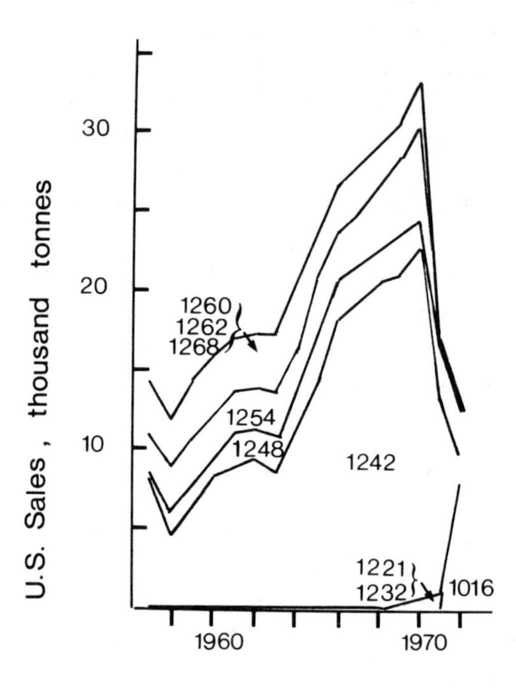

Figure 9.1: Annual production of Aroclors by Monsanto in North America. The production of each mixture is the area between lines on the graph. Note that in the last years of production, an effort was made to reduce the proportion of chlorine in the mixtures, thereby facilitating oxidation, by introducing Aroclors 1221 and 1016 (a chlorinated naphthalene, C_{10}).

of the material into the environment is likely, as opposed to "closed" uses in transformers and capacitors, where the PCB's are sealed into the unit, so that with proper disposal techniques, there should be no loss into the environment.

In North America, production of PCB's was drastically curtailed in 1972 and halted completely in 1977, although provision was made for PCB's already in use in the electrical industry to remain in service. In Canada, the provisions of the Environmental Contaminants Act (1980) permit the continued use of PCB's in electrical equipment such as electromagnets and transformers, provided that the equipment are not used to handle food or animal feed. The PCB's may be drained from the equipment, filtered, and replaced, but no new PCB fluid may be added. PCB fluids are classified as high strength (> 5000 ppm), which can comprise pure PCB's or mixtures of PCB's with chlorobenzenes or with mineral oils, and low strength (50–5000 ppm). Materials containing less than 50 ppm of PCB's do not fall under the provisions of the PCB regulations.

The reduction in the use of PCB's—and of other chlorinated substances such as DDT, chlordane etc—has led to improvements in environmental quality, as seen in data on the levels of these compounds in Ontario sport fish. For example, Coho salmon from Lake Ontario showed a reduction of PCB levels from 10 ppm in 1972 to 1 ppm in 1988; DDT levels in Lake Simcoe lake trout declined from 16 ppm in 1967 to < 1 ppm in 1983; mirex in Lake Ontario rainbow trout declined from 0.26 ppm in 1976 to 0.06 ppm in 1988[14].

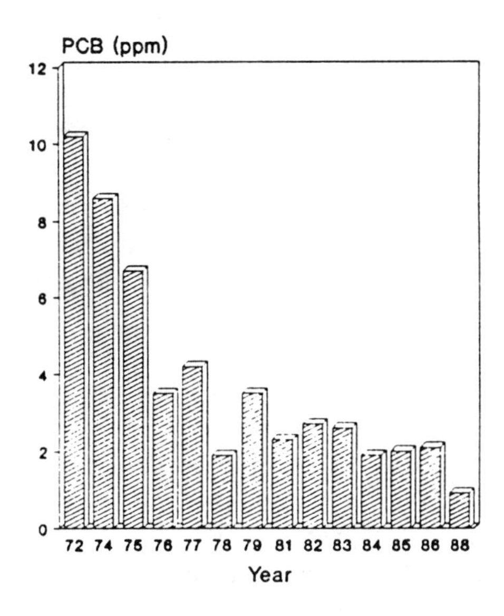

Figure 9.2: Mean concentrations of PCB's in Lake Ontario Coho salmon caught from the Credit River between 1972 and 1988. Reproduced from Reference 14.

9.2.3 *Toxicology of PCB's*

The toxicity of PCB's has been the subject of much controversy[15]. Not all congeners are equally hazardous; the work of Safe *et al* has shown that those congeners substituted in several of the 3,3′,4,4′,5, and 5′ positions are the most toxic[16]. Furthermore, distinction must be drawn between the toxicity of PCB's themselves, and that of the impurities which they may contain.

[14] C. Cox and J. Ralston "A reference manual of chemical contaminants in Ontario sport fish," Ontario Ministry of Environment, 1990.

[15] For an excellent summary of PCB toxicity, see K.L. Idler, *PCBs—the current situation,* Canadian Centre for Occupational Health and Safety, Hamilton, Ontario, Publication P86–3E, 1986.

[16] S.Safe, "Polychlorinated biphenyls (PCBs) and polybrominated biphenyls (PBBs): biochemistry, toxicology, and mechanism of action," CRC critical reviews in toxicology, CRC Press, Boca Raton, Florida, **1984,** 13, 319–395.

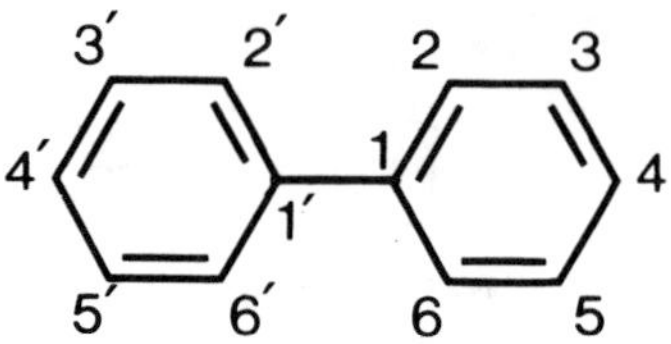

The acute toxicity of PCB mixtures is relatively low: oral LD_{50} values in rodents are near 1 g per kg body weight (not greatly different from Aspirin). At the present time few methods exist to assess the toxicology of complex mixtures such as PCB's or, worse, mixtures of PCB's and other chlorinated aromatic compounds. Recent evidence seems to suggest that the "non-ortho-chlorinated" congeners described above interact with cells by different mechanisms from the other congeners, so that even within the PCB family, the total toxicity may not simply be the sum of the contributions of the components.

Chloracne is one of the most characteristic toxic responses to chlorinated compounds in humans. The condition, which is a very persistent acne-like rash affecting principally the face and upper back, is very disfiguring and lasts months to years. This is presumably because chloro compounds are lipophilic, and the material dissolved in the adipose tissue thus provides a reservoir to perpetuate the symptoms even in the absence of new exposure. Other symptoms of PCB exposure include abnormal skin pigmentation, generalized feelings of fatigue, headaches, and joint pain (compare the symptoms of DDT poisoning, discussed above). At the present time, there is insufficient evidence to conclude that PCB's are carcinogenic in humans[15]. Liver cancers develop in some rodents after long term exposure to very high levels of PCB's. TLV's have been established for both Aroclor 1242 (1 mg m^{-3} (time weighted) and 2 mg m^{-3} (15 minute short term exposure)) and Aroclor 1254 (0.5 and 1 mg m^{-3} respectively).

Human poisonings from consumption of PCB mixtures have occurred, the best known incident of this type being the Yusho rice oil poisoning in Japan (1968), where PCB fluids became mixed inadvertently with rice oil used for cooking. The actual level of contamination of the rice oil was low ($\approx 0.2\%$). Ten years later, a very similar poisoning occurred in Yu-Cheng, Taiwan, and about 2000 people were affected. The chief symptom in both incidents was chloracne. Adipose tissue from Yusho patients was found to contain up to 75 ppm of PCB's. Children born to "Yusho mothers" were undersized and also showed unusual skin pigmentation, but they eventually recovered. Autopsies performed on Yusho patients who subsequently died revealed a variety of symptoms, including cancers, but these could not conclusively be attributed to the Yusho poisoning[17].

[17] M. Kikuchi, "Autopsy of patients with Yusho," *Am J Indust Med,* **1984,** 5, 19–30.

It is now widely recognized that the toxicity of these PCB-contaminated rice oils was not primarily due to the PCB's themselves, but to the traces of polychlorinated dibenzofurans (PCDF's, see below) which they contained[18]. The Yusho rice oil contained unusually high levels of PCDF's, and the adverse health effects among those affected were much more severe than those among workers industrially exposed to PCB's, even though the industrial workers had far higher blood levels of PCB's[15].

For many years, the news media have reacted with great alarm to examples of PCB contamination in the environment. During the 1980's, most scientists became convinced that the principal danger comes from PCDF's present in PCB mixtures rather than from the PCB's themselves, and this view now seems to be penetrating the popular news media too. The following sentences attempt to encapsulate the voluminous literature on the subject[19].

Studies with laboratory rodents have indicated that PCB's are weakly carcinogenic—i.e., they are animal carcinogens, and therefore suspected of being human carcinogens also. A very important point to note in connection with animal studies is that unrealistically high doses of the test chemical are administered, and that the genetically-inbred rodents—usually rats or mice—studied are genetically predisposed towards the end-point (cancer in this case) being examined. This experimental strategy is necessary in order to keep the number of animals per experiment within manageable bounds—imagine doing an experiment where the statistical incidence of additional cancers was 1 in a million! The usual positive finding of carcinogenicity is therefore an increase in the incidence of cancer in the animals treated with the chemical under investigation compared with the untreated controls. Assessment of the potential risk to humans requires two unknown extrapolations: from the genetically-susceptible rodent strain to humans, and from high dose to low dose.

Co-administration of PCB's and known carcinogens to animals produced fewer cancers than treatment with the known carcinogens alone: the PCB's appeared to exert some sort of "protective effect." Epidemiological studies on workers exposed to PCB's over long periods seem to show reduced—rather than greater—incidence of cancer, compared with unexposed workers, and no other detectable adverse symptoms apart from mild chloracne at high exposure levels.

---

[18] H. Kuroki, K. Haraguchi, and Y. Masuda, "Polychlorinated dibenzofuran congeners in the tissues of patients with Yusho and normal Japanese," *Chemosphere,* **1987**, 16, 2039–2046.

[19] I am indebted to Dr. M.A. Hayes of the University of Guelph for making available a copy of his unpublished paper entitled "PCB's and your health" on which this summary is based.

9.2.4 Contaminants in PCB's

PCDF congeners are formed from PCB's through oxidation at a free ortho position. This reaction requires heat and oxygen, so that PCDF contamination is a common result of PCB fires.

(3) [O], 300 - 600 °C

Similar processees can also occur in heat-transfer fluids if air is not rigorously excluded—this is probably what happened in the Yusho and Yu-Cheng incidents. These reactions occur at lower temperatures[20] ($\approx$ 600–700°C) than the extreme conditions which lead to the formation of polychlorinated dibenzo-*p*-dioxins (PCDD's) and PCDF's in incinerators, and so most of the PCDF congeners thus formed are structurally related to their PCB precursors.

Although we cannot yet say so categorically, the following comment on the toxicity of PCB mixtures seems likely to be correct. PCB mixtures themselves probably have only slight toxicity. Toxicity increases upon exposure to heat and air, due to the formation of traces of PCDF's. Because PCDF's are intrinsically much more toxic than PCB's, even trace amounts can substantially increase the overall toxicity of a PCB mixture[21].

A fire which began in a basement transformer of an office building in Binghamton, New York (1981) led to PCB's, PCDD's, and PCDF's being distributed through the building by way of the ventilation system[22]. The transformer fluids in this incident contained Aroclor 1254 (65%) and a mixture of tri- and tetra-chlorobenzenes (35%); soot deposited in the fire was

[20] M.D. Erickson, S.E. Swanson, J.D. Flora, and G.D. Hinshaw, "Polychlorinated dibenzofurans and other thermal combustion products from dielectric fluids containing polychlorinated biphenyls," *Environ Sci Technol,* **1989,** 23, 462–470.

[21] H.R. Konow and F.S. Bradley, "PCBs: myth and reality" *IEEE Canadian Review,* December **1988,** 5–6.

[22] A. Schechter, "The Binghamton State Office Building PCB, dioxin, and dibenzofuran electrical fire incident 1981–1986," *Chemosphere,* **1986,** 15, 1273–1280. The following articles in the same issue of Chemosphere discuss other PCB fires.

reported to contain total PCDD, 20 ppm and total PCDF, 2200 ppm, although the very toxic 2,3,7,8–congeners (see later) were both only minor constituents. Complete abandonment of the building due to contamination throughout the structure was required and the clean-up, which occupied several years, cost over $40 million. In 1988, a fire in a warehouse where PCB's were being stored caused the dispersal of PCB's and their oxidation products over a wide area, and the evacuation of the community of St Basile-le-Grand, Quebec for several days.

9.2.5 Ultimate destruction of PCB's

The incidents just mentioned bring into focus an important political issue: should PCB's be destroyed when they are taken out of service? Public consern about their toxicity strongly indicates destruction, even though their toxicity is now recognized to be less than originally suspected. Another compelling argument for destruction, whatever their toxicity, is their persistence and ubiquitous presence in the environment. The following questions arise:

- How completely can they be destroyed and how much will escape into the environment during the process?
- Where should the destruction facility be located?

The first of these questions is technical, and current technology allows destruction at the "six nines" level i.e., 99.9999%, which among many regulators is considered to be an acceptable standard of destruction efficiency. Nevertheless, many citizens fear the release of **any** PCB to the environment. The second question raises the "Not in my backyard" (NIMBY) syndrome. These concerns have led to political inaction over PCB destruction, without proper recognition that storage carries its own hazards. Actually, it is a simpler problem to dispose of PCB's than, say, toxic metals because the elements carbon, hydrogen and chlorine which constitute PCB's are not inherently toxic, and so PCB's can in principle be transformed chemically into innocuous byproducts. This is not possible with toxic metals, which are more or less toxic whatever their chemical form (*cf.* Chapters 8 and 10).

PCB destruction involves political as well as chemical considerations. The choice of technology, the siting of the facility, and the associated need for environmental hearings make for delay in getting a facility commissioned. Meanwhile, the PCB's must be transported to storage sites, allowing opportunities for human error in handling, and leakage from the stored containers. Opinion now favours the construction of mobile destruction facilities which can be moved to the PCB storage site, rather than a central PCB processing facility to which the waste must be hauled long distances. Mobile destruction units—incinerators—are in place in Canada at Swan Hills, Alberta, and Goose Bay, Labrador, and are more commonly used in the United States and Europe. Note in this context that "mobile" means capable of being moved with considerable effort, and not a unit that can be mounted on the back of a truck.

Incineration

Incineration is the leading contender for an ultimate disposal method for high strength PCB's at the "six-nines" level. PCB's themselves do not burn, which is one of the reasons for their use as dielectric fluids, and incineration requires them to be oxidized in the presence of a large excess of fuel oil (Problem 7). PCB's contain chlorine, hence if the combustion temperature is insufficiently high, traces of PCDD's and PCDF's may result. The formation of PCDD's and PCDF's has been a major impediment to the acceptance of incineration as a disposal method for PCB's.

It is particularly difficult to evaluate the possible threat posed by the emissions of PCDD's and PCDF's, which are not single compounds but complex mixtures of congeners. In the public mind, "dioxin" is linked to the very toxic congener 2,3,7,8–tetrachlorodibenzo-*p*-dioxin (TCDD). Many of the other congeners may be less toxic by a factor of millions or more, besides which, the toxicity of "dioxins" appears to be very species-dependent (Section 9.3.2). Incineration affords mainly octachlorodibenzo-*p*-dioxin, with less than 5% of the highly toxic TCDD, but the total amounts are exceedingly small. The formation of PCDD's and PCDF's from structurally unrelated precursors is at a maximum at a combustion temperature of 700–800°C, and even then it is only $\approx 10^{-10}$% of the theoretical yield. Under incineration conditions of at least 1000°C and a contact time of the material of more than 2 s in this hottest zone, the formation of these substances is almost zero[23]. U.S. regulations for waste incineration require 3% excess O_2, and a minimum residence time of 2 s at 1200°C; in the U.K., 6% excess O_2, and 1.5 s at 1000°C are demanded.

In the context of municipal solid waste incinerators, Choudhry and Hutzinger[24] have pointed out that there are several possibilities for the presence of PCDD's and PCDF's.

1. They may be present in the original garbage, and not burned.
2. They may arise from aromatic chloro compounds such as chlorophenols, PCB's etc.
3. They may arise through pyrolysis of other chlorinated organics such as polyvinyl chloride.
4. They may be formed by high temperature reactions between unchlorinated organics and inorganic chlorides ($Cl^-$and/or HCl).

Possibility (1) seems the least likely, but none of the routes may be categorically excluded. Pathway (2) can be demonstrated in the laboratory as in the formation of PCDF congeners from PCB's and of PCDD's from

[23] W.M. Shaub and W. Tsang, "Dioxin formation in incinerators," *Environ Sci Technol,* **1983,** 17, 721–730.

[24] G.G. Choudhry and O. Hutzinger, *Mechanistic aspects of the thermal formation of halogenated organic compounds including polychlorinated dibenzo-p-dioxins,* Gordon and Breach Science Publishers, London, 1983: see particularly Chapter 6.

chlorinated phenols. Regarding the conversion of PCB's to PCDF's, two distinct reactions can be discerned. Under relatively mild conditions (300–600°C) little molecular rearrangement occurs, as seen in the example below (the pathway by which PCDF's are formed in PCB fires)[25]. More complex rearrangements occur at higher temperatures.

Incomplete oxidation of organics, reaction (3), is well known to afford polycyclic aromatic compounds (think of soot, and the formation of polycyclic aromatic compounds over barbecues), likewise, chlorinated aliphatics pyrolyze in part to chlorinated aromatics. An example is the combustion of polyvinylidene chloride, used as plastic wrapping film, which yields chlorinated naphthalenes, biphenyls, and dibenzofurans in small amounts[26]. Regarding possibility (4), Mahle and Whiting[27] found that highly chlorinated PCDD congeners could be formed when coal was burned in the presence of NaCl or, better, HCl.

An early and imaginative use of incineration to destroy PCB's was developed in Canada by the St. Lawrence Cement Co. in the mid 1970's. PCB waste was mixed with the oil used to provide the heat needed to make cement; since the clinker is highly basic (CaO), it served to trap the HCl byproduct formed when the PCB's burned, and thus assist in the production of a high-chloride cement. The company received an environmental award for this process; ironically, the news of the award raised public awareness of the fact that PCB's were being transported to the site, and the resulting outcry led to the process being proscribed. Now, over a decade later, this technology is being used in Europe, and is being reconsidered for possible use in Canada.

Plasma technology is another high temperature method which has been proposed for destroying chlorine-containing wastes. The waste is introduced into a high temperature plasma; under these conditions the complex organochlorines are literally torn apart into atoms and small fragments, which recombine into stable small molecules such as CO_2, HCl, and water on leaving the plasma zone[28].

A milder proposed alternative to incineration is the catalytic "wet oxidation" process involving atmospheric oxygen as the oxidant, but at much lower temperatures than those needed for combustion. Impressive destruction rates for PCB mixtures and for TCDD have been reported at temperatures in the range 200–250°C. One combination that has been investigated is the bromide/nitrate catalytic system. This mixture produces reactive substances such as HNO_3—which can initiate oxidation—and NOBr—which

---

[25] H.R. Buser and C. Rappe, "The formation of polychlorinated dibenzofurans (PCDFs) from the pyrolysis of individual PCB isomers," *Chemosphere,* **1979,** 8, 157.

[26] A. Yasuhara and M. Morita, "Formation of chlorinated aromatic hydrocarbons by thermal decomposition of vinylidene chloride polymer," *Environ Sci Technol,* **1988,** 22, 646–650.

[27] N.H. Mahle and L.F. Whiting, "The formation of chlorodibenzo-*p*-dioxins by air oxidation and chlorination of bituminous coal," *Chemosphere,* **1980,** 9, 693.

[28] *Chem Eng News,* March 20, **1989,** 20.

can initiate free radical reactions, but the detailed reaction mechanisms are not yet known[29].

Other chemical methods of PCB destruction[30]

Because of the environmental significance of the PCB disposal problem, a great deal of research has been conducted into different possible technologies for their destruction. Some examples follow.

1. Reaction with active metals. This is a variant of the Wurtz reaction, by which alkyl halides are converted to alkanes.

$$2\mathrm{R{-}Cl} + 2\mathrm{Na} \longrightarrow \mathrm{R{-}R} + 2\mathrm{NaCl} \qquad (4)$$

If a PCB is used in place of the alkyl halide, the reaction has the effect of joining the aromatic rings together. Because a PCB contains several chlorine atoms per molecule, each molecule reacts more than once; whereas the Wurtz reaction on a monochloralkane gives the dimeric hydrocarbon, the corresponding reaction with a PCB gives a polymer. The result is that the volatility, solubility, and toxicity of the PCB are all greatly reduced and it is claimed that the residuum may be safely landfilled.

(5)

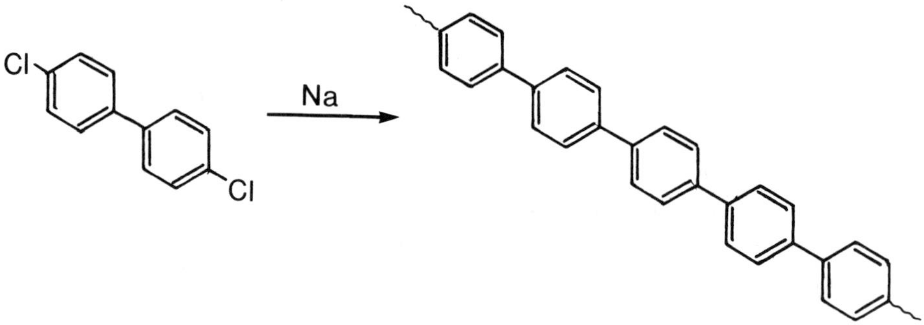

The reaction is illustrated for the simple example of 4,4′-dichlorobiphenyl. Reaction with matallic sodium is a commercially technology for removing the PCB's from low strength PCB fluids. The mineral oil can be separated from the sludge of NaCl and NaOH and reused.

---

[29] R.A. Miller and R.D. Fox, "Catalyzed wet oxidation of hazardous waste" Chapter 13 of *Detoxication of hazardous waste,* Reference 30.

[30] The following text is useful reading: *Destruction and disposal of PCBs by thermal and non-thermal methods,* D.G. Ackerman, L.L. Scinto, P.S. Bakshi, R.G. Delumyea, R.J. Johnson, G. Richard, A.M. Takata, and E.M. Sworzyn, Noyes Data Corp., Park Ridge, New Jersey, 1983. There are also several useful chapters (6–14) in *Detoxication of hazardous waste,* Ed. J.H. Exner, Ann Arbor Science, Ann Arbor, Michigan, 1982. The report "The evaluation of mobile and stationary facilities for the destruction of PCBs" Environment Canada, Report No, 46–10791 classifies documented PCB destruction methods into Commercial (in use > 1 yr), Near Commercial (permit for commercial operation applied for), and Promising (laboratory or pilot scale only) as of 1987.

The direct reaction between the reactants occurs efficiently only at temperatures > 200°C, but a different reaction pathway is possible in the presence of aprotic, cation-complexing solvents such as tetrahydrofuran. Under these conditions, the active metal dissolves to give the radical anion of the aromatic component.

$$\text{(6)} \qquad Na + ArH \longrightarrow Na^{+} + ArH^{-\bullet}$$

Chlorinated aromatic radical anions can expel Cl^{-}, and the aryl radical thus formed can go on to participate in typical free radical reactions, such as abstraction of hydrogen from the solvent or arylation of another molecule of the aromatic compound.

$$\text{(7)} \qquad ArCl^{-\bullet} \longrightarrow Ar\bullet + Cl^{-}$$

One way of bringing about this chemistry is to use sodium and naphthalene. Sodium naphthalenide $Na^{+}C_{10}H_8^{-\bullet}$ is formed in the aprotic solvent, and when the PCB (written here as ArCl) is added, the naphthalenide ion reduces the PCB molecules.

$$\text{(8)} \qquad C_{10}H_8^{-\bullet} + ArCl \longrightarrow C_{10}H_8 + ArCl^{-\bullet} \longrightarrow Ar\bullet + Cl^{-}$$

2. Reactions with base.
Two different approaches have been investigated.

1. Nucleophilic substitution has been studied by researchers at General Electric (Schenectady, N.Y.)[31]. In the presence of a high boiling additive such as polyethylene glycol, hydroxide ion will replace chlorine in the aryl chloride (i.e., the PCB) by nucleophilic substitution. KOH is used as the source of OH^{-}, the polyethylene glycol serving to solubilize the KOH by solvating the potassium ions in the aprotic medium. It does this by "wrapping" them in the polyether, much as a crown ether solvates cations in its cavity.

   This reaction is more suited to destroying low strength mixtures of PCB's and mineral oil (e.g., Askarels) rather than pure PCB's. It is carried out near 100°C, using a blanket of nitrogen, so that the oil will not oxidize and hence can be reused. The substitution product(s) precipitate upon cooling the reaction mixture to 40°C, and are removed by filtration. This reaction has also been applied recently to the destruction of 2,3,7,8–tetrachlorodibenzo-*p*-dioxin (TCDD).

2. Japanese researchers have proposed a free radical chain process in which isopropyl alcohol is used to reduce the PCB's. The chem-

[31] D.J. Brunelle, A.K. Mendlratta, and D.A. Singleton, "Reaction/removal of polychlorinated biphenyls from transformer oil: treatment of contaminated oil with poly(ethylene gylcol)/KOH," *Environ Sci Technol,* **1985**, 19, 740–746.

istry here has similarities to reduction using active metals in aprotic solvents, which was discussed above.

$$\begin{aligned} ArCl^{-\bullet} &\longrightarrow Ar\bullet + Cl^- \\ Ar\bullet + (CH_3)_2CHOH &\longrightarrow ArH + (CH_3)_2C(OH)\bullet \\ (CH_3)_2C(OH)\bullet + OH^- &\longrightarrow (CH_3)_2C\text{–}O^{-\bullet} + H_2O \\ (CH_3)_2C\text{–}O^{-\bullet} + ArCl &\longrightarrow (CH_3)_2C{=}O + ArCl^{-\bullet} \end{aligned}$$

The NaOH is needed for the deprotonation of the radical $(CH_3)_2C(OH)\bullet$, whose approximate pK_a is 12. Water is a byproduct of this reaction, hence its success depends on maintaining anhydrous conditions.

3. Hydrogenation reactions (Problem 8).

1. At temperatures around 900°C, hydrogen both cleaves the C–Cl bonds of a PCB (reductive dechlorination) and also splits the biphenyl nucleus into two benzene residues[32]. No catalyst is required.
2. Reductive dechlorination, also called hydrogenolysis, can be effected at much lower temperatures (ca. 100°C) in the presence of a mixed copper oxide/chromium oxide catalyst[33]. Under these conditions, the biphenyl nucleus is not affected, and the reaction product is biphenyl.

4. Photochemical reactions (Problems 9–11). Biphenyl derivatives absorb radiation in the deep ultraviolet, with maximal absorption usually close to 250 nm. The tail of the absorption band often extends out to just about 300 nm, the extreme short wavelength of the tropospheric solar spectrum. Following excitation, cleavage of the C–Cl bond can occur, but the quantum yield is often low[34].

$$Ar\text{–}Cl \xrightarrow{h\nu} Ar\bullet + Cl\bullet$$

The fate of the aryl radicals depends on the reaction medium; oxidation occurs in air, and hydrogen abstraction takes place in solvents bearing abstractable hydrogen atoms. Other research has been directed towards increasing photefficiency in the presence of additives such as amines and semiconductor suspensions, but at the present time none of these approaches is close to commercialization. Another proposal is a photooxidative technology involving deep-UV photolysis combined with ozonation.

---

[32] J.A. Manion, P. Mulder, and R. Louw, "Gas-phase hydrogenolysis of polychlorobiphenyls" *Environ Sci Technol,* **1985,** 19, 280–282.

[33] D.R. Hedden, R.W. Johnson, K.J. Youtsey, L. Hiffman, and T.N. Kaines, "Catalytic hydrogenation of waste oils," presentation at 80th Air Pollution Control Association annual meeting, New York, June 1987.

[34] For mechanistic discussion, see N.J. Bunce, "Photodechlorination of PCB's: current status," *Chemosphere,* **1982,** 11, 701–714.

In the atmosphere, PCB's and related chloroaromatic compounds break down naturally by two competing pathways: direct solar photolysis and attack by hydroxyl radicals. Direct photolysis is the minor pathway for most congeners because of the poor spectral overlap between PCB absorption and the tropospheric solar spectrum[35]. The reaction with hydroxyl radicals is indirectly a solar-driven process also, because OH is produced by photolysis of ozone: see Chapter 3. Reaction in the atmosphere is relatively slow, as evident from the transport of PCB's over long distances through the atmosphere, as discussed earlier.

9.2.6 Microbial transformation of PCB's

The metabolism of PCB's in mammals is oxidative, and leads to hydroxylated metabolites. The formation of ortho diols is thought to involve arene epoxides similar to the intermediates in the metabolic oxidation of aromatic hydrocarbons such as benzo(a)pyrene.

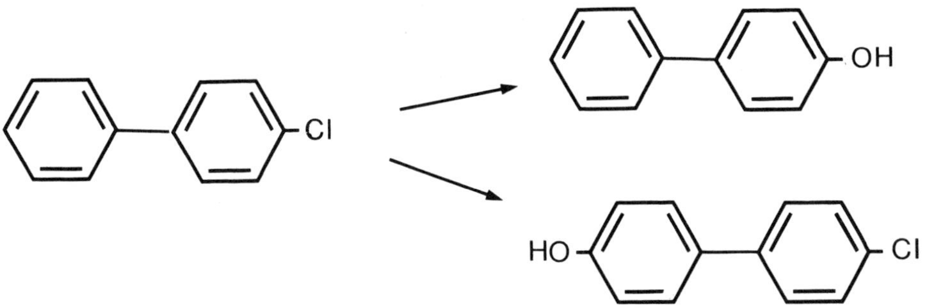

Microbial transformation of PCB's occurs both by the oxidative route, and also reductively under anaerobic conditions. The latter reaction leads to dechlorination: replacement of Cl by H.

A number of claims has been made for "super microorganisms" capable of degrading PCB's. Such microorganisms have been selected from "wild" bacteria by growing cultures on PCB's as their sole carbon source. They do indeed utilize PCB's, but since they metabolize other available carbon sources, they have not been useful as "biological catalysts" for the cleanup of PCBs from contaminated sites. More interesting are the observations by the group at General Electric at Schenectady, New York which has studied sediments in the Hudson River[36]. Company records show that Aroclor

[35] N.J. Bunce, J.P. Landers, J. Langshaw, and J.S. Nakai, "An assessment of the importance of direct solar degradation of some simple chlorinated benzenes and biphenyls in the vapor phase" *Environ Sci Technol,* **1989,** 23, 213–218; R. Atkinson, "Estimation of OH radical reaction rate constants and atmospheric lifetimes for polychlorobiphenyls, dibenzo-*p*-dioxins, and dibenzofurans" *Environ Sci Technol,* **1987,** 21, 305–307.

[36] J.F. Brown, D.L. Bedard, M.J. Brennan, J.C. Carnahan, H. Feng, and R.E. Wagner, "Polychlorinated biphenyl dechlorination in aqueous sediments" *Science,* **1987,** 236, 709–

1242 was the main formulation lost into the river before the 1970's. Analysis of the sediments in the 1980's showed congener patterns in the PCB's obtained from sediments that were different from that of Aroclor 1242. From the same sediments were isolated microorganisms which were able to transform PCB's, some strains degrading preferentially the least chlorinated congeners and others preferring heavily chlorinated PCB's. The inference is that these naturally-occurring bacteria are responsible for changing the PCB pattern by selectively removing certain of the congeners. Presumably, with time, these processes will result in clearance of PCB contamination in the environment.

9.3 PCDD's and PCDF's[37]

These compounds (usual abbreviations PCDD's and PCDF's) have been much in the news, mainly because of the extreme toxicity of the congener 2,3,7,8–tetrachlorodibenzo-*p*-dioxin (TCDD), usually called "dioxin" in the news media. The environmental chemistry and toxicology of PCDF's have much in common with those of PCDD's.

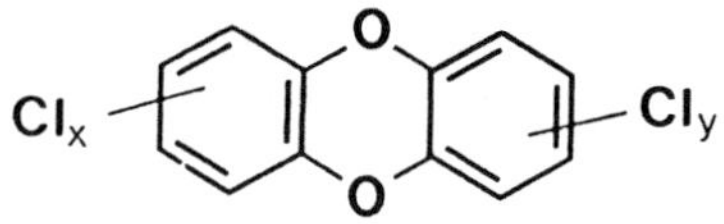

Cl_x O O Cl_y Cl_x O Cl_y

PCDD structure 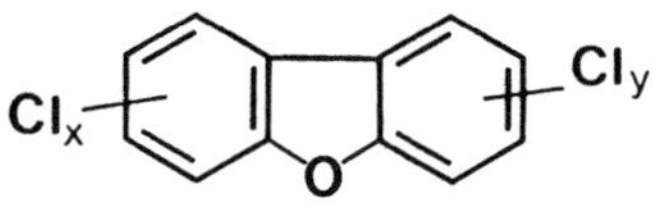 PCDF structure

Like the PCB's, PCDD's and PCDF's are families with many congeners: there are 75 PCDD's and 135 PCDF's. Most are colourless solids of moderate to high melting point and low volatility (Problem 12). Unlike the substances discussed so far in this chapter, the PCDD's and PCDF's have no known uses, and have never been manufactured deliberately[38]. Instead, they are formed as trace contaminants in other processes, some of which we have met already (incineration, and the low-temperature oxidation of PCB's and chlorinated diphenyl ethers to PCDF's). None of these reactions supplies pure individual PCDD or PCDF congeners in chemically useful yield.

712; for subsequent correspondence, see *Science,* **1988,** 240, 1674–1676. See also a news item in *Chem Eng News,* November 13, **1989,** 21.

[37]H. Fiedler, O. Hutzinger and C.W. Timms, *Toxicol Environ Chem,* **1990,** 29, 157–234.

[38]Considerable misinformation exists on this point, with TCDD often being labelled the "active ingredient" of herbicide formulations in news reports.

The pattern of PCDD congeners in air and sediment samples can be used to identify the source of the PCDD's. Incineration yields predominantly octachlorodibenzo-*p*-dioxin, and the very toxic TCDD amounts to only a few percent of the mixture[39]. Incinerator emissions of PCDD/PCDF's are mostly adsorbed onto fly ash particles and hence immobilized. Although most PCDD/PCDF congeners are less volatile than PCB's, atmospheric transport still occurs[40]. By contrast, the much more toxic TCDD is a major component of the PCDD fraction when the source of contamination is either the manufacture or use of 2,4,5-trichlorophenol, or the bleaching of paper.

9.3.1 Environmental chemistry of TCDD

2,3,7,8–Tetrachlorodibenzo-*p*-dioxin—TCDD—first came to scientific and public attention[41] when the herbicide 2,4,5-T (2,4,5-trichlorophenoxyacetic acid and its salts and esters) was suspected of being teratogenic (capable of causing birth defects)[42] in the 1970's. This was significant because 2,4,5-T was used as a brush killer, for example to clear railway and hydro rights of way, and it was also employed in very large quantities as the chemical defoliant "Agent Orange" [43] by the U.S. army in the Viet Nam War. Research showed that the actual toxic agent in formulations of 2,4,5-T was TCDD. Improved control during manufacture permitted the concentration of TCDD in 2,4,5-T to be kept below 0.1 ppm. However, the manufacture of the herbicide seemed inevitably to produce traces of TCDD, and the registration of 2,4,5-T was ultimately withdrawn in North America as a result of public concern about health hazards from its use, especially in forest spraying programs.

When TCDD was first discovered in the environment, there was controversy about its source, with herbicide manufacturers postulating a "trace chemistry of fire" hypothesis to explain its presence. Such processes are now linked to the PCDD's formed upon incineration, but there seems to be

[39]The PCDD congener pattern has been used to distinguish the product of incineration (mostly octachloro–) from herbicide contamination (2,3,7,8–TCDD): J.M. Czuczwa and R.A. Hites, "Airborne dioxins and dibenzofurans: sources and fates," *Environ Sci Technol,* **1986,** 20, 195. See also D.C. Ayres, "Organochlorine waste disposal–cremation or burial?" *Chem in Britain,* January **1987,** 41–43.

[40]B.D. Eltzer and R.A. Hites, "Atmospheric transport and deposition of polychlorinated dibenzo-*p*-dioxins and dibenzofurans" *Environ Sci Technol,* **1989,** 23, 1396–1401.

[41]The June 6, 1983 issue of Chemical and Engineering News was a "Special issue" devoted to the dioxin problem. The articles therein are an excellent source of additional information. Other background articles: F.H. Tschirley, "Dioxin," *Sci Am,* **1986,** 254, 29–35; G.H. Eduljee, "Dioxins in the environment" *Chem in Britain,* December **1988,** 1223–1226.

[42]For an introductory article, which includes data on PCB's and PCDD's, see B.S. Shane, "Human reproductive hazards" *Environ Sci Technol,* **1989,** 23, 1187–1195.

[43]Agent Orange was a 1:1 mixture of the butyl esters of 2,4-dichloro– and 2,4,5-trichloro-phenoxyacetic acids: the mean concentration of TCDD in Agent Orange was 10 ppm.

little evidence of PCDD formation from "clean" (i.e., non-organochlorine) precursors. For example, tissue from 2800 year old mummies showed no contamination by PCDD's and PCDF's[44].

The manufacture of 2,4,5-T from tetrachlorobenzene is outlined below. TCDD is a byproduct of the high temperature hydrolysis of tetrachlorobenzene to 2,4,5-trichlorophenol[45]. This reaction is usually carried out at temperatures in the range 140–170°C, in a polar, high-boiling solvent such as ethylene glycol. The higher the reaction temperature, the greater the amount of TCDD formed. Note that even at 10 ppm, the actual yield of TCDD in this reaction is very small[46] (Problem 13).

NaOH; $ClCH_2CO_2H$; O^-Na^+; OCH_2CO_2H

2,4,5-T

Byproduct:

ONa; NaO

TCDD

The hydrolysis of tetrachlorobenzene to 2,4,5-trichlorophenol is exothermic, and no additional heat needs to be applied once the reaction is under way. The normal protocol calls for gradual distillation of the water from the ethylene glycol solution, followed by quenching the reaction with water and allowing the mixture to cool. In the accident at Seveso, Italy in 1976 this protocol was disregarded[47]; the Saturday shift ended and the reactor

[44] W.V. Ligon, S.B. Dorn, R.J. May, and M.J. Allison, "Chlorodibenzofuran and chlorodibenzo-*p*-dioxin levels in Chilean mummies dated to about 2800 years before the present," *Environ Sci Technol,* **1989,** 23, 1286–1290.

[45] Pyrolysis of 2,4,5-trichlorophenol can also give 2,3,7,8–TCDD. Very recently, an enzymatic route has also been discovered, but whether it is important in the environment is not yet known: A. Svenson, L. Kjeller, and C. Rappe, "Enzyme-mediated formation of 2,3,7,8–tetrasubstituted chlorinated dibenzodioxins and dibenzofurans," *Environ Sci Technol,* **1989,** 23, 900–902.

[46] Recently a new route to 2,4,5-trichlorophenol has been proposed, and is claimed to eliminate TCDD formation completely..

H^+, $CH_3CH{=}CH_2$; $CH(CH_3)_2$; 1. O_2 2. H_3O^+; OH; Cl_2

[47] J. Sambeth, "The Seveso accident" *Chemosphere,* **1983,** 12, 681–686. See also Reference 40, and a news item in *Chem Eng News,* March 28, **1988,** 5–6.

was shut down before distillation was complete. Hydrolysis continued, and the build-up of heat led to the violent release of the contents of the reaction vessel, contaminating the neighbourhood with 2,4,5-trichlorophenol and with TCDD. This incident is discussed further below in connection with the toxicology of TCDD.

Human error was also the culprit in the contamination of the town of Times Beach, Missouri by TCDD (Problem 14). In this incident, a hauler of waste oil obtained a contract to remove still bottoms from a factory in Verona, Missouri. This plant was engaged in the production of 2,4,5-trichlorophenol, and the still bottoms were contaminated with TCDD (average concentration 33 ppm). The hauler mixed the TCDD-contaminated oil with other oils, and used the mixture to spray country roads and horse arenas for dust control. The contamination came to light following the deaths of a number of race horses, when the presence of TCDD in the arenas was discovered. Further enquiry showed that the whole town was contaminated and it was eventually purchased by the U.S. government, the residents relocated, and the site abandoned.

TCDD was again in the news in 1987, when it was revealed that chlorine-bleached paper contains traces of the chemical. Although the exact mechanism of formation of TCDD in paper is not known, it is supposed to involve chlorination of phenols present in lignin, the major non-cellulosic constituent of wood pulp. As a result, concern was expressed about exposing the population to TCDD in products such as disposable tissues, diapers, sanitary pads, and coffee filters (Problem 15). The actual amounts are very small (generally less than 1 pg/g), and opinion now suggests that there is no significant health hazard[48]. Products made with unbleached and ozone-bleached pulp are now available for some uses, and contain no chlorinated contaminants[49]. One area of concern is that milk is apparently able to leach TCDD from waxed paper containers; this is explicable in that TCDD is a very non-polar substance, and high solubility in milk-fats may be anticipated. TCDD is present in detectable amounts in many foodstuffs, but the amounts are very small[50]. The point at issue is whether the amounts represent any health threat to the population, raising once again the difficulty of extrapolating risk from experiments carried out at high doses with experimental animals to the exposure of humans and wildlife to the much lower concentrations found in the environment. Other sources of PCDD's in the environment include the use of dichloroethane as an additive to leaded gaso-

---

[48] R.E. Keenan and M.J. Sullivan, "Assessing potential health risks of dioxin in paper products," *Environ Sci Technol,* **1989,** 23, 643–644.

[49] S.E. Swanson, C. Rappe, J. Malmstrom, and K.P. Kringstad, "Emissions of PCDDs and PCDFs from the pulp industry," *Chemosphere,* **1988,** 17, 681–691. See also correspondence on this issue in *Chem in Britain,* September **1989,** 871.

[50] Y. Takizawa and H. Muto, "PCDDs and PCDFs carried to the human body from the diet," *Chemosphere,* **1987,** 16, 1971–1975; K. Davies, "Concentrations and dietary intake of selected organochlorines, including PCBs, PCDDs and PCDFs in fresh food composites grown in Ontario, Canada," *Chemosphere,* **1988,** 17, 263–276.

line (Chapter 10), and the use of chlorocompounds to activate the catalysts used in petroleum refining.

A gap appears to exist between the known rate of release of TCDD into the environment, and analytical data on environmental TCDD. Experimentally, there appears to be more TCDD in the environment than known release rates would predict, according to mathematical models which have been used to compare TCDD with other chloroaromatic compounds such as PCB's and chlordane, whose release rates and environmental concentrations are both known. This means either that the mathematical models which apply to PCB's and chlordane are inappropriate for TCDD, or that there are as yet undiscovered inputs of TCDD to the environment[51].

9.3.2 Toxicology of PCDD's and PCDF's[52]

As mentioned previously, TCDD came to public attention in connection with the suspected teratogenicity of the pesticide 2,4,5-T. Particular controversy resulted in Oregon, where a cluster of birth defects was alleged to be linked to forest spraying. Successive statistical examinations of the medical data first confirmed, and later discounted, a link between forest spraying and an elevated risk of birth defects[53]. The current situation is that there is no confirmed evidence that the use of 2,4,5-T is linked to birth defects in humans, even though 2,4,5-T formulations do contain traces of TCDD, and TCDD is a confirmed teratogen in laboratory animals.

TCDD has been by far the most studied of the PCDD or PCDF congeners. Although it is clearly very toxic, the frequent claim that it is "the most deadly substance known" is a considerable misrepresentation. In immature male guinea pigs, the LD_{50} of TCDD is about 1 μg/kg. While this is a very low value, it is still orders of magnitude less toxic than many natural toxins such as botulinum toxin. Besides this, TCDD has the interesting property that its LD_{50} is very species-dependent (Table 9.1); the hamster, for example, is 5000 times less susceptible to TCDD than the guinea pig.

The available evidence suggests that humans are probably at the less sensitive end of the toxicity spectrum as far as TCDD is concerned. At the time of the Seveso accident, there were fears that TCDD-related deaths would occur, based on the exposure of the population near the factory. Although no deaths have been linked to the accident, several residents suffered from chloracne, the severity of which correlated with concentrations of TCDD in

---

[51] C.C. Travers, presentation at the International Conference on Dioxins, Toronto, September 1989.

[52] *Dioxin in the environment: its effect on human health,* American council on Science and Health, New York, 1988. E.K. Silbergeld and T.A. Gasiewicz, *Am J Industrial Health,* **1989,** 16, 455–474.

[53] A parallel controversy has arisen over alleged reproductive and other medical problems among U.S. and Australian veterans of the Viet Nam War. However, elevated levels of TCDD have not been found in Viet Nam veterans compared with the general population: *Chem & Eng News,* April 15, **1991,** 12–13.

Table 9.1: Oral LD_{50} values for TCDD in laboratory animals (Reference 40)

Species	LD_{50}, $\mu g/kg$	Species	LD_{50}, $\mu g/kg$
guinea pig (m)	0.6	guinea pig (f)	2.1
rat (m)	22.	rat (f)	50–500.
rabbit	115.	monkey (f)	70.
hamster	1100.	frog	1000.

their blood. Long term follow-up of the population had shown, by 1990, no evidence for excess incidence of cancer, miscarriage, or birth defect that could be associated with the accident. This finding is corroborated by other data on worker exposure due to accidents in 2,4,5-trichlorophenol production. However, recent information suggests that workers who have been occupationally exposed to high levels of TCDD may be at some increased risk of soft-tissue cancers, the first time that any definite link between TCDD and human cancer has been made[54].

At sub-lethal doses, TCDD has many effects, mostly studied in laboratory animals. These include teratogenicity, carcinogenicity, reproductive failures, suppression of the immune system (shown for example by thymic atrophy), skin lesions (exemplified by chloracne), and porphyria. A particularly common effect is the so-called wasting syndrome, in which the animal gradually becomes anorexic and loses weight. The full spectrum of toxic responses is not seen in any one species, neither is any particular response seen in all species. In a study in human volunteers who had TCDD applied to the skin, doses of 3–114 ng/kg produced no observable effects (N = 60), while 107 $\mu g/kg$ caused chloracne in 8 out of 10 subjects. Notice that the latter level is considerably greater than the LD_{50} value in the guinea pig. News reports about the risks to humans of TCDD exposure are frequently couched in terms of the guinea pig data, with the implicit assumption that the guinea pig data may be extrapolated to humans.

PCDD's and PCDF's have been found in human tissues[55] and human breast milk[56]. The paper by Noren has several points of interest. First,

[54] M.A. Fingerhut *et al,* "Cancer mortality in workers exposed to 2,3,7,8-tetrachlorodibenzo-p-dioxin," *New Engl J Med,* **1991,** 324, 212–218.

[55] A. Schechter, J.D. Constable, S. Arghestani, H. Yong, and M.L. Gross, "Elevated levels of 2,3,7,8-tetrachlorodibenzodioxin in adipose tissue of certain U.S. veterans of the Vietnam war," *Chemosphere,* **1987,** 16, 1997–2002.

[56] A. Schechter, J.J. Ryan, and J.D. Constable, "Polychlorinated dibenzo-*p*-dioxin and polychlorinated dibenzofuran levels in human breast milk from Vietnam compared with cow's milk and human breast milk from the North American continent," *Chemosphere,* **1987,** 16, 2003–2016; H. Beck *et al,* "Levels of PCDFs and PCDDs in samples of human origin and food in the Federal Republic of Germany," *Chemosphere,* **1987,** 16, 1977–1987; K. Noren, "Changes in the levels of organochlorine pesticides, polychlorinated biphenyls, dibenzo-*p*-dioxins, and dibenzofurans in human milk from Stockholm, 1972–1985" *Chemosphere,* **1988,** 17, 39–49.

average levels of a wide range of organochlorines in Swedish breast milk have declined over the period 1972–1985. Second, for a given mother having successive children, the organochlorine levels in the milk decline with the number of children i.e., transport of organochlorines into the milk substantially depletes the reservoirs in the mother's body. The first two points indicate that dietary consumption of organochlorines has declined over the study period, at least in Sweden. Third, for a given mother, the concentrations of organochlorines in the milk correlate linearly with the total fat content, reflecting the high partition coefficient of the non-polar TCDD between lipid and aqueous phases.

The toxic effects of TCDD are initiated by its association with an intracellular "receptor protein." This protein, which is known as the Ah receptor, has molar mass ca. 90,000 g mol^{-1} in its denatured form, but is probably oligomeric[57]. It binds remarkably strongly to TCDD ($K_{ass} \approx 10^{11}$ L mol^{-1}). The TCDD-receptor complex then migrates to the nucleus, where it associates with DNA binding sites, thereby triggering the production of a series of messenger RNA molecules, and ultimately, the toxic responses (Problem 16).

The Ah receptor is present in many—maybe all—mammals. There seems to be little difference in either the amounts of receptor, or its affinity for TCDD, between responsive species such as the guinea pig, and less responsive species such as the hamster. A curious observation is that no natural ligand which binds to the Ah receptor protein is known, yet TCDD is not found in nature. Several lines of evidence have shown that the Ah receptor contains a hydrophobic binding site of well defined dimensions which optimally accommodates TCDD.

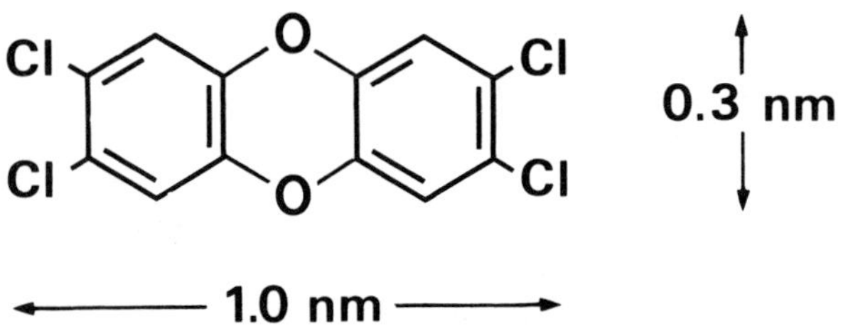

Cl O Cl
Cl O Cl
0.3 nm
1.0 nm

Among other PCDD and PCDF congeners (which are all planar, non-polar molecules) receptor affinity and toxicity are reduced if chlorine atoms are removed from the lateral (2, 3, 7, or 8 positions) and/or are present at positions 1, 4, 6, or 9. Such structural changes typically reduce both toxicity and receptor affinity by 3, 6, or more orders of magnitude (Table 9.2).

[57]R.D. Prokipcak and A.B. Okey, "Physicochemical characterization of the nuclear form of the Ah recptor from mouse hepatoma cells exposed in culture to 2,3,7,8–tetrachlorodibenzo-*p*-dioxin," *Arch Biochem Biophys,* **1988,** 267, 811–828..

Table 9.2: LD_{50} values for administration of a single dose of PCDD congeners to guinea pig and mouse[58]

congener	LD_{50}, $\mu g/kg$	
	guinea pig	mouse
unsubstituted		$> 50,000$
2,8–Cl_2	$> 3 \times 10^5$	$> 8 \times 10^8$
2,3,7–Cl_3	29,000	$> 3,000$
1,3,6,8–Cl_4	$> 15 \times 10^6$	$> 3 \times 10^6$
2,3,7,8–Cl_4	1	200
1,2,3,7,8–Cl_5	3	300
1,2,4,7,8–Cl_5	1,100	$> 5,000$
1,2,3,4,7,8–Cl_6	73	830
Cl_7	> 600	
Cl_8		$> 4 \times 10^6$

The considerations concerning the relative toxicities of PCDD congeners also apply to PCDF's. 2,3,7,8–Tetrachlorodibenzofuran (TCDF) is a molecule whose dimensions are closely similar to those of TCDD: it has comparable affinity for the protein receptor, and is also comparably toxic. The term **isosteric** is used to describe molecules having similar dimensions, and this concept has been found useful in predicting both toxicity and affinity for the Ah receptor protein. Other compounds that are approximately isosteric with TCDD are 3,3′,4,4′-tetrachloroazobenzene and its azoxy analog; these substances are important environmentally as contaminants in herbicides based on 3,4-dichloroaniline as a starting material. Like TCDD, they are persistent in the environment, bind strongly to the Ah receptor, and show similar toxic responses in mammals.

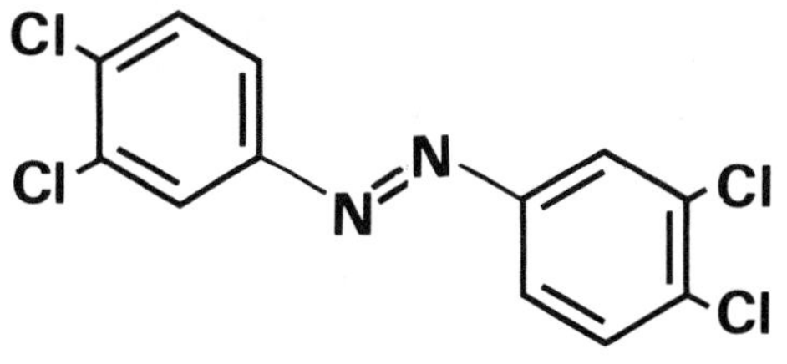

TCAB 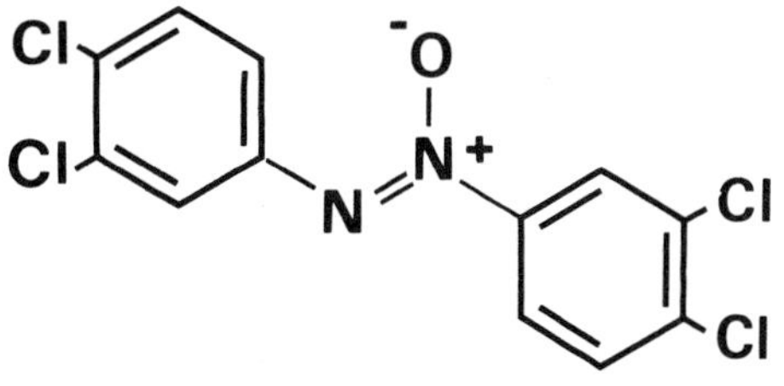 **TCAOB**

[58] R.J. Kociba and O. Cabey, "Comparative toxicity and biologic activity of chlorinated dibenzo-*p*-dioxins and furans relative to 2,3,7,8–tetrachlorodibenzo-*p*-dioxin (TCDD)," *Chemosphere,* **1985,** 14, 649–660. Where a range was given in the paper, the approximate midpoint has been selected.

The difficulty of describing the potential hazard of materials contaminated with PCDD's and PCDF's is that a large number of congeners is involved, each with its characteristic toxicity. The toxicities vary greatly, as we have seen, and in addition, it is not certain that the mixtures behave additively. A recent attempt to address this problem is through "toxic equivalences," such as the International Toxic Equivalency Factors (I-TEF's) see Table 9.3. The idea behind I-TEF's is that the toxicity of any PCDD or PCDF congener is related, using animal data, to the amount of TCDD having equal toxicity. Only those congeners substitued in all of the 2,3,7 and 8 positions are considered to be of concern.

Table 9.3: International Toxicity Equivalency Factors for PCDDs and PCDFs.

Congener	PCDD series	PCDF series
2,3,7,8	1 (defined)	0.1
1,2,3,7,8	0.5	0.05
2,3,4,7,8		0.5
1,2,3,4,7,8	0.1[a]	0.1[b]
1,2,3,4,6,7,8	0.01	0.01[c]
octachloro	0.001	0.001

[a] same value for 1,2,3,6,7,8– and 1,2,3,7,8,9– congeners
[b] same value for 1,2,3,6,7,8–, 1,2,3,7,8,9–, and 2,3,4,6,7,8– congeners
[c] same value for 1,2,3,4,7,8,9– congener

The following example shows how TCDD equivalents are calculated. 2,3,7,8–TCDD has I-TEF = 1 (by definition); 2,3,4,7,8–pentachlorodibenzofuran (pentaCDF) has I-TEF = 0.5. Therefore a mixture of 2 ng TCDD + 6 ng pentaCDF has an assumed toxicity equivalent to 2 ng TCDD + (6 × 0.5) ng TCDD, or 5 ng TCDD altogether.

Estimates of human exposures to PCDDs and PCDFs in the environment are ca. 0.1 ng TCDD equivalents per person per day, with the principal source being food. There is much public concern about PCDD/PCDF emissions from municipal solid waste incinerators; however, the best estimate is that persons in the immediate vicinity of such a facility would suffer an additional exposure of less than 1 pg/person/day, or less than 1% of background. Residential wood stoves are actually believed to contribute more to environmental levels of these pollutants than municipal or chemical waste incinerators.

9.3.3 *Treatment of dioxin wastes*

Dioxin waste-treatment is generally further from commercialization than PCB treatment, but the technologies under consideration are similar. A

"Project summary" by the U.S. EPA[59] summarized the situation as of 1987: incineration technology was under consideration, and test runs had been performed; plasma arcs were at the pilot stage, as was the KOH/polyethylene glycol method. The products of this last reaction have since been shown to have lost their toxicity towards guinea pigs[60]. One technology which has been used in practice is photochemical destruction, as shown by the following case study.

Photochemical destruction of TCDD residues in Verona, Missouri[61]

Like PCB's, TCDD absorbs UV radiation which is capable of causing cleavage of a C-Cl bond, to afford an aryl free radical. If a hydrogen atom donor is available, hydrogen abstraction completes the process of reductive dechlorination.

$$\begin{aligned} \mathrm{ArCl} &\xrightarrow{h\nu} \mathrm{Ar{\bullet}} + \mathrm{Cl{\bullet}} \\ \mathrm{Ar{\bullet}} + \mathrm{R{-}H} &\longrightarrow \mathrm{ArH} + \mathrm{R{\bullet}} \\ \mathrm{Cl{\bullet}} + \mathrm{R{-}H} &\longrightarrow \mathrm{HCl} + \mathrm{R{\bullet}} \end{aligned}$$

Reductive photodechlorination was used to cleanup a chemical production facility in Verona, Missouri which had been used for the manufacture of 2,4,5-trichlorophenol. When the plant changed owners in the early 1970's some 17,000 liters of distillation still bottoms containing nearly 350 ppm of TCDD were discovered on the site; the new owner was unable to remove them because no licensed incinerator existed in Missouri, and U.S. federal law prohibited transport of the waste across the state boundary.

The sludge was treated in the following steps.

- Extraction with NaOH(aq) and hexane removed the phenols into the aqueous phase, and transferred the TCDD into the hexane.
- Isopropyl alcohol (an excellent hydrogen atom donor) was added, and the yellow solution was irradiated using 10 kW mercury arc lamps in a 7000 L reactor. Sequential loss of chlorine atoms was followed by cleavage of the dioxin nucleus, to give a solution which was no longer considered hazardous. The progress of the photolysis was monitored by analysing for the chloride ion released, and eventually 99.94% of the dioxins were destroyed (343 ppm to less than 0.2 ppm).

The cost of this operation—excluding the salaries of company personnel—was $0.5 million, all to destroy 7 kg of TCDD. While this may be an extreme

[59] M. Breton *et al,* "Technical Resource Document: treatment technologies for dioxin-containing waste" EPA/600/S2–86/096.

[60] D.M. DeMarini and J.E. Simmons, "Toxicological evaluation of by-products from chemically dechlorinated 2,3,7,8–TCDD," *Chemosphere,* **1989,** 18, 2293–2301.

[61] See chapters 15–18 of *Detoxication of hazardous waste,* Ed. J.H. Exner, Ann Arbor Science, Ann Arbor, Michigan, 1982, and especially Chapter 17: J.H. Exner, J.D. Johnson, O.D. Ivins, M.N. Wass, and R.A. Miller "Process for destroying tetrachlorodibenzo-*p*-dioxin in a hazardous waste."

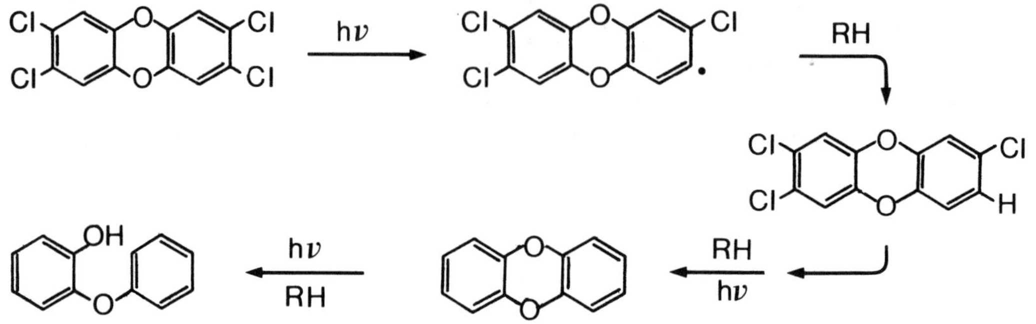

example, it does emphasise that the treatment of hazardous wastes can be a very expensive proposition.

9.4 Chlorinated phenols

9.4.1 Chlorinated phenols in the environment

Chlorinated phenols were first synthesized in the 19th century, and were found to have antiseptic properties. Of the 19 chlorinated phenols, the most important congeners industrially are 2,4-dichlorophenol, 2,4,5-trichlorophenol, and pentachlorophenol (PCP). The first of these continues to be commercially important in the production of chlorinated phenoxyacetic acid herbicides; 2,4,5-trichlorophenol is also used as the precursor of the antiseptic hexachlorophene through condensation with formaldehyde. The problem of TCDD as a contaminant of 2,4,5-trichlorophenol has already been discussed, and similarly, octachlorodibenzo-*p*-dioxin (OCDD) is found as a contaminant in PCP.

2,4-Dichlorophenol and 2,4,5-trichlorophenol enter the environment at the point of manufacture or conversion to other products. Non-point sources are mainly agricultural, since the phenoxy herbicides are hydrolyzed back to the phenols with a lifetime of about a week near 20°C. A minor local source of chlorophenols is chlorination of raw drinking water which is contaminated with phenol: see Chapter 7.

The most important chlorinated phenol is the pentachloro congener, which accounts for almost half of the total world production of some 200,000 tonnes per year. Pentachlorophenol (PCP) is used largely as a wood preservative, with smaller amounts in various pesticidal applications[62]. PCP formulations are rarely pure; substantial amounts (20% and more)

---

[62] Data in this section were obtained from:

(a) World Health Organization, "Environmental Health Criteria 71: Pentachlorophenol," U.N. Environment Program, 1987.

(b) K.R. Rao (Editor), *Pentachlorophenol: chemistry, pharmacology, and environmental toxicology,* Plenum Press, New York, 1978.

of 2,3,4,6-tetrachlorophenol are sometimes present. Nilsson *et al*[63], report the following ranges for minor impurities: PCDD's (mostly octachloro–), 0–2000 ppm; PCDF's, 50–200 ppm; chlorinated diphenyl ethers, 100–1000 ppm; chlorinated phenoxyphenols, ≈1%. The phenoxyphenols represent the halfway stage in the condensation/dimerization of PCP to the PCDD's.

Chlorinated phenols are weak acids, the acidity increasing with the number of chlorine substituents. Recalling that when pH = pK_a, a weak acid is half dissociated, we see that in waters of pH 6–8, monochlorophenols will be almost completely undissociated, di– and tri–chlorophenols will be partly dissociated, and tetrachlorophenols and PCP will be completely dissociated (Problem 17): see Table 9.4.

Table 9.4: Physical properties of some chlorophenols

Chlorophenol Congener	Temperature at which vapour pressure = 1 torr, °C	pK_a
2–Cl	12	8.5
4–Cl	50	9.2
2,4–Cl_2	53	7.7
2,4,5–Cl_3	72	7.4
2,3,4,6–Cl_4	100	5.4
Cl_5	see Problem 18	4.9

Chlorophenols also possess moderate volatility, but only the undissociated form will volatilize easily. Even though PCP will be extensively ionized in contact with water (pK_a = 4.9), volatilization of PCP into the atmosphere is an important environmental route for this compound. About half of PCP applied to brush or dip treated coniferous wood is lost by volatilization within 12 months. Disposal of waste water from wood preservation sites is often handled by allowing the water to evaporate, a practice which leads to air pollution by PCP.

Measurements of ambient tropospheric concentrations of PCP, while few in number, range from < 1 ng m^{-3} in remote areas to 5–10 ng m^{-3} in urban localities. Chlorophenols have been detected in rain and snow in a number of locations. Once in the atmosphere, PCP and other chlorophenols can undergo both direct solar photolysis and attack by hydroxyl radicals, as already discussed for PCB's. The relative proportions of these two processes vary from congener to congener, with photolysis estimated to be more important for PCP, and hydroxyl attack the major sink for 2,4-dichloro- and

[63] C.A. Nilsson, A. Norstrom, K. Andersson, and C. Rappe, "Impurities in commercial products related to pentachlorophenol," in Reference 61b, 313–324.

2,4,5-trichloro-phenols[64]. Lifetimes with respect to such chemical transformation are of the order of days, but vary with season and with geographical location because of variation in the flux of solar photons.

9.4.2 Toxicology of pentachlorophenol

As noted above, PCP formulations usually contain other, more toxic compounds in minor quantities, and so it is not always easy to distinguish the toxic effects of PCP from those of the contaminants. In rats, purified PCP has LD_{50} of about 150 mg/kg. Purified PCP samples have been tested for carcinogenicity in rats, but none has been found. Reduced reproductive capacity has been observed[65] upon long term administration of PCP to female rats at 30 mg kg^{-1} day^{-1} . PCP is also generated *in vivo* by hydroxylation of either penta– or hexa–chlorobenzenes. PCP is cleared fairly quickly from the body, in rats, fish, and several invertebrates, by conversion to a water-soluble conjugate (sulfate and/or glucuronide) in the liver. In rats, dechlorination to tetra– and tri-chlorohydroquinones and their conjugates also occurs, the reaction being effected by liver microsomal enzymes. It is reasonable to attribute the relatively low acute toxicity of PCP in rats to the rapid rate of clearance from the body. Note that at physiological pH, PCP will be almost completely in the anionic form, thus favouring excretion rather than bioconcentration.

9.5 Problems

1. The following series of problems all relate to a "one-compartment" toxicological model for the uptake of a toxic substance from water by an aquatic organism.

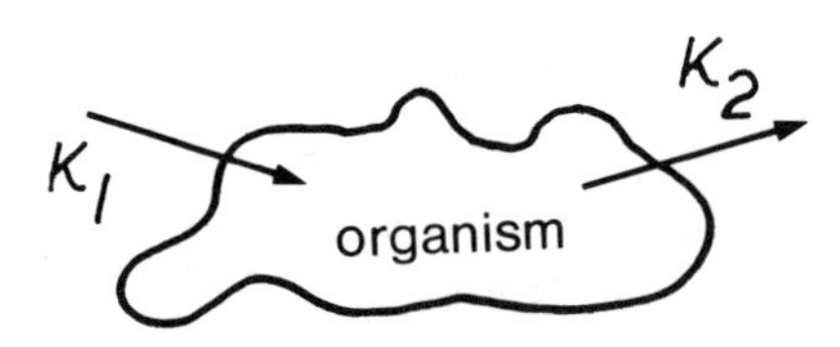

[64]N.J. Bunce, J.S. Nakai, and M. Yawching, "Estimates of the tropospheric lifetimes of short- and long-lived atmospheric pollutants," *J Photochem Photobiol (A)*, **1991**, 57, 429–440.

[65]B.A. Schwetz *et al*, "Results of two-year toxicity and reproduction studies on pentachlorophenol in rats" in Reference 62b, 301–309.

First-order rate constants k_1 and k_2 are respectively for uptake from the water and loss from the organism (metabolism and excretion back to the water). Numerical starting conditions for the problems are c_O (initial concentration of toxicant in water) = 0.020 ppm; half life for clearance of the toxicant from the organism = 3 days; K_{OW} = octanol: water partition coefficient = 3×10^5; steady state concentration of the toxicant in the organism = $(0.05 \times K_{OW} \times c_{aq})$.

(a) Calculate the steady state concentration of toxicant in the organism. What is the relationship between k_1 and k_2?

(b) Derive the equation showing the variation of the concentration of the toxic substance in the organism with time, following the initial placement of the organism in the water. Assume the water to be an infinite reservoir of the toxicant.

(c) Calculate the time required for the concentration of the toxicant in the organism to reach 5 ppm.

(d) Repeat the derivation of part (b) for the case where the reservoir of the toxicant is limited. The aqueous phase has volume V_1 liters and the organism V_2 liters.

(e) Repeat the calculation of part (c) for the case of a minnow (V_2 = 2.5 cm^3) in each of three tanks of capacity 5 L, 50 L, and 500 L.

(f) Repeat the calculation of part (e) for a large tank ($V_1 \rightarrow \infty$) but considering the size of the organism. You must take k_1 as a composite rate constant $k_1 = k_{\circ}A_1$ where A is the surface area of the organism and $k_{\circ}$ has units L area^{-1} time^{-1}. For simplicity, assume spherical organisms of radius 0.1 cm, 1 cm and 10 cm.

(g) A minnow is placed in a large ($V_1 \rightarrow \infty$) tank under the conditions of part (b) and left there for 4 days. It is then transferred to a large tank of clean water. What is the concentration of the toxicant in its tissues after a further 4 days ?

2. **(a)** Calculate the mass of DDT present in a 3 oz feeding of mother's milk contaminated by 130 ppb of DDT.

(b) The detection limit of many chlorinated aromatics such as DDT, PCB's and dioxins is of the order of 10 pg (10^{-11} g) introduced on the gas chromatography column. A sample of human adipose tissue contains 85 parts per trillion of DDT. Assuming an extraction procedure with a 35% efficiency, what is the minimum mass of adipose tissue that must be processed in order to detect the DDT by gc-ms?

3. **(a)** What is the average composition of the molecules in a commercial sample of Aroclor 1254?

(b) Explain why gc-ms analysis of a sample containing Aroclor 1254 would give a large number of "peaks" of different gas chromatographic retention time, and why many of these would give rise to different combinations of ions in the mass spectrum.

4. (a) Remembering that chlorine has two isotopes ^{35}Cl and ^{37}Cl (natural abundances 75% and 25% respectively), calculate the appearance of the "molecular ion cluster" (that is, the unfragmented molecule) 2,3,7,8–TCDD. Assume that C,H, and O consist only of their common isotopes ^{12}C, ^{1}H and ^{16}O.

   (b) Would there be any problems in doing a quantitative analysis of a mixture of TCDD, DDE ($C_{14}H_8Cl_4$) and a pentachlorobiphenyl by mass spectrometry? Explain.

5. Highly lipophilic substances are frequently assigned a "maximum" K_{OW} of 10^6. Assuming one compartment, and 5% total lipid, calculate the steady state body burden of hexachlorobiphenyl in a 1.0 kg experimental trout maintained in water containing 15 ppb of hexachlorobiphenyl.

6. Pentachlorobiphenyl $C_{12}H_5Cl_5$ (s) has vapour pressure 1.1×10^{-3} Pa at 298K. Calculate

   (a) the equilibrium vapour pressure in ppm

   (b) $\Delta G°$ for the vaporization of $C_{12}H_5Cl_5$ at 298K

   (c) The air over a landfill site (assume 298K) is found to contain PCB's (assume $C_{12}H_5Cl_5$) at a concentration of 0.022 $\mu g\ m^{-3}$. Calculate ΔG for the process:

$$C_{12}H_5Cl_5(s) \longrightarrow C_{12}H_5Cl_5(g)$$

   and conclude whether or not equilibrium has been reached.

7. A test burn of PCB in mineral oil by the Florida Power and Light Co. involved burning 34 L per hour of PCB and 91,200 L h^{-1} of fuel oil. PCB emissions were 0.0003% of intake.

   Calculate the PCB concentration in the stack gas in ppb making each of the following assumptions

   - stack gas at 800K
   - fuel oil is $C_{15}H_{30}$, density = 0.80 kg L^{-1}
   - stoichiometric amount of O_2
   - air is 20% O_2 by volume
   - molar mass of PCB = 320 g mol^{-1}; density = 1.2 kg L^{-1}.

   Neglect the PCB in considering the combustion process.

8. Examine the paper by Manion *et al*, (*Environ Sci Technol*, **1985**, 19, 280) on the use of the high temperature reaction with $H_2(g)$ as a possible method for destroying chlorinated aromatic compounds. Look up any necessary thermodynamic data and calculate K_p for the reactions below, at 1000 K.

(a) $C_6H_5Cl(g) + H_2(g) \longrightarrow C_6H_6(g) + HCl(g)$

(b) biphenyl(g) $+ H_2(g) \longrightarrow 2C_6H_6(g)$

9. Upon photoexcitation chlorobenzene undergoes decomposition from a triplet state which is formed in 64% yield from the initially formed singlet excited state.

$$PhCl^* \xrightarrow{k_r} Ph\bullet + Cl\bullet \longrightarrow \text{final products}$$

The lifetime of ^{3}PhCl is known to be 0.5 μs, and the overall quantum yield of decomposition is 0.40.

(a) What is the value of the rate constant k_r?

(b) What is the efficiency of decomposition of ^{3}PhCl?

(c) A low pressure mercury arc is used to decompose chlorobenzene. What is the rate of chlorobenzene decomposition if (i) the lamp converts electrical energy to 254 nm photons with an efficiency of 35% (ii) Only 254 nm photons are produced (iii) all these photons are absorbed by chlorobenzene (iv) the electrical rating of the lamp is 15W?

10. PCB's volatilise into the atmosphere where they are subject to attack by hydroxyl radicals. With 3-chlorobiphenyl the rate constant for the reaction with OH is 5.4×10^{-12} cm^3 molecule^{-1} s^{-1} at 295 K.

(a) If [OH] is present at a steady state concentration of 5.8×10^5 radicals per cm^3, and the 3-chlorobiphenyl concentration is 1.5×10^5 molecules cm^{-3}, calculate (i) the half life of 3-chlorobiphenyl in the troposphere, assuming that it reacts only with OH, and (ii) the initial rate of reaction in the units mol L^{-1} s^{-1}.

(b) Using your answer in part (a) explain whether or not you think that PCB photolysis is an important source of chlorine atoms in the stratosphere.

11. (a) Examine the paper by Zepp and Cline, *Environ Sci Technol,* **1977,** 11, 359. Calculate from Table II of this paper the total moles of photons having $\lambda \leq 317.5$ nm that fall on 1 cm^2 of water surface in one hour using the "summer" values.

(b) Miller *et al, J Agr Food Chem,* **1980,** 28, 1053 report that 3,4-dichloroaniline (DCA) decomposes photochemically in water with a quantum yield 0.052. The experiments were done in open dishes 12 cm diameter, 6.2 cm deep. Use Figure 9.4 of this paper to estimate the first order rate constant for the disappearance of DCA under summer conditions. (Hint: determine k_1 in the equation: rate $= k_1$[DCA].

(c) Why should the rate in this experiment depend on the DCA concentration? (Recall that in Problem 9 the rate depended only on the light intensity.)

(d) Combine the results from (a) and (b) to estimate, over the range $300 < \lambda < 317.5$ nm, the fraction of the light absorbed by DCA in the experiment by Miller *et al.*

12. Dublin *et al*, *Environ Sci Technol,* **1986,** 20, 72–77, measured the vapour pressure of TCDD at 25°C by passing dry N_2 over solid radiolabelled TCDD at a rate of 4.60 mL min^{-1}. The TCDD was collected in a trap and assayed by means of its radioactivity. After 2880 min, the mass of TCDD collected was 1.78×10^{-10} g. Calculate the vapour pressure of TCDD at 25°C.

13. (a) Calculate the chemical yield of TCDD if a sample of 2,4,5-trichlorophenol (no solvent) is contaminated by 9.4 ppm of TCDD.

(b) Calculate the mass of octachlorodibenzo-*p*-dioxin formed upon incineration of 100 kg of trichlorophenol if the chemical yield, based on chlorine, is 10^{-10}%.

14. In the "Times Beach" incident, a waste oil hauler removed 18,500 US gallons of oil contaminated by 33 ppm of "dioxin" from a 2,4,5-T manufacturing plant.

(a) What mass of dioxin was involved? (Assume the oil had a density of 1.0 g cm^{-3}).

(b) Some of the horse arenas that were sprayed with this oil had soil/solid matter dioxin concentration of 1750 ppb. What mass of this solid matter need be ingested by a 25 g mouse to reach the LD_{50} of 114 μg per kg?

15. Suppose a coffee filter contains 5 pg/g of TCDD, and that a lethal dose of TCDD in the guinea pig is 1.0 μg/kg. What mass of this paper would a 250 g guinea pig need to consume in order to ingest a lethal dose of TCDD?

16. Taking K_{ass} for the interaction between TCDD and the Ah receptor to be 10^{11} L mol^{-1}, and an estimated intracellular receptor concentration of 15 pmol L^{-1}, calculate the intracellular concentration of TCDD required to occupy 85% of all the binding sites for TCDD on the receptor.

17. (a) Calculate the percent dissociation of (i) pentachlorophenol (ii) 2,4,5-trichlorophenol in body fluids at pH 7.4.

(b) Plot the speciation of pentachlorophenol over the pH interval 4–7.

18. The vapour pressure of pentachlorophenol is given below[66]:

---

[66] R.A. McDonald, S.A. Shrader, and D.R. Stull, *J Chem Eng Data,* **1959,** 4, 311.

Temperature	Vapour Pressure
200.66°C	4.133 kPa
215.51	6.759
233.87	12.279

Determine:

(a) the temperature at which the vapour pressure of pentachlorophenol is 1 torr;

(b) the vapour pressure of pentachlorophenol at ambient temperatures (say, 20°C);

(c) the TLV for pentachlorophenol is 0.5 mg m^{-3}. Will the TLV be exceeded if the air in a room at 20°C becomes equilibrated with solid pentachlorophenol?

Chapter 10

Metals in the Environment

Introduction

In this chapter we will discuss two of the many metallic elements which cause environmental pollution: mercury and lead. This selection is arbitrary, since there are many other metallic elements which cause environmental contamination. Mercury pollution was a serious environmental concern of the period near 1970; lead is of greater anxiety today. The occurrence of these elements in drinking water was discussed in Chapter 7.

As stated in Chapter 8, for metallic pollutants it is the element itself which is toxic. Hence there is no "ultimate destruction" treatment for metals and their compounds corresponding, for example, to incineration of PCB's. Indiscriminate release of toxic metals into the environment should be avoided, and their disposal should involve in chemical forms which are relatively immobile in order to minimize their dispersal and consequent contamination of air and groundwater.

10.1 Speciation

The speciation of metallic elements affects their impact upon the environment. To take a simple example from medicine: barium compounds are very toxic, yet "barium meals" are routinely given to patients with gastrointestinal disorders without ill effects. (Barium blocks x-rays very effectively, allowing the soft tissues of the gastrointestinal tract to be x-rayed.) The patient suffers no toxicity, because the barium is administered as barium sulfate—which is very insoluble—and is not absorbed. Hence the toxic element barium is physically unavailable (Problem 1).

Chemical speciation is important when different chemical forms of an element differ in toxicity. As we shall see below, inorganic mercury is much less toxic than organomercurials, especially methylmercury derivatives. Arsenic exemplifies the opposite situation, where the inorganic forms are the more hazardous. Arsenic enters the environment through burning coal and oil, in which it is a trace element, from mining operations, and from smelting, especially of copper. It is usually released as arsenic(V), which predominates under aerobic conditions. In sediments however, arsenic(V) is reduced microbially to arsenic(III) which is about 50 times more toxic. Other microbial reactions lead to methylated arsenic oxyacids, which are less toxic than arsenic(V) compounds. Finally, fish and shellfish store arsenic as arsenolipids, which are almost non-toxic[1], to the extent that moderate quantities of these fish can safely be eaten. The point is that it is not the total quantity of the toxic element which determines the degree of environmental hazard, but the chemical form. Analytical methods for the assay of these elements in the environment must therefore be able to differentiate between the various chemical forms of the element: it is not sufficient simply to measure the stoichiometric concentration by some technique such as atomic absorption spectroscopy.

10.2 Mercury[2]

10.2.1 *Industrial uses*

Mercury poses an environmental problem because of **industrial release** coupled with **high toxicity**. Mercury is the only metal which is a liquid at ordinary temperatures; its boiling point is relatively low (357°C), and its vapour pressure is significant even at room temperature (Problems 2–4). The toxicity of mercury has long been known. It is a neurotoxin, which causes symptoms such as quarrelsome behaviour, headache and depression, and muscle tremors. The expression "mad as a hatter" derives from the use of mercury(II) nitrate in making felt for hats, a use which continued until about 1940. Workers in the industry suffered long term exposure to mercury salts, hence their symptoms.

Other occupations which formerly led to mercury exposure were gilding and mirror making. Gilding was the method by which objects were plated with gold and silver (before electroplating was developed in the mid-nineteenth century). An amalgam[3] of 10 parts of mercury to 1 part of gold was painted on to the object to be gilded, and then the mercury was evaporated away—without proper ventilation. Until the end of the last century, mirrors were made by applying a mercury/tin amalgam to glass and then evaporating the mercury[4]. Mercury was also an important ingredient in

[1] G.M.P. Morrison, G.E. Batley, and T.M. Florence, "Metal speciation and toxicity," *Chem in Britain,* August **1989,** 791–796.

[2] Unreferenced material in this section is mostly taken from J.O. Nriagu, *The biogeochemistry of mercury in the environment,* Elsevier/North-Holland Press, Amsterdam, 1979.

[3] An amalgam is the term used to describe either a solution of another metal in mercury or, if the product is solid, an alloy of another metal with mercury.

[4] The modern silver mirror is made using chemistry which resembles Tollens' test in organic chemistry. Silver nitrate is reduced using a weak reducing agent such as glucose; if the surface of the glass is clean, a thin layer of silver adheres to the glass, and a protective finish is applied on top of the silver layer.

alchemy; it has been suggested that the peculiar behaviour of Isaac Newton around 1690 was occasioned by his interest in alchemy at the time.

Modern occupations having the potential for exposure include thermometer manufacture (the end is closed by hand after the mercury is introduced), mining and refining of mercury (see below), workers in the chlor-alkali industry (see below), and laboratory and dental workers. The use of mercury diffusion pumps and similar equipment leads to serious risks of exceeding the TLV of mercury (0.05 mg m^{-3}) in case of spills (Problem 5), and it is not unknown for puddles of mercury to be found beneath the floors when old laboratories are renovated! Dental amalgams are prepared from silver (70%), tin (26%), copper (2%), and zinc (2%), and to this mixture is added 45% by weight of mercury. Dentists and their assistants may be occupationally exposed to mercury if the room in which the amalgam is made up is poorly ventilated; however, there appears to be little risk to the patient from having a mouth full of amalgam, presumably because the vapour pressure of mercury is greatly reduced upon amalgamation. Mercury is used in the electrical industry for switches, and in the manufacture of batteries; organomercurials are used in agriculture as fungicides, e.g., for seed dressings.

Mercury is a relatively rare element; it occurs mainly as HgS, cinnabar[5]. In some places a proportion occurs as the free element: in the mercury mines in Sicily, where the mercury occurs in shales, the miners are exposed to mercury vapour, levels of which in the air may reach $\approx$ 5 mg m^{-3}. Figure 10.1 is an example of the handwriting of a Sicilian miner suffering from the tremors of mercury poisoning. Even where free mercury is absent, care has to be taken to minimize the workers' exposure to dust.

Mercury refining involves heating the sulfide in air:

$$HgS(s) + O_2(g) \xrightarrow{700°C} Hg(g) + SO_2(g) \tag{1}$$

Significant worker exposure is possible during furnace operation and cleaning. The raw mercury is condensed in a water-cooled condenser, and redistilled for sale in traditional 76 lb "flasks."

The electrical uses of mercury include its application as a seal to exclude air when tungsten light bulb filaments are manufactured; fluorescent light tubes and mercury arc lamps—used for street lighting and as germicidal lamps—also contain mercury. The property of mercury being a liquid metal is exploited in certain electrical switching gear; the mercury is sealed into a glass container with electrical contacts at one end. The assembly is balanced so that under conditions of load the mercury completes the circuit; if the load is removed, the mercury runs away to the other end of the container and the electrical circuit is broken.

---

[5] Mercury sulfide comes in two forms: cinnabar, which is black, and vermillion, which has for centuries been used as a pigment for oil based paint. Mercury poisoning among artists has occurred as a result of licking the brush to get a fine point.

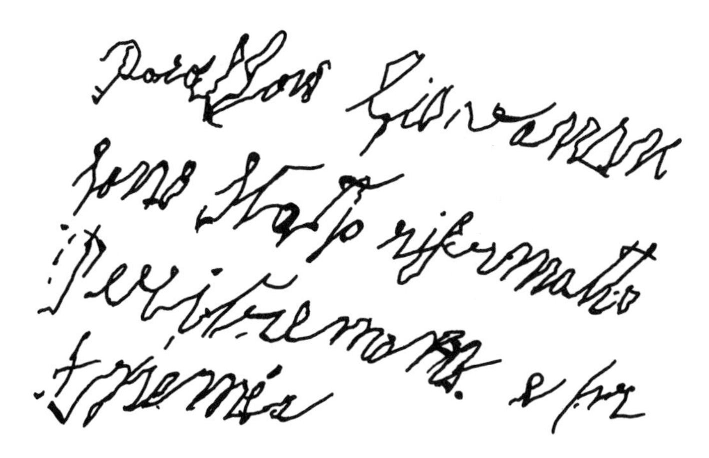

Figure 10.1: Handwriting of an Italian who worked in a mercury mine.

Mercury batteries are used widely in everyday life, in applications such as cameras, hearing aids, and heart pacemakers. About 30% of U.S. production of mercury is used in this way, the reason being the constancy of the voltage of the mercury battery, almost to the point of complete discharge. This is understandable from thermodynamic considerations.

The mercury cell is represented as follows:

$$\mathrm{Steel : Zn(s) : ZnO(s) : 40\%\ KOH : HgO(s) : Hg(l) : Steel}$$

The cell reaction is:

$$\mathrm{Zn(s) + HgO(s) \longrightarrow ZnO(s) + Hg(l)} \qquad (2)$$

For this reaction ΔG is given by:

$$\Delta G = \Delta G^\circ + RT \ln Q = \Delta G^\circ + RT \ln\left(\frac{a(\mathrm{Hg,l})\bullet a(\mathrm{ZnO,s})}{a(\mathrm{Zn,s})\bullet a(\mathrm{HgO,s})}\right)$$

In terms of the Nernst Equation:

$$E_{\mathrm{cell}} = E^\circ - \left(\frac{RT}{n\mathcal{F}}\right) \ln\left(\frac{a(\mathrm{Hg,l})\ a(\mathrm{ZnO,s})}{a(\mathrm{Zn,s})\ a(\mathrm{HgO,s})}\right)$$

Since the chemical substances in the ln Q term are all pure solids or elements, their activities are unity, and E = E° (1.35 V) throughout the life of the battery.

10.2.2 Mercury and the chlor-alkali industry

This is a story of serious environmental pollution, recognition of the problem, and successful measures to rectify the situation. In 1970, pollution by mercury was the top environmental issue; today, it is mainly of historical interest. Let us trace the development of the story.

The chlor-alkali process is the electrolysis of brine to produce chlorine and sodium hydroxide.

$$(3)\quad 2NaCl(aq) + 2H_2O(l) \xrightarrow{\text{electrolysis}} 2NaOH(aq) + H_2(g) + Cl_2(g)$$

The sodium hydroxide solution is evaporated to give the solid. From the stoichiometry of this equation, we see that approximately equal masses of Cl_2 and NaOH are produced (Problem 6)[6].

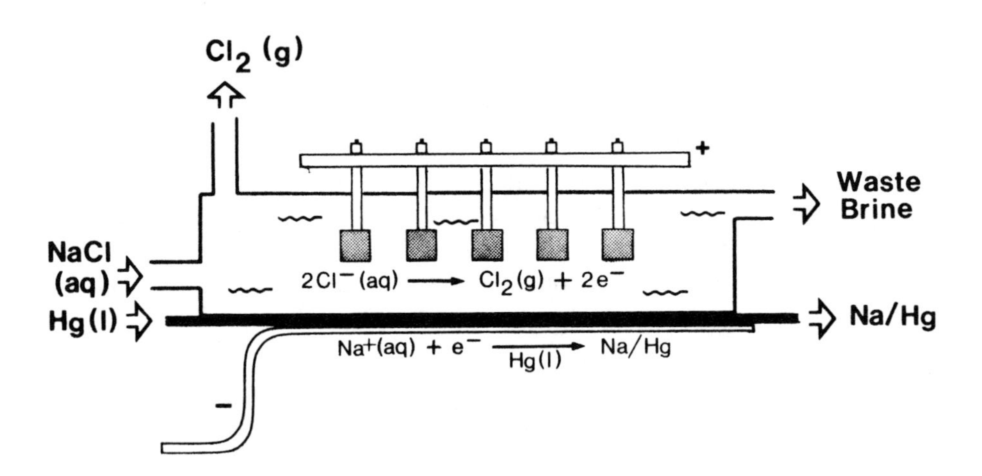

Figure 10.2: Flowing mercury cell for electrolyzing brine.

In the traditional flowing mercury cell, Figure 10.2, the anodes are made of graphite or of platinum-coated titanium. Chlorine is formed at the anode:

$$(4)\quad Cl^-(aq) \longrightarrow \tfrac{1}{2}Cl_2(g) + e^-$$

[6]This has an interesting economic consequence. If the demand for one of these two commodities increases, its price rises and production increases. This causes an oversupply of the other product, whose price then falls. In early 1989, the price of NaOH was $255–355 per tonne, depending on the grade of product. Chlorine at that time was in oversupply; its list price was $200 per tonne, but it was actually selling at only half that: *Chem Eng News,* March 13, **1989,** 9–10.

The cathode is a layer of liquid mercury which flows slowly through the cell. The cathode reaction leads to the formation of Na/Hg amalgam.

$$Na^+(aq) + e^- \longrightarrow Na/Hg \qquad (5)$$

Although sodium is much more electropositive than hydrogen, it is $Na^+(aq)$ rather than $H^+(aq)$ that is reduced because of: (i) the large overvoltage for reduction of H^+ at a mercury cathode; (ii) the lowering of the free energy for sodium formation through formation of the amalgam rather than the free metal.

The importance of the amalgam to the process is that it removes the sodium from the aqueous solution with which it would react. The dilute sodium amalgam is led into a separate chamber and reacted with water.

$$2Na/Hg + 2H_2O(l) \longrightarrow 2NaOH(aq) + H_2(g) + 2Hg(l) \qquad (6)$$

The sodium hydroxide is formed as a concentrated solution and the mercury is recycled. Further evaporation gives pure solid NaOH. An advantage of the mercury cell is that hydrogen and chlorine are produced in separate chambers, thus minimizing the risk of explosion from reaction between gaseous hydrogen and chlorine.

In the late 1960's it was discovered that chlor-alkali plants were leaking large amounts of elemental mercury into the environment. The need for abundant cooling water necessitates siting these plants on rivers; poor maintenance and inventory control led to losses of mercury, causing the river sediments downstream to become contaminated with liquid mercury. After pollution by the chloralkali industry was recognized, better housekeeping and government regulation greatly reduced losses of mercury to the environment, as shown in Table 10.1[7].

Table 10.1: Canadian losses (tonnes) of mercury to the environment, 1968–1976

Year	Hg Production (tonnes)	Losses from chloralkali plants (tonnes)
1968	195	149
1970	841	154
1972	505	52
1974	483	38
1976	400	26

Other releases of mercury in Canada at that time were from coal burning (2–6 t yr^{-1}), oil burning (18 t yr^{-1}), and the smelting of copper, lead, and

[7] Canadian data in this section are taken from the following sources: (a) I.G. Sherbin, *Mercury in the Canadian environment,* (Canadian) Environmental Protection Service, Report EPS-3-EC-79-6, 1979; (b) National Research Council of Canada, *Effects of mercury in the Canadian environment,* NRCC Publication 16739, 1979.

zinc (8 t yr^{-1}). An interesting sidelight is that in 1970, the cost of mercury was about $14 per pound; two years later, this dropped to $3.50 mainly because the chloralkali industry had been the major consumer.

Losses of mercury by Swedish chlor-alkali plants using flowing mercury cells have been reported as follows: 1970, 200 g per tonne of chlorine; 1975, 1–2 g; 1980, 0.15 g. Losses occurred at the following places: volatilization with the Cl_2 (1 g) and the hydrogen (2–10 g); trapped in the NaOH (2–12 g); carried off with the waste brine (5 ppm), and unknown amounts in the ventilation air, the cooling water, and the sludges from the electrochemical cells.

Since 1970, mercury pollution from flowing mercury cells has thus been sharply reduced, but in addition, new mercury-free technology has installed. Actually, the first mercury-free system to be developed—the asbestos diaphragm cell, Figure 10.3—was introduced by the Hooker Chemical Company in Niagara Falls, New York almost half a century ago.

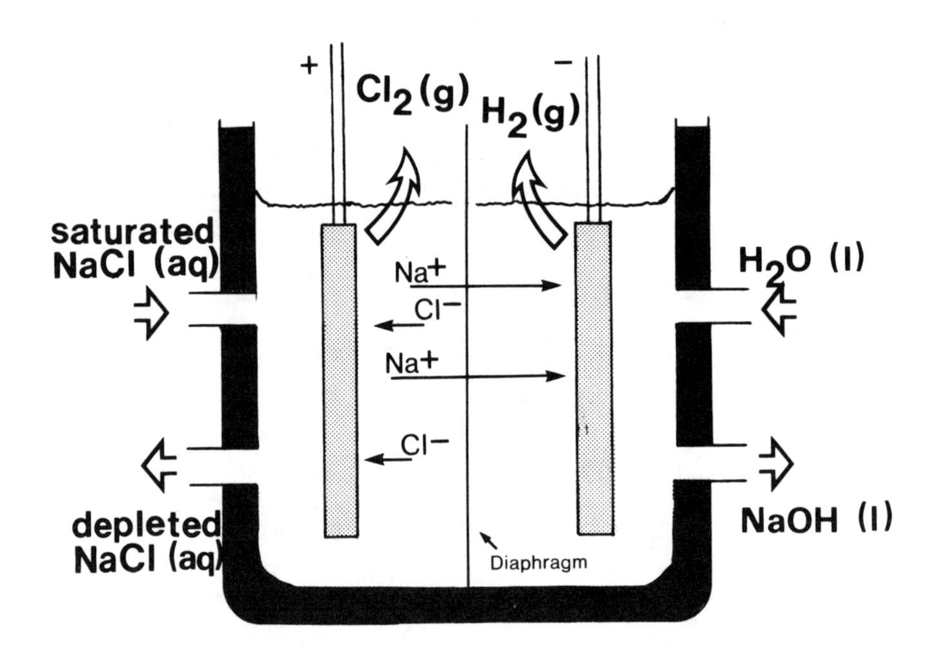

Figure 10.3: Schematic of a diaphragm or membrane cell.

The electrochemistry of the diaphragm cell is different from that of the mercury cathode cell in that H^+ rather than Na^+ is reduced at the cathode.

$$\text{(4)} \quad \text{anode:} \quad Cl^-(aq) \longrightarrow \tfrac{1}{2}Cl_2(g) + e^-$$

$$\text{(7)} \quad \text{cathode:} \quad H^+(aq) + e^- \longrightarrow \tfrac{1}{2}H_2(g)$$

Recognizing that removal of H^+ from aqueous solution leaves OH^- behind, the overall electrochemical reaction is:

$$\text{(8)} \quad H_2O(l) + Cl^-(aq) \longrightarrow \tfrac{1}{2}H_2(g) + \tfrac{1}{2}Cl_2(g) + OH^-(aq)$$

In terms of actual reactants:

$$(3) \qquad H_2O(l) + NaCl(aq) \longrightarrow \tfrac{1}{2}H_2(g) + \tfrac{1}{2}Cl_2(g) + NaOH(aq)$$

In the diaphragm cell, the anode and cathode compartments are separated by an asbestos barrier, whose functions are (i) to minimize mixing of the contents of the two electrode compartments while still allowing brine to pass through the barrier; (ii) to prevent mixing—and explosive reaction—of the hydrogen and chlorine liberated in the separate compartments. The quality of the NaOH is lower than that produced in the mercury cathode cell, because it is inevitably contaminated by unreacted NaCl.

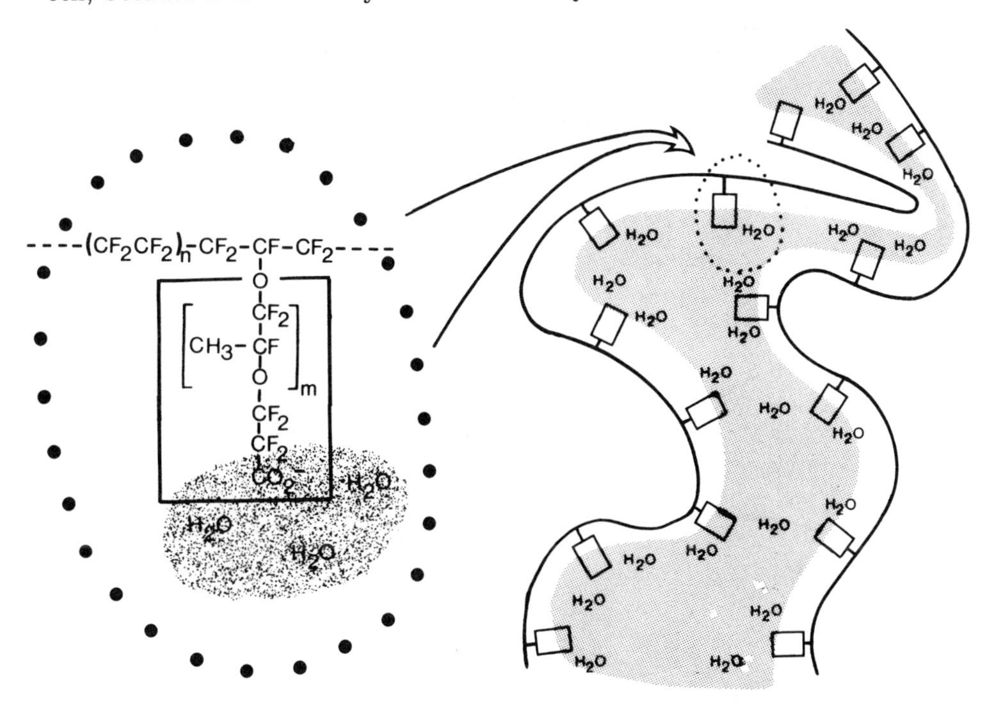

Figure 10.4: Structure of Nafion showing (a) the chemical constitution (b) the channels through the structure.

The introduction of the membrane cell, which uses a fluorocarbon-based membrane in place of the asbestos diaphragm, allows the production of high quality NaOH. The best known membrane material is Nafion, developed by DuPont. Nafion is the tradename for a range of polymers composed of a fluorocarbon backbone to which are attached fluorinated side chains terminating in various functional groups. The Nafion most often used in chloralkali cells terminates in a carboxylate group; it is a cation exchange material, which contains channels through which cations can pass. Anions are excluded from the channels because of the high negative charge in the channels due to the flanking carboxylate anions[8], see Figure 10.4.

[8] R.F. Brady, "Fluorpolymers," *Chem in Brit,* May **1990,** 427–430.

The anode and cathode reactions in the membrane and the diaphragm cells are the same. When H^+ is discharged at the cathode, OH^- is left behind and Na^+ migrates through the membrane to restore the charge balance. A pure grade of NaOH is obtained from the membrane cell because chloride anions cannot pass from the anodic to the cathodic compartment, hence the NaOH is uncontaminated by NaCl. All chlor-alkali plants commissioned since the early 1970's have employed this newer technology; there are still mercury cathode cells in use, but they are less polluting than formerly, and the last of them will probably be taken out of service shortly after the year 2000.

Abatement of mercury discharges from chlor-alkali plants have resulted in substantial improvements in environmental quality. For example, fish taken from the St. Clair River (Ontario)—a site of serious mercury pollution in the 1960's—have shown a progressive reduction in mercury levels, and are now considered safe to eat, see Figure 10.5.

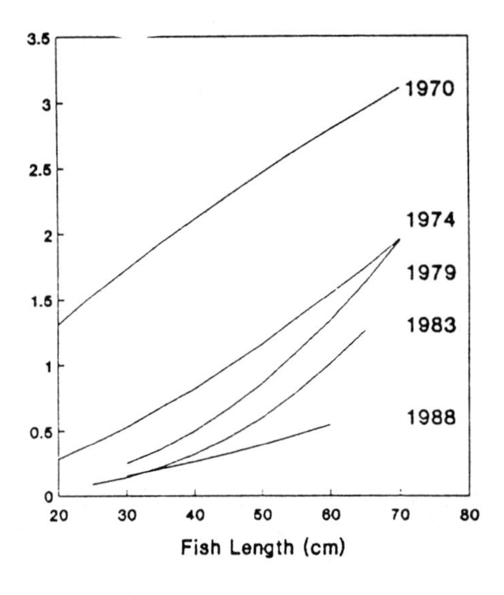

Figure 10.5: Concentration of mercury in Walleye Pike between 1970 and 1988. Note that in each year, the concentration of mercury increases with the size of the fish, showing that mercury accumulates with age. Reproduced from C. Cox and J. Ralston "A reference manual of chemical contaminants in Ontario sport fish," Ontario Ministry of Environment, 1990.

A major outlet for both Cl_2 and NaOH is the bleaching of wood pulp to make paper. Many pulp mills in remote locations (e.g., Northern Canada) have their own chlor-alkali plants. Concerns arise from the use of chlorine as a bleach since chlorine is a chlorinating agent. Pulp mills discharge large amounts of chlorinated byproducts into rivers, and their toxicological effects are largely unknown. An issue which has recently emerged for the chlor-alkali industry is the presence of PCDD's in paper products: see Chapter 9. There will probably be great pressure on the industry to replace chlorine bleach by oxygenated bleaches such as ClO_2, peroxides, or ozone[9]. This would have a considerable impact on the chlor-alkali operations associated with pulp mills, both in terms of the amount of Cl_2 required and also the relative demand for Cl_2 and NaOH.

10.2.3 Mercury poisoning

The Reed Paper controversy

The Reed paper company operated a pulp mill at Dryden, Ontario, together with a chlor-alkali plant to supply the chemicals needed to bleach the pulp. Between 1962 and 1970 it is estimated that about 10 tonnes of mercury were lost into the Wabigoon-English River systems. Use of mercury was limited after 1970 and discontinued in 1975, but minor losses from the plant continue even today, because of the large amount of mercury still dispersed on the site.

A serious situation developed because two native Indian (USA: "Amerindian") bands used the waters of the Wabigoon-English River systems for fishing. The fish caught there were said to comprise a major part of the diet of the bands as well as their principal source of livelihood[10]. Tests taken on the fish showed them to be contaminated by mercury far in excess of the 0.5 ppm which had been set as the standard for human consumption. Some members of the Indian bands were found to have tissue levels of up to 600 ppb of mercury, which is at the lower end of the range of clinical mercury poisoning. After years of legal wrangling, the bands were eventually compensated to the extent of $8 million in 1985.

What should be done about rehabilitating the river in a case like this? In this as in other similar ones, the losses of mercury were so great that pools of liquid mercury may be found in the sediments. Dredging has been rejected as an option, since it would be prohibitively expensive and would disturb the river sediments. Stirring up the mud would pose a short term threat to aquatic life and would resuspend the mercury in the biotic zone instead of letting it become ever more deeply buried in the sediment. A point often forgotten in discussions about environmental contamination by

---

[9]Note that even today, chlorine is not the only bleach used to whiten the pulp. Bleaching is a complex technology carried out in several steps, using alternately chlorine and chlorine dioxide, and the steps require different pH levels. The latter explains why pulp manufacture also employs large amounts of NaOH. For more on the major uses of Cl_2 and NaOH, see *Chem & Eng News,* May 21, **1990,** 19–20.

[10]There is some doubt as to whether most of the residents really were fish-eaters: see *Final report of the task force on organic mercury in the environment: Grassy Narrows and White Dog, Ontario,* Health & Welfare Canada, 1976.

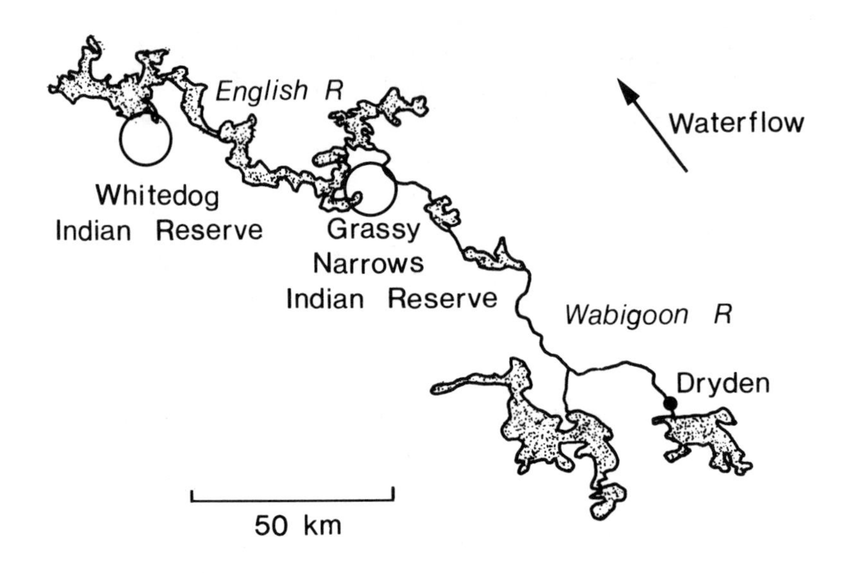

Dryden, Ontario, and area affected by mercury pollution.

metals is that these elements occur naturally in the environment; it is from the natural environment that they were extracted in the first place.

Minamata disease

The largest episode of mass poisoning due to mercury occurred in the 1950's in the Japanese fishing village of Minamata. About 1300 people were afflicted with physical signs of mercury poisoning over the period 1953–1960, and about 200 died. The patients showed signs of anorexia, irritability, and other psychiatric symptoms, but considerable research was needed to make the link to mercury.

The affected residents all ate large amounts of fish and shellfish. Cats fed on fish scraps showed similar symptoms. At first, it was believed that food poisoning was responsible, but the patients showed no fever or gastrointestinal disturbance, and all bacteriology was negative. Chemical poisoning was then considered, and a chemical plant manufacturing acetaldehyde came under suspicion because both the number of victims and the severity of their symptoms increased with proximity to the plant. The actual toxic agent was not traced for some time, but was eventually identified as mercury, used as a catalyst in the production of acetaldehyde. Mercury was detected in the fish and shellfish, the amount decreasing with distance from the acetaldehyde plant, and the methylmercury cation CH_3Hg^+ was detected in the waters of Minamata Bay.

The chemical process by which acetaldehyde was produced is shown below[11].

$$\text{(9)} \qquad H{-}C{\equiv}C{-}H \xrightarrow{Hg^{2+},\ H_3O^+} [CH_2{=}CHOH] \longrightarrow CH_3CHO$$

Minamata disease was thus caused by the loss of mercury residues from the acetaldehyde plant into Minamata Bay, where they were taken up by the fish and shellfish, and bioconcentrated in the form of lipophilic methylmercury derivatives (see below for the methylation of mercury).

Regular consumption of mercury in the diet leads to the accumulation of mercury in the body (Figure 10.6) because the rate of clearance of mercury from the body is slow (Problem 8). For this reason, mercury is described as a cumulative poison. A single meal contaminated by mercury at a specified level may cause no ill effects, but the same concentration in a steady diet can lead to sickness or death. This situation stands in contrast to a readily excreted or metabolized poison such as cyanide, which does not accumulate.

Screening of the Minamata residents was facilitated by gas chromatographic analysis of samples of their hair for CH_3HgCl, the concentration of which in hair was found to be proportional to the concentration of mercury in the patients' blood. The amount of methylmercury chloride in the hair was found to correlate linearly with the average amount of mercury in the diet. Up to 500 ppm of CH_3HgCl was detected in the hair of clinically affected patients, compared with 5 ppm among Japanese in general. Mercury levels in the blood of clinically affected patients were 70–900 μg L^{-1} (compare Figure 10.5), and 100–900 μg day^{-1} was excreted in the urine.

10.2.4 Speciation of aqueous mercury

The usual form of mercury in aqueous solution is the Hg^{2+} ion. Mercury has two oxidation states, Hg(I) and Hg(II), but the first of these—which contains the unusual ion $^{+}Hg{-}Hg^{+}$—is stable only as insoluble salts such as Hg_2Cl_2. It disproportionates in solution.

$$\text{(10)} \qquad Hg_2^{2+}(aq) \longrightarrow Hg^{2+}(aq) + Hg(l)$$

This means that reduction of Hg^{2+}—for example in sediments—gives the metal.

[11] Up to the early 1950's, acetaldehyde was made from acetylene, which in turn was derived from coal. Like the more familiar alkenes, alkynes are hydrated under acidic conditions, but unlike alkenes, alkynes require Hg^{2+} as a catalyst. The rise of the petrochemical industry during the early 1950's made oil-based organic chemicals cheaper to produce than coal-based ones. The acetylene route to acetaldehyde has been completely superceded by the Wacker process in which ethylene is the reactant and no mercury is used.

$$CH_2 = CH_2 \xrightarrow{O_2, \text{Pd catalyst}} CH_3CHO$$

Today, even acetylene is no longer made from coal, but from partial oxidation of methane.

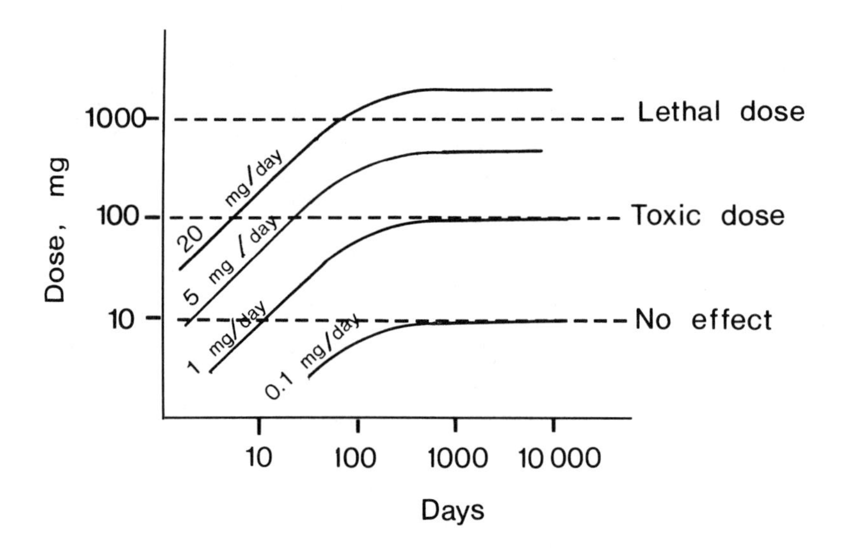

Figure 10.6: Accumulation curves for different levels of mercury in the diet.

Mercury(II) is a very "soft" Lewis acid, which forms stable complexes preferentially with soft Lewis bases such as sulfur ligands (recall that mercury is found in nature as the sulfide). Increasing the pH of an aqueous solution of Hg^{2+} leads to precipitation of HgO; HgO has finite solubility in water, and the solution may be described in terms of mercury(II) hydroxide (as an aqueous species $Hg(OH)_2^\circ$, though not as a solid phase). The relevant equilibria are as follows: species other than HgO are (aq).

$$(11)\quad HgO(s) \xrightarrow{H_2O} Hg(OH)_2^\circ \xrightarrow{+H^+,\,-H_2O} HgOH^+ \xrightarrow{+H^+,\,-H_2O} Hg^{2+}$$

The major complicating factor in the environmental chemistry of mercury is its biological methylation to CH_3Hg^+ and $(CH_3)_2Hg$, which converts inorganic mercury to forms which are both more toxic and more lipophilic. This is the subject of the next section.

10.2.5 Methylation of mercury

Mercury is responsible for neurological conditions because certain chemical forms of mercury are lipid soluble (and hence bioconcentrate) and are able to cross the blood-brain barrier. As a result, different forms of mercury represent different levels of hazard. Methylmercury derivatives are a particular danger in this regard (Problem 9).

Mercury is methylated in nature by the attack of methylcobalamin (Vitamin B_{12}) upon Hg^{2+}. Methylcobalamin contains a methyl group bonded to a central cobalt atom, making the methyl group somewhat carbanion-like. Representing methylcobalamin as $L_5Co–CH_3$, the simplified equation is shown below as attack by the electrophilic Hg^{2+} on the carbanion-like methyl group.

$$L_5Co–CH_3 + Hg^{2+} \longrightarrow L_5Co^+ + CH_3Hg^+ \tag{12}$$

CH_3Hg^+ occurs mostly as CH_3HgCl; in shellfish, CH_3HgSCH_3 is also found, since mercury has a strong affinity for sulfur. While CH_3Hg^+ derivatives predominate when mercury is methylated in sediments below pH 7, further methylation to $(CH_3)_2Hg$ becomes important as the pH rises: Figure 10.7.

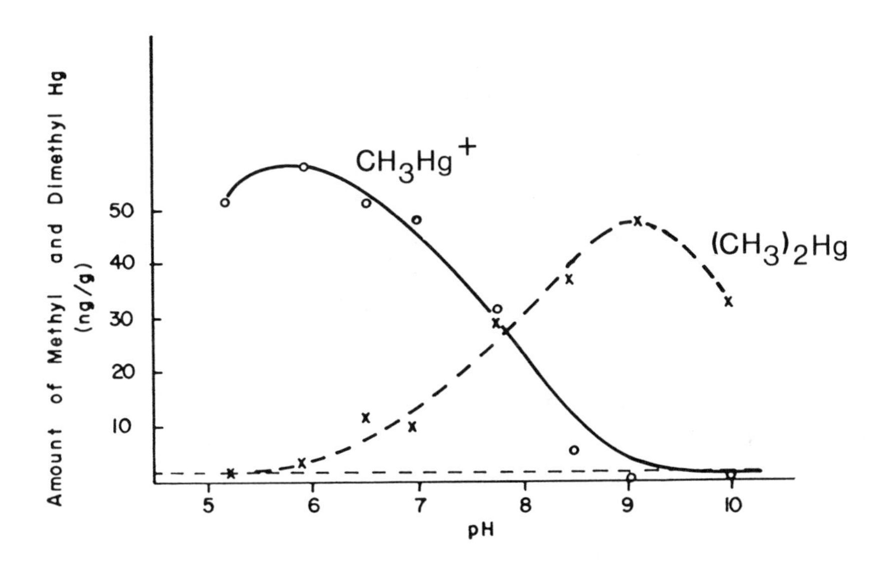

Figure 10.7: Methylation of 100 ppm of Hg^{2+} in sediments over two weeks. Reproduced from Reference 7(a), 30.

The carbon-mercury bond is intrinsically weak (about 200 kJ mol^{-1}), but is almost completely non-polar. Neither nucleophiles nor electrophiles react readily with such a centre, and so organomercurials tend to be kinetically unreactive. This contrasts with the behaviour of more familiar organometallic compounds such as Grignard reagents and organolithiums, where the carbon-metal bond is strongly polarized in the sense $C^{\delta-}$—$M^{\delta+}$, and it behaves like a nascent carbanion.

10.2.6 Environmental regulation of mercury

As noted in Chapter 7, standards for mercury in drinking water are typically 1.0 $\mu g\ L^{-1}$ (1 ppb). The Great Lakes joint water quality agreement between Canada and the United States sets a target of 0.2 μg of mercury per liter of unfiltered lake water. This agreement implies that states and provinces bordering the Great Lakes will control their discharges; Ontario's standards are 1.0 ppb of mercury in waste water and 0.1 ppb for water discharged into sanitary sewers. Fish taken from the Great Lakes should not be eaten if their mercury content exceeds 0.5 ppm. The substantial reduction in environmental pollution by mercury since about 1970 means that contamination of fish by mercury is no longer perceived to be a serious threat by the public, whose attention and concern have now shifted to the presence of PCB's, PCDD's and PCDF's in these same fish (Problem 10).

Typical workplace air quality standards are 0.05 mg m^{-3} for inorganic mercury and 0.01 mg m^{-3} for organic mercury. This is in consideration of the greater toxic potential of the lipophilic and bioconcentratable organic forms of mercury. Alkylmercurials are more toxic than arylmercurials as shown by LD_{50} data for birds: $HgCl_2$, 5000 mg kg^{-1}; C_6H_5HgOAc (seed dressing), 1000 mg kg^{-1}; C_2H_5HgCl, 20 mg kg^{-1}. Alkylmercury compounds, especially short chain alkylmercury derivatives, are able to cross the blood-brain barrier, and this explains why mercury poisoning causes mental disturbance. In studies on mice, the brain of the unborn fetus appeared to be particularly susceptible, and fetotoxicity and teratogenicity were observed at levels of CH_3HgCl in the mothers' diet below 1 ppm. At this concentration, the adults showed loss of coordination, and were able neither to swim properly nor to climb normally on their wire cages.

Mercury salts cause kidney damage, while elemental mercury—like the alkylmercurials—affects the central nervous system, giving rise to symptoms such as tremors, irritability, and sleeplessness. Elemental mercury is mainly hazardous as the vapour; there is less danger of absorbing the metal from the digestive tract.

Acute mercury poisoning is treated by administration of the antidote 2,3-dimercaptoethanol (known also as BAL: short for British AntiLewisite, because 2,3-dimercaptoethanol was also used as an antidote for a World War I poison gas known as Lewisite). Mercury has a high affinity for sulfur compounds such as BAL, see Equation 13. The complex BAL.Hg is not a monomeric chelate, since the bond angle for dicoordinated mercury is 180°. Instead it is a polymer, perhaps better represented as $(BAL.Hg)_n$.

Ambient levels of mercury in air vary widely. Clean air contains less than 10 ng m^{-3} of mercury, and in the absence of human intervention most of this probably arises from volcanic activity. A volcanic area in Hawaii showed 20 $\mu g\ m^{-3}$, while the levels near a working mercury mine in California were 1.5 $\mu g\ m^{-3}$. Ontario's Environmental Protection Act sets 2.0 $\mu g\ m^{-3}$ as the 24-hour average standard for mercury in air, with 5.0 $\mu g\ m^{-3}$ the maximum permissible in any 30 minute period. Actual ambient levels

(13) 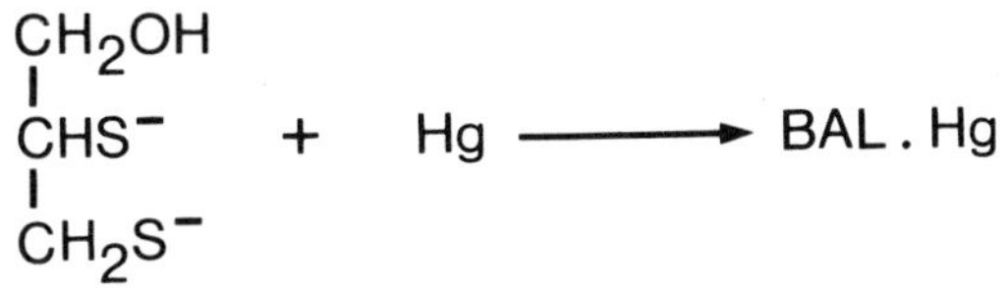 $\begin{matrix} CH_2OH \\ | \\ CHS^- \\ | \\ CH_2S^- \end{matrix} + Hg \longrightarrow BAL.Hg$ log K =25.7

$$\cdots S{-}Hg{-}S\diagdown CH_2\underset{\displaystyle CH_2OH}{\underset{|}{CH}}\diagup S{-}Hg{-}S\diagdown CH_2\underset{\displaystyle CH_2OH}{\underset{|}{CH}}\diagup S{-}Hg{-}S\cdots$$

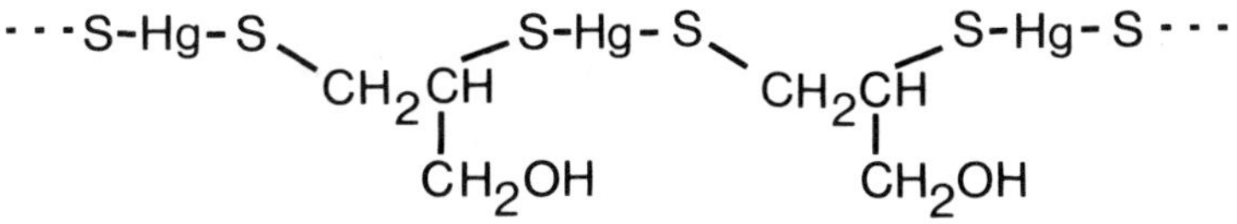

sampled have been much lower, less than 2 ng m^{-3}. The widespread use of mercury in disposable batteries leads to significant emissions of mercury from municipal solid waste incinerators, on account of its high volatility. High levels of mercury emissions led to temporary closure of the municipal waste incinerator in Detroit, Michigan, in 1988.

10.3 Lead[12]

10.3.1 Uses of lead

Lead has been used as a metal at least since the times of the Egyptians and the Babylonians. The Romans employed lead extensively for conveying water, and the elaborate water distribution systems and bathing arrangements of that civilization depended upon the easy working and bending of the soft metal lead. The Latin word plumbum gives us the expression plumbing and the chemical symbol Pb for the element. There has long been speculation that a contributing factor to the decline of the Roman Empire was subclinical or sublethal lead poisoning among the ruling class. Through the Middle

[12]Unreferenced material in this section is taken from J.O. Nriagu, *The biogeochemistry of lead,* Elsevier/North Holland Biomedical Press, Amsterdam, 1978, and from *Pathways, cycling and transformation of lead in the environment,* Ed. P.M. Stokes, Royal Society of Canada, 1986.

Ages and beyond, the malleability of lead encouraged its use as a roofing material for important public buildings such as the great cathedrals of Europe. Present day production of lead is in the millions of tonnes annually (Figure 10.8); uses of lead include the familiar lead-acid storage battery and the addition of organolead compounds to gasoline as an anti-knock agent.

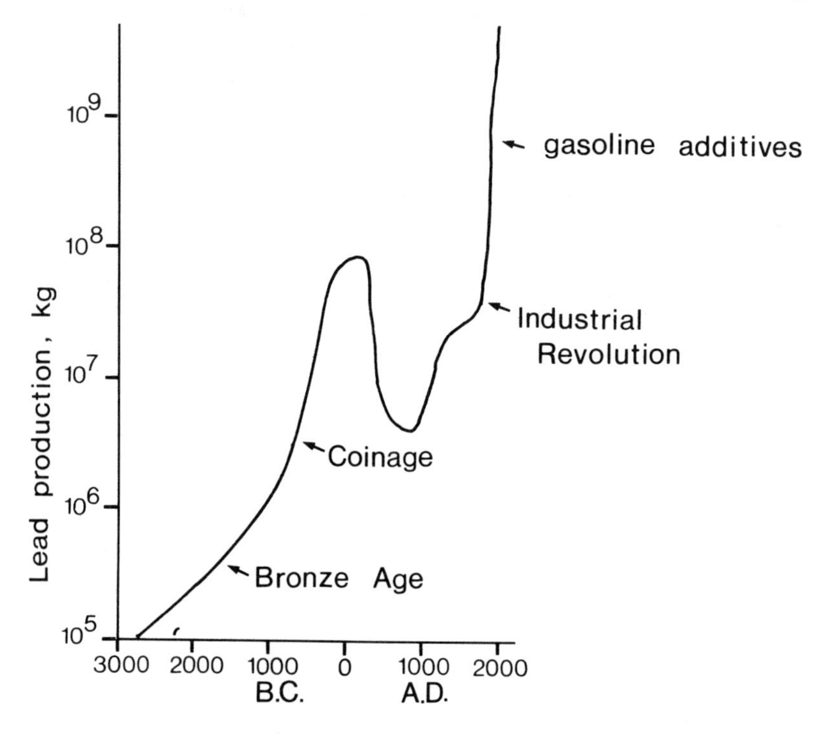

Figure 10.8: Historical production and consumption of lead. (Reference 12a)

Smaller scale uses of lead are in solder, leaded glass for ornamental purposes, and as a shielding material for radioactive sources. Compounds of lead are widely used for their ability to coat other surfaces. The best known example is "red lead" (Pb_3O_4), used to undercoat steel. Other lead-based covering materials are lead chromate ($PbCrO_4$), familiar to North Americans as the yellow pigment used in the paint of school buses, and "white lead," a basic lead carbonate of formula $2PbCO_3.Pb(OH)_2$, which until recently was used as the pigment base for paints. White lead discolours easily: for example in sulfurous atmospheres it blackens with the formation of lead sulfide, and this explains why interior decorations of a century ago favoured such drab colours—lighter shades would darken quickly, and frequent redecoration would be required. Today, white lead has been replaced as the pigment base by zinc oxide and titanium dioxide, compounds which do not darken.

10.3.2 Lead in the environment

The varied uses of lead explain why this element should be so widely dispersed in the environment. The question arises as to what is the "natural" background level of lead. This has been a question of some controversy. Lead levels in modern people are frequently 10% of the toxic levels; some analyses of ancient bones and ancient ice cores seems to suggest that this situation is not new: i.e., that relatively high levels of lead have always existed in the environment and hence that life evolved in the presence of this toxic element.

C.C. Patterson of the California Institute of Technology has challenged this view, claiming that these high lead analyses in ancient samples are the result of inadvertent contamination of the samples during their collection and analysis. He argues, for example, that the ice cores are contaminated by lead from the drilling equipment. His data on Greenland ice cores show the trend in Figure 10.9.

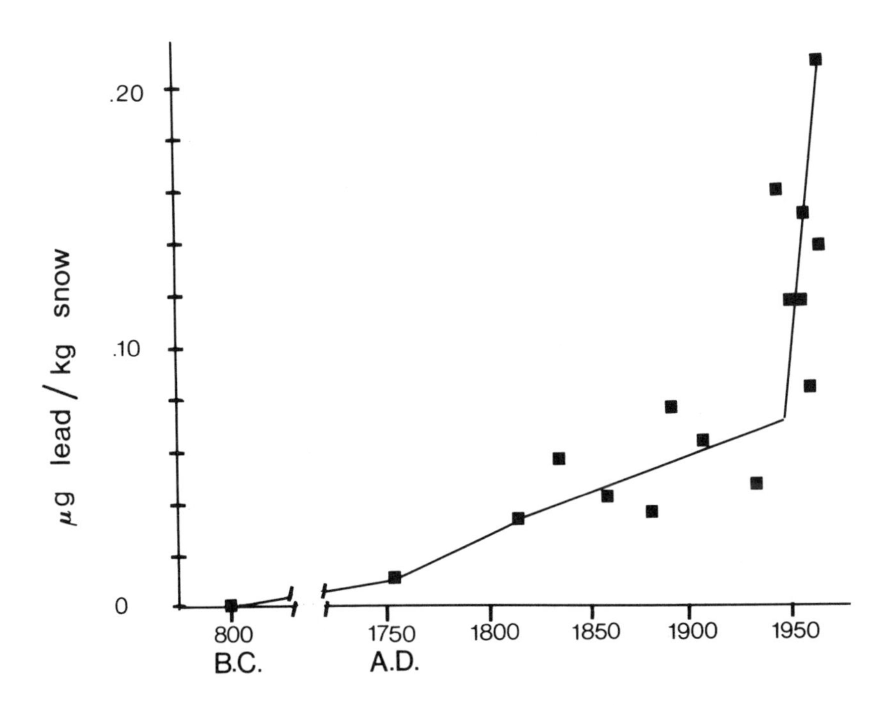

Figure 10.9: Increase of lead in Greenland snow, 800 BC to the present. Reproduced from Reference 12, p. 189.

Patterson, who maintains a laboratory which is carefully protected against accidental lead contamination, also reports that meticulously preserved old skeletons contain 0.01 to 0.001 times as much lead as contemporary skeletons, and that in nature, organisms high in the food chain accumulate calcium to the exclusion of other divalent cations, including lead[13]. A different perspective is provided in a recent analysis of preindustrial and contemporary Alaskan sea otter skeletons. The total concentrations of lead

[13] S. Budiansky, "Lead: the debate goes on but not over science" *Environ Sci Technol*, **1981,** 15, 243–246.

in the two groups of skeletons were similar, but their isotopic compositions were different. The pre-industrial skeletons contained lead with an isotopic ratio corresponding to natural deposits in the region, while the ratio in the contemporary ones was characteristic of industrial lead from elsewhere[14].

10.3.3 Toxicology of lead

The idea of a competition between Pb^{2+} and Ca^{2+}, which are of similar ionic radii, has also been raised in the context of the effects on small children who chew objects covered with old lead paint (Problem 11). The Needleman study (1979) indicated that high body burdens of lead, especially in children, are associated with mental retardation and hyperactivity. Poor children seem to be more at risk, one hypothesis being that the diet of poor children contains less calcium (for example in milk) and hence that similar amounts of lead ingested will affect poor children more than those from wealthier families, an application of Le Chatelier's Principle (Problem 12).

The half-life of lead in humans is estimated to be about 6 yr (whole body) and about 15–20 yr (skeletal). Movement out of the skeleton is thus very slow and lead, like mercury, is a cumulative poison. However, it is not true to say, as appears sometimes in the news media, that lead is accumulated through the lifetime and never eliminated. What has been shown in several studies is that skeletal burdens of lead increase almost linearly with age; this suggests that the steady state with respect to lead is not normally reached. For patients clinically affected, chelation of Pb^{2+} with ethylenediaminetetraacetic acid (EDTA) has been found beneficial in reducing body burdens of lead.

Lead, like mercury, causes neurological problems. Children can suffer mental retardation, lower performance on I.Q. tests, and hyperactivity. Severe exposure in adults causes irritability, sleeplessness, and irrational behaviour. The appetite is depressed, which can lead to emaciation, and death can ensue due to starvation. Again like mercury, organolead compounds are more toxic than simple lead salts because they are non-polar, lipid-soluble, and can more readily cross the blood-brain barrier.

An interesting example of lead poisoning overtook the ill-fated Franklin expedition, which left England in May 1845 in a bid to discover the elusive Northwest Passage through the Canadian Arctic. It was never heard from again[15]. Ultimately, the graves of many of the crew members were located, and in 1984 and 1986, the bodies of three crew members were subjected to autopsy, having lain perfectly preserved for 130 years in the permafrost. The emaciation of the bodies, plus high levels of lead in bone and hair,

---

[14] D.R. Smith, S. Niemeyer, J.A. Estes, and A.R. Flegal, "Stable lead isotopes evidence anthropogenic contamination in Alaskan sea otters," *Environ Sci Technol,* **1990**, 24, 1517–1521.

[15] O. Beattie and J. Geiger, *Frozen in time,* Western Producer Prairie Books, Saskatoon, Saskatchewan, 1987.

pointed to lead poisoning. The source of the lead? Lead solder, which was used to seal the tin cans which contained the crew's provisions.

10.3.4 Chemistry of lead

Lead occurs in nature mostly as the sulfide, the principal useful ore today being the mineral galena, PbS. Lead and zinc sulfides are commonly found and mined together. Silver is an economically important minor component of these ores; in ancient times, lead was considered a byproduct of silver manufacture. The various sulfides are separated by flotation prior to smelting. Lead is more electropositive than mercury, and roasting the sulfide in air produces lead oxide rather than the free metal.

$$2PbS(s) + 3O_2(g) \longrightarrow 2PbO(s) + 2SO_2(g) \qquad (14)$$

The oxide is then reduced to the metal with coke, and the impure metal is refined further by electrolysis (impure lead anode, pure lead cathode, H_2SiF_6 electrolyte).

Lead lies immediately above hydrogen in the electromotive series of the elements, and the reaction $Pb^{2+}(aq) + 2e^- \longrightarrow Pb(s)$ has $E^\circ = -0.13V$. The structural and decorative uses of the metal depend upon its resistance to corrosion, since in the atmosphere, the metal surface quickly becomes covered with a barrier of oxide and carbonate, inhibiting further attack. In the absence of such a coating, lead will dissolve slowly in water, especially at acidic pH: this has been mentioned in Chapter 7 as a problem when lead plumbing is used. If the water is not highly oxygenated and is soft (recall that soft water supplies tend to be acidic and to contain only low concentrations of carbonate), it is the metal itself that is exposed to the water, and slow dissolution will occur. Hard water tends to coat the pipes with a protective layer of calcium carbonate.

The common compounds of lead derive from the +2 oxidation state. As a member of the periodic group IV-A, lead also forms tetravalent compounds, which are covalent. Of these, the most important commercially are the tetraalkylleads, which are used as gasoline additives. The Pb–C bond is very non-polar, and organolead compounds tend to be kinetically inert, like organomercuries.

The speciation of lead(II) in aqueous solution involves several polymeric hydroxo-complexes[16]. Below pH 5.5, $Pb^{2+}(aq)$ predominates, but as the pH increases, $Pb_4(OH)_4^{4+}$, $Pb_6(OH)_8^{4+}$, and $Pb_3(OH)_4^{2+}$ appear in turn, prior to the precipitation of $Pb(OH)_2(s)$.

10.3.5 The lead-acid battery

This device is an example of a storage cell, meaning that the battery can be discharged and recharged over a large number of cycles unlike, say, a

[16] V.L. Snoeyink and D. Jenkins, *Water Chemistry*, John Wiley and Sons, New York, 1980, 214–215.

flashlight battery, which is thrown away when it is spent. The lead-acid cell is familiar as the battery in your car. The overall chemistry follows (Problem 13).

$$\text{(15)} \quad Pb(s) + PbO_2(s) + 2H_2SO_4(aq) \longrightarrow 2PbSO_4(s) + 2H_2O(l)$$

Charging reverses this reaction. The charge in the battery is monitored through the density of the electrolyte, since sulfuric acid (charged) is denser than water (discharged). During charging, the following reactions occur.

cathode: $E^\circ = -0.13V$

$$\text{(16)} \quad Pb^{2+}(aq) + 2e^- \longrightarrow Pb(s)$$

anode: $E^\circ = -1.46V$

$$\text{(17)} \quad Pb^{2+}(aq) + 2H_2O(l) \longrightarrow PbO_2(s) + 4H^+(aq) + 2e^-$$

An important disadvantage of the lead-acid battery is its high mass, on account of the high density of lead. The lead-acid cell is described as having a low energy density, meaning that the ratio of the extractable energy to the mass of the battery is low. Much research has been devoted to finding alternatives which would employ lighter metals such as lithium or aluminum. The successful development of the all-electric car seems to depend upon producing a lightweight power source which can either be recharged simply, or whose spent electrode can be easily and cheaply replaced.

Used car batteries distribute a lot of lead into the environment; despite recycling, they are the major source of lead in municipal waste[17]. Left lying about, for example on the farm, they are the cause of numerous poisonings, especially of inquisitive cattle. Recycled, they result in local lead pollution when the old electrodes are redistilled (lead has m.p. 327°C and b.p. 1740°C) to recover the metal. Lead recycling plants in Canada fall under the definition of secondary lead smelters, emission standards for which limit the amount of lead emitted to 46 mg m^{-3}.

10.3.6 Lead in gasoline

Simple distillation of crude petroleum oil affords a "straight run" gasoline, which is a poor fuel for the modern automobile because under compression it burns erratically in a series of small explosions. These explosions are known as "knocking" or "pinging." They represent premature detonation of the gasoline-air mixture before the piston is ready to move upwards in the cylinder, thus greatly reducing the efficiency of the engine. In the 1920's Thomas Midgeley of General Motors' research laboratories discovered that the addition of small amounts of organolead compounds to gasoline greatly reduced the tendency to knocking, and hence improved fuel efficiency.

---

[17] E.A. Korzun and H.H. Heck, "Sources and fates of lead and cadmium in municipal solid waste," *J Air Waste Management Assoc*, **1990**, 40, 1220–6.

Tetraethyllead (TEL) has been the most favoured gasoline additive. It is prepared from sodium-lead alloy.

$$(18) \quad 4Na/Pb + 4C_2H_5Cl \xrightarrow[\text{acetone}]{75^\circ C,\ 4\ atm} Pb(C_2H_5)_4 + 4NaCl + 3Pb$$

TEL promotes smooth combustion of gasoline by decomposing evenly into ethyl free radicals, which act as free radical initiators.

$$(19) \quad Pb(C_2H_5)_4 \xrightarrow{\text{heat}} Pb(g) + 4C_2H_5\bullet$$

This reaction serves to introduce a steady source of free radicals into the combustion mixture, promoting smoother combustion through free radical chain oxidation. Environmental consequences of adding TEL to gasoline are of two kinds: emissions at the point of manufacture, and emissions from the cars using leaded gasoline.

During manufacture of TEL, up to 2 g of lead per kg of TEL produced may be released to the atmosphere when the TEL (b.p. $\approx$ 200°C at 1 atm) is distilled, together with a further 30 g of lead per kg TEL when the lead which is left over from the Na/Pb alloy is recovered by distillation (Problems 14 and 15). (These amounts can be greatly reduced by careful emission control.)

When TEL decomposes in the car engine, lead atoms are released, initially into the gas phase. However, if the lead were not removed, it would condense on cooler engine parts such as valves, causing them to seize. For this reason, dichloroethane and dibromoethane are also added to the gasoline, and the lead is ultimately emitted from the exhaust as $PbCl_2$, PbBrCl, and $PbBr_2$, which are gaseous at the temperature of the exhaust gases[18].

Lead emissions from vehicles are a major source of environmental contamination by lead, and the nature of the source is such that the wide dispersion of the pollutant is inevitable. Numerous studies, in North America and in Europe, have shown that the soil close to freeways contains high levels of lead, and that the concentration of lead increases both with proximity to the roadway and with traffic volume. In urban areas, airborne lead is highest near busy intersections, and its concentration decreases with increasing elevation, even over 10–20 meters. Canadian reports indicate that the percentage of lead emissions to the environment due to leaded gasoline peaked in 1973 at 70%; the introduction of unleaded gasoline reduced this to 60% in the early 1980's. Concern over lead pollution has caused many governments to phase out leaded gasoline completely: in Canada, the date for the complete removal of lead from gasoline was December 1, 1990. This

[18]Recent research has shown that the chlorine thus incorporated into the combustion mixture is a significant source of TCDD (Chapter 9) in the environment, as a consequence of incomplete combustion. Replacement of leaded by unleaded fuel will minimize this source of environmental contamination, along with the primary objective of reducing contamination by lead.

measure is expected to reduce lead emissions to the Canadian environment to one third of their 1980 levels[19].

The move to remove lead from gasoline originated in the 1970's with the thrust to reduce air pollution from unburned hydrocarbons and nitrogen oxides. As discussed in Chapter 3, these substances are ingredients of photochemical smog. California was the first jurisdiction to legislate emission controls on automobiles. The tie-in to lead is that in order to limit hydrocarbon and NO_X emissions, it is necessary to use emission control devices (catalytic converters). The precious metal catalysts used in these converters are rendered inactive ("poisoned") by lead, and consequently the use of leaded gasoline is incompatible with controlling emissions by means of catalytic converters. The present catalytic converters are capable of eliminating 96% of unburned hydrocarbons and CO, and 76% of NO_X emissions from exhaust gases[20]. These devices have become increasingly efficient since their introduction in 1975, and in addition, older cars are gradually being replaced. It is expected that by the year 2000, total emissions of RH, CO, and NO_X will decline 57, 73, and 46% respectively from 1989 levels, despite the anticipated increase of 1.8% annually in the size of the fleet.

In the 1980's, the move against lead in gasoline has been directed mainly towards eliminating contamination by lead for its own sake, rather than as a prerequisite for limiting air pollution by automobile combustion products. In Europe this movement was initially retarded by a EEC rule that mandated minimum levels of organolead additives in gasoline.

Automobile emissions illustrate an interesting problem in stoichiometry. The hydrocarbon fuel (RH) is to be burned completely to CO_2 and water. At the same time NO_X is produced, because the oxidant is air rather than pure oxygen. A "rich" mixture in the carburetor means a slight stoichiometric excess of RH, and so hydrocarbon emission is inevitable. Conversely, a "lean" mixture is deficient in RH, so there is an excess of oxygen, and this can react with N_2 in the engine to give NO_X. "Rich" operation minimizes NO at the expense of RH emissions, while "lean" operation has the opposite consequence. A further complication is that the endothermic formation of NO_X from the elements is favoured by raising the temperature at which the engine operates, while the thermodynamic efficiency of any "heat engine" such as a car is maximized by running the engine at the highest possible temperature. Reducing the emission of NO_X is therefore in direct competition with fuel efficiency.

The first generation of catalytic converters, using oxidation catalysts, was introduced with the 1975 automobile model year. The active catalyst was a noble metal (e.g., Pd, Pt) or a metal oxide mixture such as Fe_2O_3 or CoO/Cr_2O_3 on an inert support. The carburation used with this system was

[19] A 1986 report (*Environ Sci Technol*, **1986,** 20, 171) attributed a decline in wet deposition of lead in Minnesota to reducing the average concentration of lead in gasoline.

[20] J. Haggin, "Catalyst gains from auto emission work," *Chem Eng News,* May 15, **1989,** 23.

a rich mixture, so that unburned RH would pass out of the primary combustion chamber. Extra air was then admitted, and the excess RH oxidized at a lower temperature over the catalyst, thus minimizing the production of NO_X. Problems with this approach included lower fuel efficiency due to the use of the rich mixture, deposits of carbon on the spark plugs, and excessive emissions of RH when the catalyst lost its effectiveness. These problems caused many motorists to (illegally) remove the emission control device.

Reduction catalysts, employing rhodium on an inert support, were introduced in the 1981 model year. Here the strategy is to use a lean mixture, and then reduce catalytically any NO that formed, using CO in the exhaust gases as the reducing agent.

$$CO + NO \longrightarrow CO_2 + \tfrac{1}{2}N_2 \tag{20}$$

A disadvantage of this system is that the exhaust gases contain small amounts of H_2—from the reaction of alkanes with steam formed during combustion. This reacts with N_2 catalytically (recall the Haber process) to form ammonia, which is emitted instead of NO_X. The most recent development is a dual system, operating as follows on an almost stoichiometric fuel/air ratio:

1. reduction catalyst (reduce NO_X to N_2);
2. air injection;
3. oxidation catalyst to oxidize residual RH and CO, and any NH_3 produced at step (1).

Given that organolead compounds offer such improvements in combustion efficiency, how can they be replaced? Research has focussed on developing different additives, changes in engine design, and changes in petroleum refining technology.

The search for free radical initiators other than TEL has not been successful, with fouling of the engine parts and/or excessive engine wear being unwelcome side effects. Changes in engine design and changes in petroleum refining are in fact related problems, and to understand them we must digress briefly to examine how crude oil is refined.

Crude petroleum is a mixture of many thousands of components, encompassing a full spectrum of molar masses. The process of refining converts this complex, highly viscous mixture into the familiar range of useful petroleum products seen in Table 10.2. These various fractions are obtained when crude oil is fractionally distilled. However, the composition of a barrel of oil matches the demand for the fractions very poorly; in particular, the demand for the gasoline fraction far exceeds the supply of straight run gasoline. In addition, straight run gasoline is a rather poor fuel as previously mentioned; it is composed principally of linear alkanes, which have a strong tendency to knock. Branched chain alkanes are superior fuels from this point of view. Modern refinery technology includes two important processes: isomerization, by which linear alkanes are isomerized into branched chain ones; and cracking, by which the more abundant longer chain alkanes are broken catalytically into smaller fragments.

Table 10.2: Fractions of crude oil

Number of Carbon Atoms	Name and/or use
1	methane, natural gas
2	ethane, used to make ethylene
3,4	propane and butane, compressed gases
5–7	naphtha, solvent
7–12	gasoline
10–15	kerosene, aviation fuel, home heating fuel
15–30	lubricating oils, heavy fuel oil
30–40	paraffin waxes
> 40	asphalt

Octane ratings are a means of classifying the knocking tendencies of a given gasoline. The branched alkane 2,2,4-trimethylpentane (known, incorrectly, in commerce as isooctane) is assigned an octane rating of 100 and heptane, a poor fuel, is given an octane rating of zero. A fuel with an octane rating of 90 (for example) burns under test conditions exactly as efficiently as a mixture of 90 parts of isooctane and 10 parts of heptane.

Octane ratings are improved by the use of additives (such as TEL) and by increasing the amounts of branched, aromatic, or "oxygenates" (oxygen-containing organic compounds) in the gasoline. Therefore, in order to manufacture a lead-free gasoline, changes in refinery operation are needed so as to change the composition of the blend.

Among oxygenates, the lower alcohols have long been known as effective fuels and have been used to power racing cars. In the 1970's much interest centred around alcohol/gasoline mixtures ("gasohol"), not only from the point of view of raising octane ratings but also because of (i) interest in replacing part of the fossil fuel gasoline by a renewable, biomass-derived fuel, and (ii) pressure from farmers in corn producing U.S. states to reduce the large corn surplus of the time. Both these pressures have now abated substantially, and with them, also the interest in gasohol in most of North America. Both gasoline/alcohol blends and pure ethanol are used as the principal automotive fuels in Brazil, however.

A serious obstacle to the use of methanol or ethanol blends with gasoline is their tendency to separate into immiscible layers in the presence of even small amounts of water. This is because the hydrophilic lower alcohols are more soluble in water than in the hydrophobic alkanes of gasoline.

For all these reasons, there is more interest today in boosting octane ratings by means of methyl (and ethyl) tert–butyl ethers, formation of which has the bonus of using surplus capacity in the production of isobutylene. These compounds have octane ratings in the range 108–110; they are added at the rate of 1–2% of the blend, so that it is debatable whether they should be termed additives or fuel components.

Along with changes in gasoline manufacture, car engines have been redesigned to operate at lower compression ratios since about 1970. This change means that fuels of lower octane number can be used, because premature ignition (knocking) is more prevalent when the fuel/air mixture is at high pressure. Through negotiations with the major automobile manufacturers in North America, successive reductions in the production of cars that operate on leaded fuel have seen the virtual elimination of leaded gasoline since 1970. Unleaded fuel is somewhat more expensive to produce than "regular" gasoline; an important contribution to minimizing lead emissions prior to the removal of leaded gasoline from the marketplace was an additional tax on regular gasoline which was introduced in Ontario in 1988, thereby removing the price differential in favour of leaded fuel.

The introduction of oxygenates has prompted considerable research into the effects of changes in fuel composition upon automobile emissions. In studies involving gasoline-methanol blends, substituting methanol for gasoline in the blend caused the replacement of gasoline hydrocarbon emissions by formaldehyde. Oxygenates appear to reduce CO emissions, particularly from older vehicles, which are the ones with the dirtiest emissions.

10.4 Problems

1. A patient takes a barium meal containing 25 g of $BaSO_4$, whose K_{sp} is 1.0×10^{-10} (mol $L^{-1})^2$. If this were to become equilibrated with the 8 L of blood in the patient's body, what would be the body burden of Ba^{2+}?

2. Show by calculation the difference in mg L^{-1} between 1.0 ppm of mercury vapour in the air, and 1.0 ppm of mercury in water, both at 20°C.

3. The boiling point of mercury at 1 atm is 356.9°C, and its enthalpy of vaporization is 272 J g^{-1}. Estimate its vapour pressure at 25°C, then look up the experimental value in the *Handbook of Chemistry and Physics* for comparison.

4. **(a)** Andren and Nriagu[2] presented the following model for mercury cycling in the atmosphere: steady state Hg in atmosphere = 1.2×10^9 g; rates of input (all in g yr^{-1}): volcanoes 2×10^7; decomposition of biomass 4×10^7; continental degassing 1.8×10^{10}; ocean volatilization 2.9×10^9; anthropogenic 1.0×10^{10}. What is the residence time of mercury in the atmosphere according to this model?

(b) The variation of mercury with altitude in the atmosphere has been deduced to follow the relationship

$$c_h = c_o\, e^{-0.001h}$$

where h is the altitude in meters. At zero elevation over land Hg concentrations are ca. 4.0 ng m^{-3}. Above what elevation is the Hg concentration less than 0.1 ng m^{-3}?

5. The TLV of elemental mercury is 0.05 mg m^{-3}. A laboratory worker spills mercury on the floor of a room of dimensions 8 m × 6 m × 3 m high, and retrieves all but 0.5 mL. At equilibrium, is the TLV exceeded?

6. Calculate the masses of all products formed by the complete electrolysis of 1.00 tonne of NaCl in the chlor-alkali process. What is the ratio by mass of NaOH to Cl_2?

7. The maximum acceptable concentration of mercury in water is 0.001 ppm (Chapter 7).

(a) Calculate this concentration in mol L^{-1}.

(b) Is the MAC likely to be exceeded by dissolution of HgS ($K_{sp} = 1 \times 10^{-56}$ (mol $L^{-1})^2$ or HgO (solubility = 5.3 mg per 100 mL) in water? If not, why is contamination of drinking water by mercury a possible problem?

8. From Figure 10.5, deduce the steady state value for the concentration of mercury in the tissues for a dietary intake of 0.1 mg of mercury per day. What is the half-life for excretion of mercury from the body according to these data?

9. Certain microorganisms can degrade methylmercury compounds to elemental mercury.

(a) Look up the appropriate data in the *Handbook of Chemistry and Physics* (65th or later edition) to calculate ΔG° for each of these possible degradation pathways.

(i) $(CH_3)_2Hg \longrightarrow C_2H_6 + Hg$
(ii) $(CH_3)_2Hg + 2H^+ \longrightarrow 2CH_4 + Hg^{2+}$
(iii) $(CH_3)_2Hg + CH_3OH \longrightarrow 2CH_4 + Hg + CH_2{=}O$

In reaction (iii) CH_3OH is used as an example (only) of an oxidizable organic substance. In calculating the ΔG°'s, choose appropriate standard states for the compounds as best you can.

(b) Would you expect the actual demethylation pathway to be the one with the most negative ΔG°? Explain.

10. **(a)** The World Health Organization sets a standard of 0.2 mg for each 60 kg person per week as an acceptable mercury intake. In Canada, fish from the Great Lakes are considered edible if their mercury content is ≤ 0.5 ppm. Are these values compatible?

(b) Calculate the masses of the following pollutants in a 1.5 kg lake trout:

(i) 0.5 ppm of Hg^{2+}
(ii) 35 ppt of TCDD (Chapter 9).

11. A "low lead" paint contains 0.5% lead by weight and loses 60% of its weight upon drying. An 11 kg child chews on an object painted with this paint. What mass of dried paint needs to be ingested for the child to take up the World Health Organization's recommended daily lead intake of no more than 6 μg kg^{-1}?

12. The movement of lead in the blood of adult males may be summarized:

Blood lead = 140 μ g L^{-1} Net transfer to bone = 7.5 μg day^{-1}
Blood volume = 4.8 L Net excretion rate = 24 μg day^{-1}

Calculate the residence time of lead in the blood.

13. On the basis of concentrations rather than activities, plot the concentration of H_2SO_4 in a lead-acid battery as the cell voltage falls from 2.0 to 1.5 V.

14. In summer (assume 20°C), tetraethyllead (TEL) in the atmosphere is destroyed mainly by reaction with OH radicals.

(a) Write out the reactions by which OH radicals are formed in the lower atmosphere.

(b) In the vicinity of a TEL manufacturing facility, atmospheric TEL levels are 13 ppb, and the steady state concentration of OH is 8.2×10^6 molecules cm^{-3}. The half life for TEL under these conditions is 1.2 h. Calculate both the rate constant and the initial rate of the reaction below.

$$\mathrm{TEL} + \mathrm{OH} \xrightarrow{k} \text{products}$$

15. A lead recycling plant begins operation on the shores of a hitherto clean lake of capacity 3.0×10^6 m^3. It discharges into the lake 12 m^3 per hour of waste containing 15 ppm of Pb^{2+}. The other inflow and outflow of the lake is a river with a flow rate 8400 m^3 h^{-1}.

(a) Calculate the steady state concentration of Pb^{2+} in the lake, which is well mixed, and has no other source or sink for Pb^{2+}.

(b) Calculate the residence time of Pb^{2+} in the lake at the steady state.

(c) How long does it take for the Pb^{2+} level to reach 50%?, 90%?, 99%? of its steady state value?

Index

About the Author

Nigel J. Bunce was born in Sutton Coldfield, England. He earned B. A. (1964) and D. Phil. (1967) degrees from Oxford University. He moved to Canada in 1967, and held a Killam Memorial fellowship at the University of Alberta for two years. Since 1969 he has been with the Department of Chemistry of the University of Guelph. Dr. Bunce's researches are in environmental chemistry generally, and his specific areas of interest at present are the toxicology of dioxins, and organic photochemistry. Dr. Bunce is also the co-editor of a weekly science column in the Guelph *Mercury*.

Back Cover

The Great Smoky Mountains, Tennessee, USA. Their attractive bluish haze is caused by the oxidation of hydrocarbons emitted by pine trees, and is chemically closely related to photochemical (Los Angeles type) smog. It differs in that nitrogen oxides, a key ingredient of photochemical smog, are not present at elevated levels. See Chapter 3.